SOCIAL STUDIES OF SCIENCE AND TECHNOLOGY: LOOKING BACK, AHEAD

Sociology of the Sciences

VOLUME XXIII

The titles published in this series are listed at the end of this volume.

SOCIAL STUDIES OF SCIENCE AND TECHNOLOGY: LOOKING BACK, AHEAD

Edited by

BERNWARD JOERGES
Wissenschaftszentrum Berlin,
Berlin, Germany

and

HELGA NOWOTNY
ETH Zürich,
Zürich, Switzerland

KLUWER ACADEMIC PUBLISHERS
DORDRECHT / BOSTON / LONDON

A C.I.P. Catalogue record for this book is available from the Library of Congress.

ISBN 1-4020-1481-3

Published by Kluwer Academic Publishers,
P.O. Box 17, 3300 AA Dordrecht, The Netherlands.

Sold and distributed in North, Central and South America
by Kluwer Academic Publishers,
101 Philip Drive, Norwell, MA 02061, U.S.A.

In all other countries, sold and distributed
by Kluwer Academic Publishers,
P.O. Box 322, 3300 AH Dordrecht, The Netherlands.

Printed on acid-free paper

Printed in the Netherlands.

TABLE OF CONTENTS

Postscript

PREFACE

BERNWARD JOERGES AND HELGA NOWOTNY

YET ANOTHER TURN

The thing that doesn't fit is the thing that's most interesting.
Richard Feynman

This volume was originally conceived as a contribution to yet another 'turn', not captured by one of the many adjectives that have served to describe the collective meandering of the scholarly community in search of direction. It was meant to mark the millennial turn, a seemingly purely chronological event, but one in search for great meanings and invested with loaded significances. The editors wanted to seize the opportunity of the moment in order to pause and reflect on the sociology and history of social studies of science and technology. The moment came and went and the new millennium, barely nine months later, thrust its historical marker upon the world through a horrendous and cruel shock in an unforeseen and unforeseeable way. Since then, the world appears more vulnerable and volatile, fragmented and fraught with uncertainty. The universal values as bequeathed by the Enlightenment are either refused or appear refuted. The dream of a universal civilization which has accompanied the unfolding of the existing multiple modernities in their historically unique trajectories, has been discarded and its promises in tatters. The belief in a process of progressive rationalization and in the ongoing, although repeatedly disrupted, validation of universal cultural forms, which have been hailed as providing the underlying strength in the relation between science and society (whatever is understood by these terms), clearly is in need of revision.

At this conjuncture of humanity and its future, is it sufficient to forge ahead at the frontiers of knowledge, exciting as they are, while social institutions and human agency need to be re-thought on a global scale? What is the role of science in an age of uncertainty, given the diversity of its contemporary institutional forms and its multiple complex layers of practices, driven by the dynamics of scientific ideas and intertwined with new technologies and instrumentation? There can be no doubt that science is a powerful institution, still working in close alliance with industry and the state, whose innovative potential fuels the economic competitiveness and military prowess of nations. Much has been said about the highly differentiated nature of scientific practices, representations and beliefs and their assumed malleability, subject to cultural influences, which are in turn based upon material forms, shaped and driven by instrumentation, technologies and techniques. Universal theories are challenged to stand up to a recalcitrant societal reality

B. Joerges and H. Nowotny (eds.), Social Studies of Science and Technology: Looking Back Ahead, xi–xvii.

that places them into many different, changing and competing contexts. But where do we go from here?

With social worlds becoming increasingly regulated, all scientific and technical 'deliverables' must conform to norms and regulations, to standardization and the expectations of citizens and consumers while also satisfying economic and technical demands. What have social studies of science and technology to contribute to these ongoing transformations – be it as a sharply analytical and/or critical public commentary, as probing exploration and explanation and as the originators of knowledge deeply rooted in a social science and historical perspective?

This volume of the *Yearbook* is the first exclusively written by the collective of its editors. Initially, taking advantage of the millennial moment, we had sought to address questions revolving around the tension between a compact global time-space and divergent cultural and intellectual tendencies. We started by looking into the connections that underpinned the history of time measurement and how to measure and assess progress. We wanted to trace the processes through which time's arrow and hence its irreversibility and with it history, had entered the sciences. We were fascinated by the recurrent tensions between the smooth surface of ever greater precision, abstraction and enhanced ability to divide and measure time in ever smaller segments and facing the phenomenon of emergence, the possibility of unique or extremely rare events, catastrophes or even, in the cultural interpretation attributed to them, the various 'end-isms'. We intended to delve into the recurrent theme of rupture vs. continuity, following the ongoing swings of the scholarly pendulum in the sciences, but also in science studies. After a period of historicizing nature with its attendant diversification of scientific narratives, we now are seemingly back on a path towards naturalizing history and society. We considered how the tendency towards global synchronization and the tendency towards local fragmentation hold up, reinforce or disrupt each other in their mutual entanglements.

The present volume considerably deviates from the one we had in mind. To be sure, all the themes set out above are present throughout, however not arranged around such headings or offering straightforward answers to these questions. One reason is that the reorganization of the editorial concept and composition – new members of the editorial board make their first appearance in this publication – happened to coincide with the millennial turn and prompted us to re-write our contributions around another event: the 25th anniversary of the *Yearbook*. Such a date may seem parochial. Nevertheless, we chose it as a timely frame of reference as the anniversary marks an intended re-orientation of the *Yearbook*. The rank of the senior editors who had started the *Yearbook* in 1977 in the youthful spirit of intellectual rebellion, has been widened and rejuvenated. Together with the new generation we intend to open up the field towards other social science disciplines. At a time where so much is steeped in uncertainty and volatility we feel a renewed intellectual responsibility to confront, to follow Feynman's dictum, 'the thing that doesn't fit' which otherwise risks to remain without voice and recognition in the stream-lining of academic consensus.

Coupled to this re-orientation, and based upon the insight that other questions need to be asked and new audiences addressed, comes a renewed emphasis on policy aspects. The range of developments in this field is vast and has far-reaching impacts. Too many of the underlying issues remain unexplored, either out of willful diffidence or buried in too narrowly conceived specialist accounts. In line with such developments, the so-called relations between science and society have assumed a public visibility, if not urgency, that call for more then mere acknowledgement. Careful intellectual scrutiny of an emerging public space is needed in which much of what is happening there may be merely 'imagined', but nevertheless is real in its consequences.

Accordingly, the volume is structured with the intention to articulate some of the issues to be tackled in the future and explore further the directions to be taken. Looking back/ahead implies to take a stand in what, following William James, can be called the specious present. It is based on the assumption that it is possible to situate oneself in the fluid temporal medium, framed in the collective endeavor of writing, while gaining a moment of stability, if not of closure. But this is not the 'closure' of settling an issue or a theoretical question, which may grant conditional finality. Rather, it is more akin to the 'closure' of the photographic act, which follows the 'exposure' to light. The reflexive eye, in this case, a collectively reflexive eye, has attempted to focus as sharply as possible on the inevitable variety in ways of seeing and to render visible what it has seized in the brief time span between exposure and closure. Like any photographic image, the picture cannot pretend to 'capture reality', but hopes to convey our way of seeing and telling by looking back/ahead.

We have chosen to arrange the remarkably different chapters that resulted from our debates around this particular *Yearbook* project – different in style, personal involvement, temporal frame, substantive focus and militancy concerning the fate of social studies of science and technology (or STS) – in three parts. We start off with wider issues regarding the place of science and technology studies in society (Part I). Moving inward, the next series of contributions focuses on trading zones between science and technology studies and other social science disciplines (Part II). We close with a series of chapters reflecting on the emergence of STS as a distinct but far from stable and 'mature' field (Part III). Needless to say that many of the chapters could find their place in either of these sections.

The volume opens with *Richard Whitley*'s reminiscence of the *Yearbook*'s own particular history. As one of the *Yearbook*'s main protagonists, he traces its impact on the field and, in turn, the effects of changing landscapes of science and technology as well as relevant state policies on the *Yearbook*.

Part I of the volume begins with *Ulrike Felt*'s examination of "Sciences, Science Studies and their Publics", speculating on possible and desirable future relations. Covering the past thirty years of research in STS, she points to a series of inconsistencies, dilemmas and unexamined issues that prevent the field from clarifying its relations with its publics inside and outside science and from writing its own history. In the future, thus her argument, public understanding of science is as much about STS understanding their publics as the public understanding science.

In "Is *that* Politics?", *Emilie Gomart* and *Maarten Hajer* turn to one very specific set of the publics of science studies: political scientists, policymakers and technocrats, and practitioners of STS who wish to render their work politically relevant. They argue that 'the empirical tactic of STS' could inform not only new kinds of policy analysis, but adaptive democratic forms themselves.

Yaron Ezrahi takes this issue considerably further in his chapter on "Science and the Postmodern Shift in Contemporary Democracies". Taking the long view back, he traces the erosion of forms where the perceived apolitical, quasi-religious authority of science could set limits to legitimate public actions and discipline the exercise of political powers. Towards the closing of the twentieth century, he argues, both politics and science have lost their foundations in encompassing normative states, the apolitical has shrunk dramatically, each public decision concerning science-based action requires open and elaborate political processes of negotiations and compromise in order to provide a stable political mandate.

Part II opens with *Bjoern Wittrock*'s wide-ranging interpretation of "Modernity and the History of Social Science", setting the scene for a series of reflections on the changing character of discursive and technological practices which made possible previously inconceivable institutional practices in modern societies. His time frame extends from the late eighteenth to the late twentieth century and, focusing on three distinctive moments, provides a platform for discussing – in view of their fit with social theory and global history – alternative modes of analysis in today's social studies of science.

Terry Shinn takes as his focal point Wittrock's third moment, criticizing 'new orthodoxies' that have come to dominate much of current science and technology studies. Scrutinizing the two concepts of the "Triple Helix" and "The New Production of Knowledge" (mode 2), he also offers a preliminary survey of their take-up in science studies, of their differences and commonalities. Both approaches, he argues, share a concern with massive contemporary 'transversality' in science and research but fail to contextualize the phenomenon historically, relating it to the emergence, over the past two hundred years at least, of what he has termed 'research-technologies' elsewhere.

Sabine Maasen then turns to a field of knowledge production and discursive practices which was, in the past, largely neglected and unaffected by science studies. In "The Conundrum of Consciousness – Changing Landscapes of Knowledge at the Turn of the Millennium", she points to the last decade of the twentieth century as the moment when consciousness became at last a subject of consciously scientific research efforts. Mapping the multitude of discourses in the field, she singles out 'the hybrid engine called consciousness studies' as a model case for science studies in view of understanding the social significances of the phenomena they study.

We conclude this part on trading zones between science and technology studies and other social science disciplines with *Sheila Jasanoff*'s essay "In a Constitutional Moment: Science and Social Order at the Millennium". How can we probe the structuring effects of science and technology on basic ordering commitments, such as changing popular conceptions of what we are and what we can demand from our ruling institutions? Diagnosing the poor fit between legal discourse and science studies, Jasanoff offers, in what she calls constructive constitutionalism and at the same time constitutionalism from below, an interpretation of emerging global issues concerning self, identity and community. And she reflects on the preconditions, both outside and inside science, for turning yet inchoate and uncodified changes in a globalizing civil society into a constitutional force that is felt in the citadels of organized power and governance.

In Part III of the volume, devoted to the emergence and future of science studies proper, *Peter Weingart* takes us back to earlier visions of the growth of science. In "Growth, Differentiation, Expansion and Change of Identity – The Future of Science", he takes as his point of departure post-World War II prophesies of the growth of science and extracts lessons as to how sociologists of science might adapt their categories and methodologies to the changing nature of knowledge systems.

Stephen Hilgartner then turns from analysis to the practicalities of "Institutionalizing Science & Technology Studies in the Academy". After contrasting impressive intellectual sophistication and empirical strength of the field with its precarious institutional standing he engages in a strategic planning exercise in view of stimulating discussions of STS institutional (and in doing so intellectual) agendas.

Michael Hagner and *Hans-Joerg Rheinberger* take us back to the intellectual lessons to be learnt from a century of science studies. In "Prolepsis. Considerations for Histories of Science after 2000" they attempt to overcome the traditional dichotomies that have dominated, in their case, the history of biology. In an attempt to overcome the basic dichotomy of structuralist versus constructivist approaches (and criticizing such neutralizing syntheses as Bourdieu's constructivist structuralism) they argue for histories of scientific cultures as artful practice 'relying on difference, not contradiction, in combination with an encompassing notion of material culture relying on repetition, not identity.'

Finally, in their "Joy in Repetition Makes the Future Disappear", *Michael Guggenheim* and *Helga Nowotny* put this advice into practice with respect to STS itself. After their critical review of the yet to be written history of STS, they set out to recommend a series of approaches which might create for STS the audiences it still misses to an alarming degree.

The volume ends with *Bernward Joerges'* musings on the power of chronologies, epitomized by the Millennium. In "Reflections on the Millennium, Calendars and the Gregorian Hegemony" he draws attention to the organizing power of periodizations implicit in most of the chapters collected in this volume and wonders why calendars and similar big time machines remain a neglected field of science and technology studies.

The contributions brought together in this volume resemble spotlights thrown on the past twenty-five years of science and technology studies. Some of them throw their light through a deliberately narrow focus on selected fields, while others open a wide angle lens to science studies at large. In their different ways they nevertheless seek to reconcile an analytic recollection of past developments in the heterogeneous fields of STS, a diagnosis of where we stand at present, and the articulation of thoughtful concerns where to go from here.

Do these contributions add up to an answer to the question 'What next?' Looking back/ahead thrives on the distinction between what has been achieved in the past and what is therefore worth to be preserved, to be built upon and to be passed on in a deepened, and hopefully enriching way. It presumes also a sense of direction which is future-oriented. The current joys of STS, as *Guggenheim and Nowotny* assert in this volume, invite the comparison with a modern Sisyphus. The mission of STS is perceived to correct the many distortions in the prevailing understanding of what science is and how it works. Yet, despite the numerous empirical case studies, painstakingly and rigorously located at the micro-level, the fog that prevents us to see clearly has not lifted or fell again to hide the view. Meanwhile, the short-lived cries of triumph, followed by another round of long-lived *lamenti*, are increasingly addressed exclusively to other members of the same scientific community, or even smaller sub-circles. All too rarely is a wider public, be it in other disciplines or beyond, even imagined to be interested in what the work of STS reveals and why this knowledge is relevant.

Why does it matter and how should this be communicated? The answers range from the necessity to build a research program which is specific and owned by STS to writing its own history in accordance with its precepts and teachings. But having and showing the courage of one's convictions is not enough. What is at stake – ahead – is also a redefinition of science, not as defined by scientists alone. A famous US lawyer is reported to having once said: law is what lawyers do. Looking at the range of the often-invoked scientific practices, their divergence and continuous intersection with other kinds of practices, institutions, and belief systems, scientists 'do science' in a variety of ways that far surpass what is dealt with within the boundaries of STS. The emergence and creation of the new – new phenomena, new artifacts and instrumentation, new understandings at new levels and in new contexts – is at the heart of this endeavor. Yet, very often there is a lack of words and meanings, since no language is available apart from the purely technical language which forms part of the 'experimental system' which would permit to capture and communicate 'what science does' and 'what scientists are up to,' using Clifford Geertz' phrase. This poses a major challenge to STS, in the sense that it is right here, where novel social forms emerge in the production of new scientific knowledge, that 'society' is and should be present, since it is here where future cultural understandings or misunderstandings, interpretations and meanings, future life-styles, identities and social forms are generated.

We would like STS to face up to the challenge we see in the process of how to incorporate, absorb and appropriate science and technology in the wider societal context. To the extent that Nature is increasingly drawn into Society and becoming socially contextualized, a process of rendering nature 'endogenous' to society has set in. It is worth to be explored further. This includes not only language and the genesis of cultural meanings and interpretations along the interfaces, but must also aim to bring in this realm the best available understanding of economics and a solid sociology of economics.

It also has become clear that the material practices of science, important as they are, do not cover the range of what we usually identify as technology (which is now treated with a curious ambiguity within STS). Thinking about technology and looking ahead, maybe the heterogeneous fields of scientific engineering can be conceived as trading zones in their own right between experimentation and production, much like the transversal research technologies mentioned above. Can the current developments in biotechnology, to cite one concrete example of a rapidly evolving field and its novel practices, be seen as the potential nucleus of a generic field of scientific engineering of the 21st century? It is a field in which technological platforms invite asking new kinds of research questions by bringing together an unprecedented standardization in the handling of and experimentation with living organisms, the processing of huge masses of data and information, and research technologies which easily spread across different disciplines and sub-fields. And if such developments adumbrate the scientific engineering of the future, which other organizational and institutional innovations will follow?

Another set of concerns to be addressed in the future is the parochial, but persistent eurocentrism of STS. If STS were to be established in a few select centers of outstanding quality in the South – which would be the leading questions to be tackled? What baggage of assumptions and profiles of (vested) interests would have to be left behind or actively discarded, before being able to re-think some of the most cherished notions and categories that simply do not find an equivalent in countries and societies ravaged by poverty, aids and violence?

Scholars socialized in the constructivist (and in many ways relativistic) research cultures that have become almost the new orthodoxy in science and technology studies over the lifetime of the *Yearbook*, find it perplexing that their counterparts in the South have difficulties to share 'symmetric' approaches to knowledge production and validation. Seemingly old-fashioned problems of upholding verifiable standards of reliable and valid observation, of constructing uncontested causal relationships or of charting unfamiliar knowledge-scapes in terms of secular vs. religious, superstitious vs. rational, gain unexpected urgency. The 'North' finds itself confronted with charges of indulging in the academic luxury trade or simply having become decadent as a result of its economic abundance. What is 'the thing that does not fit' in this context (and maybe in others as well)? It needs to be identified and may turn out to be most interesting.

It remains for us to wish the *Yearbook* – as befits any special anniversary and even more one which can conveniently be synchronized with more inclusive chronological turns – many joyful and non-repetitious returns.

May it flourish in new and unexpected places and intellectual spaces and, to remain within an organic metaphor, bring forth fruit beyond the genetically-modified zone of STS.

CHAPTER 1

RICHARD WHITLEY

THE SOCIOLOGY OF THE SCIENCES YEARBOOK: A PERSONAL RETROSPECTIVE

For this volume of the *Yearbook*, I have been asked by the editors to provide a personal account of its genesis and development as its Managing Editor for the first 20 years. This involves recalling the broader intellectual and social environment in which it was conceived and developed, and considering how this context has changed and affected the *Yearbook*. As well, then, as summarising the major episodes in the Yearbook's history, I shall briefly discuss its location in what has developed as science and technology studies (STS) in many countries, and how changing state policies have affected both the *Yearbook* and STS.

UN PEU D'HISTOIRE

The *Yearbook* grew out of a conference held in London in 1972 that was jointly sponsored by the International Sociological Association's Research Committee on the Sociology of Science and the British Sociological Association. At that meeting, many participants agreed that the sociology of science should be more concerned with the analysis of the development of scientific knowledge as a sociologically explicable process. This view contrasted quite sharply with much previous sociological work that focused on the social institution of science as a largely autonomous and distinct set of social practices rather than seeking to explain patterns of intellectual development and change. Especially in its North American variant, the postwar sociology of science had not considered how and why the sciences varied in the sorts of knowledge they produced, in how research was organised and controlled, or how controversies were generated and resolved. On the whole, analysis of intellectual structures, styles and change was left to philosophers and historians, with little comparative work being undertaken. Such a reorientation of the sociology of science not only challenged the dominant approach in that field, but also the current conceptions of scientific knowledge held by many historians and philosophers of science, especially in the Anglo-Saxon world.

The resulting book from that conference, *Social Processes of Scientific Development* (Whitley 1974), began to formulate this new approach, but some of us felt that a more extensive series of publications was required to encourage and institutionalise social sci-

B. Joerges and H. Nowotny (eds.), Social Studies of Science and Technology: Looking Back Ahead, 1–8.

entific research on the sciences as cultural phenomena, particularly Norbert Elias, Peter Weingart and myself. After a number of informal discussions over the succeeding few years with colleagues on both sides of the Atlantic Ocean, it was agreed to propose an annual series of thematic volumes that would publish multidisciplinary research on the sciences as 'a plurality of socially constructed ways of comprehending natural and social phenomena,' to quote the initial editorial statement. Two publishers, Macmillan and Reidel, showed an interest in this project, but only Reidel were willing to commit themselves to an annual publication. Accordingly, at the first meeting of the editorial group, held in December 1975 at London Business School, it was agreed to pursue negotiations with them. Everett Mendelsohn was especially helpful in these discussions.

It may be wondered why we decided to develop our research programme through an annual series of books rather than by establishing a journal. As well as being mindful of the administrative burdens involved in editing a journal appearing three or four times a year, the relatively small size of the science studies research community in Europe and North America at that time needs to be borne in mind, as does the prior foundation of *Science Studies* and *Research Policy* at the beginning of the 1970s. Although higher education – and the public funding of scientific research more generally – was expanding in many industrialised countries in the 1960s and 1970s, the number of academics working on science as their central intellectual concern was still quite limited, and the whole science policy research industry only beginning to become established. Indeed, sociology itself had become accepted as a respectable discipline relatively recently in many European universities, and so both the demand for ideas and results and the supply of high quality work in the sociology of science were thought to be too restricted to justify another English language journal. We also hoped that a series of thematically focused volumes would distinguish our venture from conventional journals and attract contributions from a wide variety of disciplinary backgrounds, as well as making a distinctive impact on the emerging field of science and technology studies.

The first three volumes (*The Social Production of Scientific Knowledge, The Dynamics of Science and Technology, Countermovements and the Sciences)* were agreed at the first full meeting of the Editorial Board held at the Zentrum fur interdiziplinares Forschung at Bielefeld University in April 1976. As might be expected, these were largely developed by members of the Editorial Board, which then consisted of Gernot Bohme, Norbert Elias, Yehuda Elkana, Roger Krohn, Wolf Krohn, Gerard Lemaine, Wolf Lepenies, Herminio Martins, Everett Mendelsohn, Helga Nowotony, Hilary Rose, Peter Weingart and myself. It is worth noting here that 10 of the 13 members were based at West European institutions, and that most were not affiliated with units specifically focused on the social study of the sciences. It was, then, a predominantly European venture, coordinated by social scientists from a variety of disciplinary backgrounds located in different institutional settings. Since there were very few established centres of STS anywhere in the world at that time, this last characteristic of the *Yearbook* is not too surprising, and highlights the low level of institutionalisation of the field in the mid-1970s.

Contributions to these early books were largely organised by post and telephone, which resulted in some hectic efforts to meet deadlines, and occasional delays in going to press. As a result, we decided to build volumes from papers presented to thematic conferences. This increased the costs associated with each volume considerably, and meant that proposers had to be able to raise substantial funds. However, it also gave editors considerable influence over dilatory authors and enabled them to elect the most significant and thematically focused chapters for their volume. Nearly all subsequent books were preceded by such conferences, with the Editorial Board taking a leading role in selecting and improving them.

In the early 1980s the publishers, Reidel, and the Editorial Board agreed to extend the *Yearbook* by adding a Monograph series. This would enable us to publish studies that fitted into the broad programme of the *Yearbook* but could not be incorporated into the annual, thematic format. The initial volume, Marc de Mey's *The Cognitive Paradigm* (1982), sold quite well, and the subsequent five books also found an audience despite their high prices. However, by the mid-1980s it became increasingly clear that the publishers' distribution system and, arguably, the general business model, was not encouraging sales to individuals, either of the Monographs or the *Yearbooks*. Additionally, Reidel underwent a reorganisation that resulted in its being taken over by Kluwer at the end of the 1980s. The paperback editions of the *Yearbook* ceased in 1989 and no Monographs were published after Olga Amsterdamska's *Schools of Thought* in 1987.

These events, together with broader changes in the structure of the sciences and their institutional environments, led to a reconsideration of the *Yearbook*'s role and functioning at a Board meeting in 1990. Originally scheduled for November 1989, this discussion had had to be postponed due to the coincidental opening of the Berlin Wall at the same time as our meeting, and the development of links with East German colleagues. These were transformed overnight, as Terry Shinn experienced at Checkpoint Charlie that momentous week. On the Wednesday, he had had to walk across no-man's land without his guide dog and at the mercy of machine gun armed guards. On Friday, he and I walked the same route with his dog to be greeted by relaxed, if not exactly smiling, guards, and celebrated the transformation by drinking Romanian champagne with colleagues from the Academy of Sciences in their canteen. Those discussions led to later collaboration on the 1995 *Yearbook* and other projects.

One of the major conclusions to emerge from our 1990 meeting was to focus the *Yearbook* more on the changing relationships between 'science' and 'society' in the late twentieth century, and in particular the relative deinstitutionalisation of science as a distinct and separate social system producing certified knowledge in particular ways. The expansion of the public and private research systems of many countries since W.W.II, coupled with their growing internal differentiation and integration of activities across what had previously been quite strongly segmented boundaries, meant that science could no longer be satisfactorily studied as a separate institutional entity in which cognitive and organisational boundaries were coterminous, but required a more differentiated approach that systematically distinguished between levels of analysis and organisation. This was further developed in subsequent Board meetings, which formulated a number of themes for future volumes. After 20 years of the *Yearbook*, we decided to restructure the Editorial Board and invited a number of younger scholars to take charge in the 21st century.

Looking back on the 22 volumes that have appeared by mid-2001, a number of points seem worth making. First, the early volumes focused more on the physical sciences than the biological or other sciences. Indeed, the first book to discuss the social sciences was not published until 1991, shortly after Bjorn Wittrock joined the Editorial Board. I suspect that this reflected both the dominant position of physics in the hierarchy of scientific disciplines in the 1960s, and the concentration of many of the 'new' studies in the sociology of scientific knowledge on the physical sciences because they were often seen as being the most prestigious and recalcitrant to sociological analysis.

Second, the broad range of topics addressed, and of intellectual approaches adopted, are quite striking. Although the *Yearbook* was intended to promote a new understanding of the sciences, and so challenge the prevailing style of work in the sociology of science as institutionalised in many universities, it has not become narrowly focused on a single research programme, or dominated by the same school of researchers. Perhaps because it was predominantly a European enterprise, with few members of the Board being located in departments of Sociology in highly departmentalised universities, or in separate STS units, the *Yearbook* was not used to establish a particular school of thought or become the outlet for a self conscious and highly organised group of researchers seeking to differentiate themselves as a sub discipline or speciality. The volumes do not, on the whole, build systematically on each other, but rather deal with a variety of themes from a number of different intellectual standpoints. Given the considerable variations between the sciences, between disciplinary ways of studying them, and between national institutional contexts, this is not, perhaps, surprising, although it might have been different if it had been primarily a North American affair.

The third notable characteristic of these volumes is their consistently international flavour. Despite the growing domination of the US science system in many areas of post W.W.II research, very few of the *Yearbooks* have been dominated by US contributors, and nearly all have papers from authors in many different countries with distinctive intellectual traditions. This undoubtedly reflects the nationally diverse nature of the Editorial Board – which has been maintained as a deliberate policy in recruiting new members – and the continued significant contributions made to our understanding of the themes covered in *Yearbook* volumes by scholars in Europe and elsewhere beyond the USA.

Another feature of this series that might surprise twenty first century observers is the lack of books on the economic role of the sciences, either historically or contemporaneously. Furthermore, only one has dealt with the relationships between states and scientific and technological development, *Science, Technology and the Military* published in 1988. While no doubt reflecting the balance of expertise and interests in the Board, these gaps represent significant lacunae in the *Yearbook* programme that should be remedied soon. This is especially so now, given the growing emphasis on technological innovation in many states' macroeconomic policies, and the close connections between some areas of public scientific research and the development of new 'science-based' industries.

Allied to this is the relative lack of attention paid to private science –whether undertaken in state or corporate organisations – in these volumes. The bulk of the contributions focus on published research, mostly the results of work undertaken in academic institutions, without considering its changing cognitive, economic and social connections to research that is not conducted for publication and gaining intellectual esteem. Relatedly, the *Yearbook* has not analysed the role of research training and the expansion of research skills in the late twentieth century in many countries, which has resulted in a general growth in knowledge production capabilities, as well as in the number of people able to make sense of current research for a variety of purposes.

THE CHANGING CONTEXT OF THE *YEARBOOK*

To a considerable extent, these omissions reflect the original purpose of the *Yearbook*, and the intellectual and social context in which it was established. In considering its history over the past 25 years and future development, it is perhaps worth recalling that context and how it has changed. To begin with, many discussions of science in the early 1970s were dominated by Kuhn's *The Structure of Scientific Revolutions*, which had been republished in 1970, and the responses of historians and philosophers to its central claims. In the English speaking countries, these focused on the 'internal-external' debates in the historiography of scientific development, and the epistemological rationality of scientific change as understood by Kuhn and the Popperians, perhaps best represented by the essays in Lakatos and Musgrave's *Criticism and the Growth of Knowledge* (1970).

In Germany, the idea of a general pattern of scientific development, and its relationship to changing science-society connections, was taken up by scholars at the Max Planck Institute for the Study of the Conditions of Life in the Scientific-technical World in Starnberg, who published a controversial paper entitled *Die Finalisierung der Wissenschaft* (Bohme, van den Daele and Krohn 1973). This paper, published in English in 1976, and subsequent contributions generated considerable debate in the scholarly literature and conservative media, with its opponents accusing the authors of seeking to plan scientific development in a Stalinist mode. This controversy probably contributed to the closure of the Starnberg institute in 1981 (Schafer 1983).

The significance of Kuhn's book lay more perhaps in its catalytic role than in the details of its analysis of scientific change. By proposing a simple model of intellectual development and linking it explicitly to social structures with overt use of political metaphors and parallels, he provided a framework and justification for the growing number of social scientists becoming dissatisfied with conventional accounts of the sciences and seeking to take their humanly constructed nature more seriously. While generating a number of superficial applications to the social sciences, this scheme had more substantial effects on the history and sociology, and perhaps philosophy, of science by proposing a political approach to intellectual change.

Together with concurrent attacks on the empiricist research programme in the philosophy of science, this seemed to legitimise – and even provided an exemplary model for – the social scientific analysis of scientific development as a socially contingent phenomenon. The difficulties of demonstrating the epistemological rationality of scientists' decisions, exemplified by the Kuhn-Lakatos-Feyerabend debates, and the apparently political nature of key episodes in the history of Physics, meant that research goals, authority relations and intellectual disputes in the physical and biological sciences became amenable to sociological explanation in a comparable manner to political and social phenomena.

The expansion of higher education in many countries, and the associated growth of university posts and intellectual innovation that it encouraged, provided a favourable environment for such challenges to conventional representations of the sciences. While the *Yearbook* was part of this desacralisation of the sciences, it never saw that goal as its central purpose and did not formulate a 'strong programme' against current orthodoxies. Rather, it has taken the socially constructed nature of scientific knowledge for granted and tried to build a sociological – in the broadest sense – understanding of scientific development, including the possibility of orientating the public sciences for social ends.

Since then, of course, the economic and political environment has become rather less accommodating to intellectual autonomy in the social sciences, and science itself is seen as a national resource for generating economic growth and prosperity more directly than seemed possible in the 1970s. This belief has been greatly reinforced by the growth of the biotechnology, electronics and information technology industries over the last 30 or so years, especially in the USA, which depended on the considerable state investment in public science since World War II (Langlois and Mowery 1996; Pavitt 2001).

The increased value being put upon formal knowledge as a key resource for innovation and economic advantage has been matched by the expansion of potential knowledge producers in many countries. Both first degree and postgraduate student numbers have grown dramatically since the 1950s, albeit to varying degrees in different subjects and societies. This has reduced some of the costs of research and increased the available labour force for investigating technological and similar problems in a systematic manner. As a result, systematic research into a wide range of problems has expanded greatly, especially in organisations outside the traditional universities.

These contextual changes have altered the organisation of universities, and the public sciences in general, in many countries. Although it is doubtful if the disciplinary structure of the sciences was ever quite as rigid and strongly bounded as some have claimed, the last three decades of the twentieth century have undoubtedly witnessed a shift in the structure and roles of universities compared to the 1950s (Clark 1995). Additionally, there has been a considerable expansion of applications-oriented published research conducted in varied organisational settings.

The expansion of state funding for long term political and social goals, such as health care improvement, in both state and university laboratories, together with the growth of non-profit and business funding of academic research, have generated both a greater pluralism of intellectual goals and more variety in the standards used to judge research excellence in the public sciences. The concomitant reduction in block grant funding through university hierarchies and the rise of project based support in many countries has reduced the ability of university institute heads to develop coherent organisational research policies around disciplinary goals and increased the autonomy of individual researchers – especially in the Anglo-Saxon university systems. Together with the increasing variety of legitimate audiences for research outcomes in the more mission-oriented fields, this has resulted in more varied hierarchies of intellectual prestige and reduced the degree of integration and cohesion of research goals.

These changes can be expected to have three major consequences for the organisation of the public sciences. First, they reduce the extent to which researchers have to depend on the results of disciplinary colleagues in other organisations to make their own competent contributions to collective intellectual goals. Rather, they become more able to use ideas and information from a wider variety of sources. Second, their dependence on the verdicts of particular disciplinary elites to gain relations that lead to rewards declines as alternative sources of reputations and different standards of intellectual merit become available. Third, this increasing variety of audiences and goals means that the reputational payoff from implementing specific research strategies is increasingly difficult to predict, and so the degree of strategic task uncertainty grows (Whitley 2000).

In the sciences most affected, such as the biomedical ones in the USA, these outcomes imply that theoretical diversity and the variety of intellectual objectives will increase as disciplinary elites lose the power to make sure that research projects contribute to the fields' theoretical goals and follow currently dominant approaches. Such growing individual and group autónomy from intellectual elites will encourage intellectual innovation as researchers develop alternative research programmes and respond to different groups' demands. The strength and stability of prestige hierarchies of research topics, problem areas and research teams will also decline in these circumstances. Depending on the level of intellectual competition for reputations, this can result in rapidly shifting research priorities, intellectual boundaries and dominant theoretical frameworks as research teams conflict over the relative merits and importance of their approaches

The expansion, differentiation and variety of organisational arrangements governing the production of formal knowledge in many industrialised countries since the 1950s have made it difficult to characterise 'science' as a coherent social institution. Rather, different kinds of knowledge are produced for various purposes in multiply structured organisations that are funded in different ways across market economies. To understand these changes and their historical antecedents, the *Yearbook* and STS in general are developing longer term research programmes that compare the dynamics of scientific and technological fields in differently organised national and international contexts. How cognitive and organisational boundaries and trajectories are constructed, reproduced and changed in contrasting environments need to be deliberately studied rather than assumed, as do the changing understandings of public and private scientific and technological research.

Just as deregulation of some economic activities at the national level is being succeeded by the construction of new institutional frameworks at varied levels of collective organisation, so too the deinstitutionalisation of the postwar public science system in some countries is leading to a new set of organising principles and practices that need to be systematically analysed. I am sure that the *Yearbook* will make significant contributions to this understanding over the next 25 years.

Manchester Business School, Manchester, UK

REFERENCES

Bohme, Gernot, Wolfgang van den Daele and Wolfgang Krohn (1973), Die Finalisierung der Wissenschaft, *Zeitschrift fur Soziologie* 2: 128–44.

Clark, Burton R. (1995), *Places of Inquiry: Research and Advanced Education in Modern Universities*, Berkeley, CA: University of California Press.

Kuhn, Thomas S. (1970), *The Structure of Scientific Revolutions*, Chicago: University of Chicago Press, second edition.

Lakatos, Imre and Alan Musgrave (eds.) (1970), *Criticism and the Growth of Knowledge*, Cambridge: Cambridge University Press.

Langlois, Richard and David Mowery (1996), The Federal Government role in the development of the U.S. software industry, in D. Mowery (ed.), *The International Computer Software Industry: A Comparative Study of Industry Evolution and Structure*, Oxford: Oxford University Press, pp. 53–85.

Mey de, Marc (1982), *The Cognitive Paradigm*, Dordrecht, The Netherlands: Reidel.

Pavitt, Keith (2001), 'Public policies to support basic research: What can the rest of the world learn from US theory and practice? (And what they should not learn)', *Industrial and Corporate Change* (forthcoming).

Schafer, Wolf (1983), 'The finalization debate', in G. Bohme, W. van den Daele, R. Hohlfeld, W. Krohn and W. Schafer (eds.), *Finalization in Science: The Social Orientation of Scientific Progress,* Dordrecht, The Netherlands: Reidel, pp. 275–99.

Whitley, Richard (ed.) (1974), *Social Processes of Scientific Development*, London: Routledge.

Whitley, Richard (2000), *The Intellectual and Social Organization of the Sciences*, Osford: Oxford University Press, second edition.

PART I

STS AND SOCIETY

CHAPTER 2

ULRIKE FELT

SCIENCES, SCIENCE STUDIES AND THEIR PUBLICS: SPECULATING ON FUTURE RELATIONS[1]

INTRODUCTORY REMARKS

Moments of commemoration such as anniversaries or the beginning of a new century/millennium are often used as occasions to do both, look back and analyse past expectations in the light of present experiences as well as speculate about future challenges to be taken up. The heterogeneous ensemble of investigations which identify themselves or are identified as part of the research field 'science and technology studies (STS)',[2] can surely look back upon a lively development throughout the last 30 years, having opened up areas of debate and introduced profound reflection on the mutual shaping processes between (techno)science and society. To varying degrees – largely depending on national contingencies and traditions, but also on the investment of individuals and networks – STS has managed to get institutionally established as a research and teaching domain and gained some visibility. Curricula allow for reproduction; regular international conferences[3] highlight the issues at stake; publication networks bear witness to the large variety of academic production;[4] academic societies try to give visibility and coherence to the rather spread out community; and finally the expertise available in the field is partly integrated on the policy level. So, everything's at best?

Yet when taking a closer look, it is quite easy to perceive that STS has also run into a number of difficulties and will in the coming years have to reposition itself in view of fundamentally changing relations between technoscience and society. While some of the tensions are due to the above mentioned changes both in academia itself as well as in the environment it is embedded in, it should nevertheless not be overlooked that part of them are due to the field's own history – or better to say its non-history. Being transdisciplinary from its very conception, this research area proved complex to get institutionally settled in ways which allow for a rather open development on the one hand, while conforming sufficiently with classical academic 'discipline' on the other hand. In a certain sense it had to confront the institutional limitations of transdisciplinary research in general as well as the hesitation towards a field that would study the 'social side of science'. As a consequence, much of STS research has never left the disciplinary structures (sociology, philosophy, history, psychology, communication science, etc.) which also meant that interaction between the disciplines working in the STS domain was left

B. Joerges and H. Nowotny (eds.), Social Studies of Science and Technology: Looking Back Ahead, 11–31.

to the individual researcher. The fragility of transdisciplinary structures becomes more concrete when looking at the fact that in some cases existing interdisciplinary structures, which have been landmarks on the STS territory for a while, have been dissolved (e.g., Amsterdam's Science Dynamics Unit). Not being established in clear structures carries the danger of allowing the flourishing of autonomous sub-areas of research which can either stage their dissent however without needing to resolve it or live ignoring completely similar research questions addressed in other settings. Looking back it seems that there have been, neither internally nor externally, very strong and lasting forces that stimulated the continuous development of an increased internal coherence in this research field. The STS community not forming a 'disciplinary structure' but an open heterogeneous forum could thus give way to both a flourishing variety of approaches and thus produce rich analyses of science/society interactions but also employ conflict avoidance strategies, leading to a lack of internal consensus building and thus coherence. This becomes partly apparent in debates around quality control in this area of research. Out of this as well as along with it grew the difficulty to discern something like a clearly delimited, defined scientific field. These issues became visible in the 'science wars' debate started by Alan Sokal and others, who used the argument of lack of internal coherence to fundamentally question the credibility of science studies at large and its legitimacy to speak about science.

Summing up, one could thus put forward the hypotheses that STS all along its development has never really left its rather ambivalent position towards the question of how much establishment is needed and aimed at and in what forms as well as how much internal structuring would be necessary to become externally recognisable as a research field.

The second kind of tensions STS is confronting has to do with the 'political paradigm' that drives the field. While there seems large agreement on a more general agenda – the mutual relationship in all its different facets between science and society is at stake –, it appears to be far more difficult to find consensus on the ways and the degree up to which this knowledge should be tied into the process of actually shaping the science/society relationship (policy making). Should science studies have a more politically oriented agenda, thus re*acting* to the changes in the relations between science and society, trying to impact on the developments taking place and showing direct involvement on the policy level? Or should it much more follow the trajectory of a purely academic field like many others, with the only difference that its object of study is science itself? Thus the motives that drove/drive members of the STS community to do their research and to pick up particular research questions cover a broad range from political/democratic engagement to a more science intrinsic interest in studying and theorising about the object/system 'science'.

And, finally, seen from this perspective, STS has increasingly also to face the fundamental question of its own publics. Who are those they want to address with their research? Are the projects carried out and the knowledge produced directed towards the own scientific community or are we aiming at engaging in debate with other scientific communities, do we look for interaction with science policy makers and position our research as important and valuable for their work, or are larger publics our potential addressees? How should the contexts of communication within the community and with the outsides be structured accordingly? And how will these acts of communicating STS research to wider circles impinge on the epistemological development of the field? To be even more explicit: How will STS position itself towards the fact that the very claim of producing relevant knowledge for advisory structures (be it in the domain of science policy, risk, etc.) will change also the kinds of questions that will be asked and the ways in which knowledge is produced and presented? What will happen to STS once being integrated into teaching structures of science on a regular and general basis? Will it adapt and not be the 'reflexive other' anymore?[5] All that has to be seen in the context of a society where fundamental rearrangements in the economic, social, cultural and power-structures are taking place, involving or being based on scientific and technological developments at large (Nowotny et al. 2001; Weingart 2001; Heidenreich 2002).

The above identified problem zones will be at the centre of the following essay. In order to trace some of the difficulties and shifts in the field of STS in a more concrete way, my focus of analysis will be on a particular research area, which could be subsumed under the heading 'public understanding of science'. This term will however be used in what follows in a rather large sense, making reference to the broad variety of studies and research traditions, which look into the complex relationships between publics and sciences. At the heart of this research area are questions on communication between sciences and the publics, on different knowledge forms, their conflicts and cohabitation in public arenas, on forms and functions of expertise, on boundary drawing mechanisms and power relationships in modern knowledge societies, and on issues of public participation in major technoscientific decisions.

Although this will evidently only offer a partial perspective on the STS problematic at the beginning of this new millennium, this choice seems valid for at least two reasons. First this research territory reflects very clearly what I think is the core set of past performances or lacks, present difficulties and future possibilities but also limitations of STS approaches. It shows the capacities of STS to drive or be driven by societal change in the areas where science, politics and publics interact. Second, the way the relationships of science with different publics have changed over the years and have been/are conceptualised in this research opens up a rich terrain to reflect on the way STS understands and structures its own relationship with its publics.

Thus the main interest in this paper will not be to study *public understanding of science* as a research territory *per se*, but to reflect STS, its past performances and future potential in the light and the framework of this field. Thus, it will be much more on understanding the logic behind the formulation of the research questions, on the way political contexts and ideological presuppositions shaped and shape the research setting, on the different national patterns that appear in this research, on how STS researchers got increasingly caught up in between their different roles of analysts of the relations on the one hand and players in shaping them on the other as well as on the capacity of being self-reflexive. When using the notion 'understanding science' it is not meant in the limited sense of being able to grasp parts of the knowledge produced. Far more it addresses an understanding of the social structures of science, of scientists' behaviour both towards their objects of investigation as well as towards the environment in which their knowledge will be embedded, of the institutional settings in which science is taking place, of the ways in which the question of responsibility is taken up and many more.

In a first step the focus will be on main shifts in the PUS field and the different elements that play an important role will be analysed. These are for example the specific relations between science, politics and publics at a given point in space/time, the ways the boundaries of science are defined and drawn, the concepts that are introduced or the general democratic agenda which is supposed to be at stake. We will see the impact of the growing contradictions between the techno-science system which is increasingly integrated into a larger socio-economic context and develops according to this rhythm, and the situations in which science is either communicated to publics or applied in a way that seems to touch the public perception.

Concretely we will distinguish four phases in the PUS debate. The account that will develop is by no means directed towards completeness, but rather aims at pointing out some key features as the debate moves on. In that sense the literature quoted is only to be seen as exemplary. While they are presented in a somehow chronological order, this does not mean that one phase is clearly separated from the next. Much more one would have to understand the new approaches as additional, opening new territories of reflection, while not closing the previously existing ones. This is particularly true – as I will argue – for the linear model, which seems to have survived all the major changes through gradual adaptation without changing the most fundamental assumptions.

Given the contemporary tensions between increasingly globalised developments and their reconfiguration in/translation into local settings, it seems interesting in a second step to reflect the impact this has both for PUS issues but also for STS at large. Thus the main focus will not be on the production of science but on those procedures that give sense to this knowledge, as well as on the mechanisms that allow for (re)interpretations and the taming of these knowledges. Thus, we will need to reconsider the omnipresent notion of the public in its locality and variations, in its cultural and geographic complexities.

DISTANCE LENDS ENCHANTMENT?! – COMMUNICATING SCIENCE TO 'THE PUBLIC'

Little STS activity and the idea of linear communication models between science and the larger public could be identified as the two features, which characterise best the first phase of studies dealing with science-public relations which lasted until the 1970s. Most of the research was carried out under the heading popularisation of science and it seemed as if STS researchers did not think that there was much to be studied in this domain. The basic explanatory model was taken over from communication studies and described scientists in their role as the producers of genuine scientific knowledge, which would then be simplified in order to become understandable, or better to say, acceptable for a wider public. The public was restricted to the role of more or less passive consumers and perceived as a rather undifferentiated ensemble of individuals. In that sense the model contains a number of hierarchies. Scientific knowledge was understood as being separable from popular knowledge and was staged as superior mainly due to its intrinsic and unique form of rationality. While scientific knowledge stood for complexity, public's knowledge could be ignored because of its alleged simplicity. Information was seen as flowing only in one direction namely from the producers to the consumers/receivers. And as a consequence scientists were the only ones, who could claim legitimately the status of experts.

The unquestionable authority to science was thus assured by the model. Communication between science and larger publics was reduced to a process of translation, based on the strong belief that the public's understanding of science was a mere question of improving the quality and density of the communication and of stimulating the public's readiness to get involved with science. As far as the public was concerned one assumed that people were basically ignorant while at the same time being keen to become knowledgeable about science. If people would reject science and technology then this was interpreted as being linked to a lack of information, which caused distance to, fear of and alienation from science. This information about science was also supposed to convey the importance and value of science in society.

These models started to be heavily criticised from the 1970s onwards, but until then most of the research done in the STS field did not profoundly question the basic assumptions behind this setting, but mainly fit their research projects within this frame. Thus we find more pedagogically oriented considerations on how to construct information on science addressing a wider public or on problems of language and translation from scientific to everyday language[6], but little is to be found about the consequences and the limitations of this communication model, on the role of situational settings in which this communication takes place and on the up-take of this knowledge by the public. Further one did not take into consideration the more implicit, symbolic forms of communicating science to larger publics through the use of images or metaphorical language. Above all, it was not reflected that this type of mono-directional communication was meant to do both: to convey the impression to give people access to science while at the same time distance was reconstructed. Thus one could say that distance could only lend enchantment if it passed through the step of creating 'imagined closeness' through science popularisation (Felt 1997, 2000).

What do scientists think that the public should know? – was another way this issue was addressed. The research trying to formulate answers were surveys on public knowledge of and attitudes towards science. They saw their early tests in the late 1950s in the United States (Wynne 1995). Also Lewenstein (1995, 352) mentions in his overview on science and the media the existence of early research in this area, addressing specific groups (high school students and politicians) and underlines in particular investigations carried out in the aftermath of the Sputnik shock.

These survey instruments were nothing like 'inscription devices'[7] which would allow the production of 'visible' and thus politically usable 'traces' of science in the public sphere. The inscriptions produced through these questionnaires were replacing the public, which could then 'disappear' in the debate and be substituted by a set of reconstructions. This methodology was thought to allow to 'see' and follow over time science's position in American society, while it did in fact nothing else than reflect the representation of science held by those commissioning and/or developing this research. It was a way to measure the degree to which *the* scientific rationality, the ideas about the institutional character of science as well as a set of standardised 'knowledge-packages' had been enforced upon the public as 'the correct answers'. Already the very fact of entering the question-answer game reflects a particular vision of science and the place it should hold in society. But at that moment in time the STS community did not seem to show much interest into what was going on in this terrain.

MISE EN SCÈNE OF SCIENCE: THE PERFORMATIVE CHARACTER OF SCIENCE COMMUNICATION

The questioning of the classical communication models began in the late 1970s. The start of this shift can be seen as rooted in a variety of changes, which characterise the period. Social movements like the environmental movements or women's movements offered contexts in which doubts about science and its impact on society could find its expression. Alternative knowledge forms started to claim their place in societal decision making and as a consequence technoscientific expertise felt gradually threatened in its authority.

In particular the work of those STS researchers studying scientific controversies and the work in the laboratories showed in a rather detailed way how highly constructed 'scientific facts' were. The process of communicating within the laboratory, the scientific community at large and partly with the societal environment should thus be seen as a place where the substance as well as the meaning of scientific knowledge was negotiated. Therefore also popularising science could not be seen as a mere simplification of knowledge anymore, but as a highly complex act of constructing a public vision of science. Far more one would need to focus on the performative nature of the discourses. In this sense one could paraphrase Jacobi and Schiele: "the very fact that this discourse exists and the framework it offers practitioners is more important than the question whether the information that is processed is right or wrong" (Jacobi and Schiele 1988, 14).

In parallel – as technoscientific change seemed to be omnipresent – there co-existed the idea of a growing necessity of educating the public with regard to science. Only if one could convince a larger segment of the public to accept the technoscientific rhetoric of progress as well as new artefacts and procedures, would scientific and technological development be able to continue its trajectory. In that sense we have to understand the survey research on public attitudes which appeared as a regular feature of US policy with regard to science. (Lewenstein 1995)

In parallel one also has to see the growth of risk perception research and the study of public controversies on science and technology.[8] In this field investigations were triggered by a clear change in the science policy arena. Increasingly the rather positive vision of science and technology was tempered by growing awareness of negative impacts. Questions of responsibility and power started to be posed. They related to the social distribution of risk, the role of citizens in decision making about science and technology as well as the access to expertise became rather central. In these studies the instrumental character of staging science became clearly visible and was addressed from different angles.

Yet these changes in context did not translate into the creation of abroad and coherent research programme on PUS during this period – the different strands managed to remain both geographically and intellectually quite separate. The classical relationship between science and the public as well as between different forms of knowledge was only questioned in a rather hesitating and indirect manner. One could say that STS research started to manifest interest in this domain, but remained focused on the science side – thus being interested in the ways in which information was constructed and distributed as well as in the 'products' through which communication took place.

A good example of this trend is the *Sociology of the Sciences Yearbook* published in 1985, under the title *Expository Science: Forms and Functions of Popularisation* (Shinn and Whitley 1985). The choice of the essays in this book, the very title and part of the introductory remarks, allow us to make a number of observations, which stand for the focus of the issues debated during that period.

The problems investigated showed a clear bias for those topics that were linked to and important for science. The way science popularisation was produced stood at the centre and the expository effects on science were investigated, while what happened once the information/the messages conveyed arrived at the different publics did not seem to be of so much interest. What was new however, was – in a similar way Ludwik Fleck (1935) already developed it – that "exposition was defined as a continuum of methods and practices utilized both within research and far beyond, for purposes of conveying science based information, whether as pure cognition, pedagogy, or in terms of social and economic problems" (Shinn and Whitley 1985, viii). There was an implicit continuity between modes of communication within science and external communication. In that sense it became clearer how multiple the possibilities and motivations were for scientists when moving to this terrain. Popularisation of science was also seen as a way to exert influence on institutional settings and society at large by imposing certain visions/images of the world around us and it had developed into an extended terrain to fight scientific controversies. Thus one could see a clear shift away from the idea that scientific knowledge was communicated here in a simplified way, towards studying the performative character of such undertakings. Through popularisation of science one produced so-called 'icons of truth' (Whitley 1985), which would then also transport the non-explicit and non-deliberative dimensions of science. Michel Cloître and Terry Shinn brought this aspect nicely to the point:

> In the case of popularization of science, the language, the reasoning and the images do not manage to elucidate the phenomenon, but quite to the contrary there is a tendency to create a conceptual incomprehension. (...) Popularization constitutes thus not an efficient instrument for the transmission of a better knowledge about the physical world. Its force and its pertinence lays in the links which it establishes between a scientific subject and the social sphere (Cloître and Shinn 1986).

Thus one could say that science studies largely regarded the communication between science and publics still as something dominantly controlled and structured by science. The public remained mainly in their classical roles of knowledge acquirer and the scientists as producers, although the former was already perceived as much more diverse, structured and guided by different interests and as having a potential impact on the way science would develop. However, it was more the projection work done in the course of popularisation which was supposed to produce internal, retroactive effects on knowledge production. But although the idea that popularisation "cannot be separated from knowledge generation and development, but needs to be considered as part of the overall process of intellectual change" (Whitley 1985, 12) was formulated, this did not mean that in the research projects that were presented the public was investigated as an acting, knowledge producing entity which could meet science on different grounds than the ones characterised by a clear hierarchy.

LOOSING SUPPORT FOR SCIENCE: *MISE EN SENSE* VERSUS *MISE EN SCÈNE* OF SCIENCE AND TECHNOLOGY[9]

Virtually in parallel with the *Sociology of the Sciences Yearbook* there also appeared in 1985 the Royal Society report with the title *Public Understanding of Science* – a report which would launch a rather broad range of initiatives both in research and communication (Royal Society 1985). The public and how to make them 'understand' science moved to the centre of interest. In a British context, largely damaged by Thatcher's science policy measures, the support of the larger public seemed to have become an essential pre-requisite for a decent survival of the research system and one wanted to gain the public as allies.

This report is interesting for a number of reasons. It was a document clearly subscribing to a linear model idea and nearly a caricature of the hopes and expectations that were put in this effort. People were supposed to have a knowledge deficit, there was a gap to be filled that separated them from science and all that was needed was to find out where the deficiencies lay and then develop programmes of wider education. Thus we find the declaration that "some basic understanding of how they (technologies) function should make the world a more interesting and less threatening place" (**reference?**). Or as people would be allowed to participate in democratic decision making it was seen as crucial that they "recognize and understand the scientific aspects of public issues" (Royal Society 1985). Like that – one postulated – lay-people would themselves be able to make the difference between competing claims and would be able to recognise and choose *the scientific claim as the right one*. 'The uninformed public is very vulnerable to misleading ideas' and thus educating them was clearly seen as the way out. To sum up: Science should better communicate its ideas, 'facts' and methods in order to become more acceptable to a wider public and to make vanish any of these 'unreasonable' fears present in the public arena. The similarity to the nineteenth century enlightenment discourse, aiming at educating the bourgeoisie and later also the working class due to the new needs caused by technoscientific change, is striking in particular if we consider that we find ourselves in the late twentieth century, in a period which would be described by analysts as reflexive modernisation (e.g., Beck 1986). It worthwhile to draw attention to the fact that one presumed to impose this framing of the issue onto social science without any question or exchange. The report as well as the money attached to it for doing research, would definitively have influence on the way the issue would be handled in the years to come.

Two extremely different lines of research managed to develop from this starting point. The first would actually build on the survey research experiences and define the notion 'understanding' as operationalisable through defining and measuring a set of 'factual' knowledge about science – which scientists would decide upon beforehand. These tests of what was called 'scientific literacy' started to become more regular in the 1980s first in the US context and later on also in Europe (Miller 1983; Durant, Evans and Thomas 1989, 1992; National Science Board 1989). In that sense this strand can be seen as a clear reinforcement of the separation of science and its publics and thus to safeguard the authority status of the former. It was presumed that there was a clear border between scientific knowledge and other forms of knowledge, that the former would hold a privileged position due to its rationality and universality. The conclusion was that a minimum scientific literacy would be required in order to allow citizens to fully appreciate technoscientific progress in contemporary societies.

Guided by the idea that science functions in a universal manner and is communicated in highly standardised ways in networks that span our world, it was assumed that one should also search for 'best practice' models for science communication and to implement them widely. Starting from this perspective much effort and money was invested in regularly observing public opinion on science and technology issues. With wide ranging surveys questioning the attitudes people have towards, but also the knowledge they would hold on scientific issues, policy experts, scientists and some social scientists thought they had found a policy tool to monitor this domain.

Underlying these surveys – into the design of which also some STS researchers[10] were involved – was/is a set of rather normative ideas about science. Scientific knowledge is equated to a number of facts and this after years of research that argued how strongly contextualised scientific knowledge is, how interpretation processes and tacit knowledge was needed to produce it and how deeply it was dependent on consensus building processes within a social community. Questions in these surveys generally allow for *one correct answer*, even if the former would be open to diverse interpretations.

Furthermore this research reproduces a number of existing prejudices e.g., concerning the gender divide with regard to science, of the north/south differences and many more. No longer than a few months ago, I had the possibility to hear a French protagonist of this method at a workshop on food and technoscientific information explain that women generally have a tendency to be more fearful than men, which explains – according to him – why they refuse GM-food to a higher percentage than men.

Criticism of this research was formulated from within the STS community, mainly on the methodological level: Respondents were taken out of their social environment when questioned, knowledge and understanding were completely decontextualized and thus answers would loose their meaning, the questions were simply reproducing certain assumptions (e.g., on 'the' scientific method) and many other weaknesses were underlined. To use the terms of Brian Wynne: "Evidence of internal coherence among survey data is not itself evidence of wider validity – only of consistency. Too often the latter is mistaken for the former" (Wynne 1995, 370).

More than debating on the methodological level, I would like to point to the power of the discourse developed around the results of these surveys on the science policy level and in the media. What are the consequences to be drawn from statements that large fractions of the population do not know 'basic' scientific facts? How do we handle results showing that "men tend to give correct answers more than women do" (INRA & Report International 1993, 60)? Or what does it mean for democratic participation if we are told that an important percentage of the public believes in a link of acid rain and nuclear power plants? This is why it might seem astonishing that there is so little debate around that and so much readiness to just ignore what is happening there.[11]

The second line of research that was triggered by the PUS debate, was more qualitatively oriented studies which focused on the ways in which people would up-take science in concrete settings. What do they do with scientific information, how do they interpret it with regard to the existing knowledge-structures and what place do they give to scientists and scientific institutions in their decision making. The shift that took place was therefore twofold. First, it was not so much the *mise en scène* which was studied, but the *mise en sense*, i.e., the way people would attribute meaning to the scientific knowledge and information they managed to obtain, or were offered. Secondly, of interest were not abstract knowledge structures, but particular settings in which people would encounter science and be forced to take decisions, which would at least involve partly technoscientific knowledge. Thus our attention was drawn to the fact that the public uptake of science depended upon previous experiences, on the social setting in which people would meet science and what expectations they would have with regard to science. These negotiations concern the degree of importance that would be attributed to scientific knowledge and expertise as compared to other forms of knowledge and expertise as well as how scientific knowledge was socially mediated and embodied when it was experienced by people (Michael 1992; Wynne 1992; Irwin and Wynne 1996). Much of the research focused on the way people framed techno-scientific problems in society and thus attributed meanings. The conclusion reached was that people would live/experience science through social relations and that the core of the problem was thus less the lack of knowledge about science, but *trust* in the science system and in scientists.

From recent research carried out on publics visiting the Austrian Science Week presentations, we could witness how far reaching the contextualization efforts from the side of the public go, how important the personal motivational background is and finally how sophisticated in key respects are the models which people develop with regard to questions of science and society. Interesting to remark was that most of our interviewed visitors still tend to stick to the 'we need more information on science' discourse, staging themselves as 'ignorants' as well as saying that they feel themselves as not sufficiently informed. In that sense it is possible to say that the 'information discourse' held by those who still adhere to an enlightenment argumentation is quite strongly taken up in the typical public self-representations. The boundary drawing between science and 'the public' thus seems to have worked in a quite efficient way. In longer and more detailed discussions, however, the multiple layers of their quite differentiated relationship with science become visible (Felt 2001). Further it is remarkable how resistant scientific and policy institutions are to getting really beyond a rather superficial and often quite behaviourist construction of 'the public'.

Now if we look at the impact these two research strands have on the policy level, it is easy to perceive the powerful position of the quantitative indicator oriented research. The more qualitatively oriented research strand was criticised and partly ignored, as the evidence produced was deeply embedded in a particular setting and thus seemed less generalisable or would not offer easy ways to compare and monitor the developments in these areas. The case studies addressed socially rather coherent knowledge networks, such as patients touched by a particular illness, sheep-farmers or Saami population who would have particular ways of confronting 'outside' knowledge compared to those who could not draw upon this collective quite organised kind of experience or knowledge (Paine 1992; Wynne 1992; Irwin and Wynne 1996). Thus is was at a first glimpse seen as less attractive on the policy level or as steering instrument and much more attention was given to the 'hard facts' produced by survey research. A further key reason for this resistance to the qualitative research was that it quite explicitly understood that the 'PUS problem' was one of scientific understanding of the public as much as of public understanding of science. Thus it represented in a certain extent a challenge to scientific culture and institutions too. The hybrid phenomenon that emerged was that for a while the more reflexive positions found their place mainly in general statements (*Sonntagsreden*) or in the preface to papers and reports, while the 'hard core' of argumentation and practice would be built on the quantitative survey results.

Yet ironically, we should draw attention to the fact that Eurobarometer research was producing quantitative results, which challenged the deficit model. For example the assumed correlation between ignorance and non-acceptance of science (e.g., in the case of GMO risks) did not correspond with the results obtained. Thus the 'hard-core' of the PUS policy was not really built on the results of such survey research, but much more on the very method of this social science, which projected and performed a standardised public with standardised science-centred worlds of meaning. It thus reaffirmed and reinforced policy assumptions that existed prior to research.

For our purpose it is also interesting to remark, that the co-existence of these two rather contradictory research approaches within the field of STS seemed not to have caused any major or regular questioning. It was neither really debated where such different methods of accessing the problem would lead and how they could coexist without apparently too high frictions. I would argue that this was possible as the 'public' this field addressed, but also those who would finance research showed a high degree of heterogeneity. Indeed one could say that the development of the field was shaped by external political interests and pressures (be it from the side of scientists or science policy makers) more than by the concern of getting some degree of internal coherence. PUS thus grew into a wide and ill-defined area involving several different disciplinary perspectives, which did not really engage with each-other (Wynne 1995, 361).

In that sense I would follow the argumentation developed by Dierkes and von Grothe at a conference devoted to this issue in 1995, who underline that "this dominance of political intentions and the analytical fuzziness of some of the concepts used in this field of inquiry are directly related" (Dierkes and von Grothe 2000).

Thus it would be legitimate to conclude that the problems that became highly visible in this period did concern as much the relationship between science and publics as it did between science studies and its publics including the science policy sphere. The procedures defining the field therefore seem strongly directed to the outside but it was less important to get internal structures/interests clarified.

REPOSITIONING: DIALOG, PARTICIPATION AND WORKING THE BOUNDARIES

If we extend the argumentative narrative developed so far, we can pose a number of questions when observing the actually quite radical apparent shift in the way science-public relationships are conceived. Both technoscientific knowledge and all kinds of artefacts have entered contemporary societies on a large variety of levels. But it is not only the artefacts and the bits of knowledge, but much more also the imagined roles attributed to them and the practices created by or based on them which build the cultural tissue into which people are wrapped.

Public understanding of science has become a matter of growing concern both for governments as well as on a supranational level (e.g., in the European Union) and for research institutions and enterprises in the technoscientific area alike (see, for example, Collins and Pinch 1993; Ashworth 1997; Fuller 1997). For governments what is at stake is the loss of science as an advisory/legitimatory device which people would trust sufficiently to accept decisions based on expert opinion (Hilgartner 2000). Scientists feel threatened both in their freedom to carry out research that seems interesting and promising to them, and in the failure to reproduce their field, as the number of science students has gone down dramatically in recent years. And finally enterprises are worried about the 'unpredictable and irrational behaviour' of consumers, who might decide at any given point in time not to accept anymore that technoscientific progress would automatically lead to societal progress appreciated by them.

Recent cases such as BSE, where scientific advisors have come to play a more than doubtful role in the public eye and where the weakness of the relation between science and politics has become highly visible, but also the debates around GMOs and human genetics have just rendered more concrete this more diffuse uneasiness (Science and Society 2000). The analysis made in the report pointed at diminishing trust in science from the side of the public although people showed interest in science and this was perceived as having major consequences for public policies in a number of ways. It called for a change in science-policy culture, thus emphasising the need for science to understand the changes in society and shift accordingly their positions.

Obviously one could no longer argue rejection of technoscientific innovations by the public as being due to a lack of information, but a new dialogue between science and the public was needed as well as to allow for public participation in decision making for sensitive issues (in particular those linked to risk and uncertainty). This debate started in many industrialised countries and clearly showed an important shift on the language level. Trust, co-operation, dialogue or participation are but a few buzz words which dominate the discourse. Examples would be the British Science and Society report by the Select Committee appointed to consider Science and Technology, the British Department of Trade and Industry White Paper, the OECD report on Promoting public understanding of science and technology, the *Memorandum: Dialog Wissenschaft und Gesellschaft* by the Stifterverband für die Deutsche Wissenschaft and finally the Report of the EU working group *Democratising Expertise and Establishing Scientific Reference Systems*. The last case the problem is phased as follows:

> 'Experts' are consulted by policy makers, the media and the public at large to explain and advise on such diverse issues as climate change, employment policy, BSE (...), and genetically modified organisms. However recent cases have shown that *expertise, while being increasingly relied upon, is also increasingly contested* (White Paper on Governance 2001).

It seems as though there is agreement that it would not be possible to merely continue to follow the 'we need more science communication' logic. What we can see from this development is the fact that the role accorded to publics is about to shift, increasingly posing a challenge also to the development of science and its integration into society. The debates seem to be torn apart between some elements of enlightenment science and the increasing necessity to integrate the public in some kind into the decision making procedures. But more important there is a clear shift in the notion itself of the public confronting science. Thus it is not only necessary to shift the notion of understanding and participation, but there is an additional element to be taken into account namely the cultural mechanisms of forming publics that (potentially) exercise power in contemporary societal structures and we have to face the temporal character of these power-balance relations.

Let me reflect about these changes from four different angles:

First it is important to understand what is happening in the light of strong tensions between ever more globalised knowledge production and distribution on the one hand and the fact that sense-giving (*mise en sense*) takes place on a local level. We are confronted with a process of ever-increasing individualisation in western industrialised nations, which also brought along the fact that those individuals develop their own positions and become increasingly aware of the power they hold. At the same time they are more bound than ever before into global socio-economic structures which counteract these development (Giddens 1991; see also Felt 1999; Nowotny et al. 2001; Weingart 2001). This means what is needed is that social theory and public understanding of science co-operate in order to find ways of better understanding and modelling these interactions. We have no stable public or several clearly separated publics, but temporary and localised configurations of individuals which form for a while a 'public' with regard to a specific problem. Yet we have to consider at the same time that even these individualised publics can rely on a background of often quite intense senses of collective solidarity which are available for more organised mobilisation, even if they are often simply passively existent as implicit forms of legitimation.

The second point to make is linked to the fact that when studying public up-take of science we also have to consider that even the very notion of science may look quite differently according to national traditions. This has so far been largely ignored as the position taken was merely from the side of science and knowledge questions or it was posed in an explicitly anglo-saxon context. If we want to understand the ways people perceive science we surely have to take this into account. Thus it is important to remark that the very notion of *Wissenschaft* – to take the German example – has a profoundly different meaning than *science*. It does not only embrace the social sciences and humanities, but it addresses also different features that were seen as central during the formation process of the science system, a fact which was also reflected in the German PUS programme (Stifterverband 1999). It is called 'Public Understanding of Science and Humanities' and expressed clearly this much more extended notion of 'Wissenschaft', which is not reduced to natural sciences and its technological applications.

Thirdly, it is interesting to remark that although we are pretty aware of the complexity of the communicative settings and the major differences in the local publics we are addressing, there is nevertheless the strong idea that there could be something like a 'best practice' in dealing with science and PUS. When moving to the science policy terrain however, there is a clear demand for such modelling work and bench marking exercises, and there is a certain danger that there will also be an offer (also from the STS side) that meets the demand. The danger is that the very aim of trying to measure the success through the establishing of clear indicators that should be followed, will lead us always back to the idea of linearity. On the policy side it seems difficult to admit that we are confronted with a fast moving and unstable terrain, where we at best can develop some basic notions on how to move in it. For a context like Europe this would mean, that it will be important to understand the rather different local setting, with their histories, structures and experiences in order to able to engage into communication about science. And effective communication takes place only if there is somebody who is ready to speak *and* to listen on both sides involved.

Finally, it would be worth while to position this debate around public understanding of science in the framework of the discussions around the new production of knowledge (see in particular Nowotny et al. 2001). The authors Nowotny et al. start the preface of their book complaining that "despite the mounting evidence of a much closer, interactive relationship between society and science, current debate still seems to turn on the need (...) to maintain a line to demarcate them." They stress the "preumption that communication flows one way – from science to society – with scant attention paid to describing the transformative effects of any reverse communication." Having outlined the problems in what is actually taking place they develop among other concepts describing the new situation of science in society, the notion of socially robust knowledge. The stronger the contextualisation of a scientific field or research domain is, i.e. the more different actors are already involved in the process of production of knowledge, "the more socially robust is the knowledge it is likely to produce" (Nowotny et al. 2001, 167). Science would thus move into a new agora of interactive forces that "invited exchanges of all kinds, and creates a context in which wishes, desires, preferences and needs can be articulated as well as demands. It is also a space in which particular forms of contestation are allowed, and even encouraged (...)" (Nowotny et al. 2001, 209).

I would like to challenge this vision by questioning the concrete ways in which participation of diverse actors will be assured. 'Social robustness describes a process' as the authors continue to explain, "that in due course, may reach a certain stability.' However they also underline that broader participation in the production process alone would not guarantee acceptability – 'there is a fine distinction between the robustness (of the knowledge) and its acceptability (by individuals, groups or societies)" (ibid.).

Thus will participation be 'organised' in specific settings – an idea that appears regularly in policy documents –, which raises a number of questions. Who will be those having access to the agora and participate in the name of 'society'? How much weight must a social group have in a societal context in order to be given a voice? In what ways more individualised positions will be considered? Will participation become a highly institutionalised and rigid construction? Could the notion of agora and the way access is envisaged not also be interpreted as a new hierarchical ordering in the public voices that should be heared? What will the 'normality' of socially robust knowledge production look like? And finally what will be the next step when acceptability is not achieved in domains which seem crucial on a political or economic level?

In that sense would it not be interesting and a valuable reflexive exercise to investigate what Nowotny and co-authors describe as one further step in the boundary work of science? Science has to confront contestation and doubt, has to rethink its boundaries to politics, industry and publics alike. At the same time science as a system has many hundred years of 'experience' of extending its territory in order to get control over the new (institutional) settings and actors that on their part are also seeking to get control over science. Thus one could ask the question whether we really observe an opening up of the science system towards societal forces or if the system is not rearranging its boundaries through integrating new players, but keeping out society at large?

CONTINUITIES IN THE PUS DEBATE

Let's make a halt and see what these seemingly different phases of the STS debate around science-public relations have in common. At least two points can be perceived with remarkable clearness. First, I could stress with Wynne (1995) that although there is no coherent paradigm that has gained sovereignty over this area, there exists a political paradigm which leads – although through the changing societal contexts appearing in different form – to a very specific framing of the problem. Indeed one could argue that a basic drive behind the different measures is still the idea that there is a particular rationality that should gain over the others and that social control in a technoscientific society would only be assured through 'assimilation of the natural order as revealed by science' by the public. Thus there is an issue of control of the system over its environment by bringing in part of the environment – which involves ordering it – into the science system. So the ones who define the terrain on which to move and argue are still scientists, however, and this is a major change, chosen in different ways.

In fact when we look at the debates in the PUS field it might also seem astonishing how certain forms of analysing and describing constellations – in our case the linear model – exhibit an amazing robustness. Even in many of the more complex models constructed in current STS literature there is a tendency to keep alive the basic hierarchy, even though space is given to more complexity and variations in the interactions that take place between science and the divers publics. Indeed many models remain somehow linear even though they try to give up the dichotomy science/popularized science. To give two examples: Hilgartner's continuum of contexts in which science is communicated which stretches between scientific knowledge production (upstream!) on the one hand and media accounts (downstream!) on the other hand; and Bucchi's continuity model which stretches between communication in the intraspecialist space and the popular space, passing through the interspecialist and the pedagogical space (e.g., Hilgartner 1990, Bucchi 1998). Both do not seem to question profoundly the established order of knowledge production. Other modelling efforts such as the one presented by John Durant (see Hargreave and Ferguson 2001) sketch a whole network of interactions between the different actors in the field such as science, general public, media and policy makers. While the model seems rather interactive and far away from a deficit/linear model a closer look reveals some interesting details. While all actors receive information from science (mainly through PR work!) and exchange amongst each other, science is staged as only giving information without being a receiver. In fact when taking a closer look it is not so astonishing at all as much of the values that underpin this model are deeply embedded in the institutional setting of science in society as well as in the power-relations that have been established. Thus questioning such models has to be understood as shaking the very foundation of the way science and partly policy makers perceives itself.

One reason for this ideological resistance can also be seen in growing closeness of STS research to the policy arenas, which leave their traces also on the epistemological level. Therefore it will be important to ask the question of 'Why studying public understanding of science?' and to discuss the impact of funding and policy structures as well as new publics for the knowledge produced on the STS domain. So we will definitely have to address the changed interrelations between science, the political sphere and different publics.

The second point, which seems interesting is that in none of the critiques of standard communication models, of the PUS programme as perceived by Royal Society or of other attempts formulated later on, did question the implicitly underlying assumption that there is an irrevocable trend that the scenarios of the future will be based on ever more science and technology. In a way science studies – at least by default – thus co-construct this narrative by not questioning it and it seems as if the question would thus be reduced to 'how do we accommodate with this reality in a reasonable way?' Thus choice would then mean that we can refuse for a certain time certain developments, but that it would be in a certain way only a question of postponing.

LOOKING AHEAD: WHEN STS MEETS ITS PUBLICS

At the beginning of the 21st century technoscientific changes have shown their far-reaching impact on our society in a number of ways. Ranging from the redefinition of the relations between the economic system, nation state and citizens/consumers, to the threat that is perceived for human identity through developments of scientific knowledge e.g., in the genetic domain, to the reorganisation of overall power relations, all that is part of the changes people live in many ways. But the change in the position of technoscience in society has evidently also crucial consequences for the place science studies has to envisage for its future.

One striking aspect in the decades of the constitution of an STS territory is the fact that there has been no creation of something like a common history as part of the identity of the field. By history I do not mean the identification of great men who shaped as pioneers this new territory, but critical examination of the societal premises, which constituted the context for the first questionings, on the social and institutional mechanisms that left their traces, on the way exchange and debate got organised, how reproduction was assured and many more. Not engaging in something like a negotiation about perspectives on the past and visions for the future, is an indicator for a rather ambivalent stance towards how much engagement and investment would be good. In the introductory remarks to the *Handbook of Science and Technology Studies*, which appeared in 1995, we certainly find traces for a reluctance to engage in a broader self-reflexive project. On the one hand STS was described as too young as a research field to start writing a history – it would be like 'constructing a map of a half-seen world.' This not very convincing argument suggests that one would only need to wait until the picture of the past has become sufficiently cleared up as if it was only the distance to the object, which would lead to the possibility of seriously writing such an account. On the other hand it was stated that "this map making (...) would necessarily entail statecraft as well as politics" (Jasanoff et al. 1995, xi) and this seems much more to be at the core of what is at stake. Drawing such a map which would be widely accepted as reflecting a common perception of the territory occupied, thus have a meaning for the community, and becoming also visible to the outside would need places where dissent can be expressed and consensus created.

The fact of these negotiations not taking place thus means that a number of the underlying implicit assumptions in this field can remain unquestioned. One such problem zone became visible when looking at the very notion of Public Understanding of Science. I do not mean it in the way Harry Collins posed the question as early as 1987, when he said that this is an ill-defined notion. In my view he did not ask the most crucial question when he wanted to know whether it was about the content or the nature of science PUS was talking (Collins 1987, 690). It is about the implicit meaning of science itself, which shows the cultural differences. While the PUS acronym has become internationally taken up and integrated into policy discourse, it was very little reflected upon that it imposed a very anglo-saxon tradition of thinking about science, and that it is hardly possible to find an adequate translation for example into German language. The German notion of *Wissenschaft* – as I outlined already – would stand for a different set of values many of them also implicitly inscribed through the centuries of its development. Thus we transfer notions, values and a debate into a context where the basic common grounds are not the same. What does that mean for the self-reflexivity of STS? It surely is part of the explanation of why the social sciences and the humanities, with the exception of a limited number of excellent pieces of research – have never really entered the field of preoccupation. Thus it was also easy to avoid the self-perception of science studies as a social science field. And in the case of PUS the problem is still one level more complex: We are not talking about science, but about science and society and thus would have to reconsider the more local contingencies of the debates.

Having made these two more general observations, I see the starting point for debates about the future development of the field from a number of dilemmas it is actually in. The first dilemma is that STS researchers are analysts and observers while at the same time they are advisors and players in the field under observation. This problem becomes particularly virulent when science studies moved into the realm of advisory science. STS analyses and hypothesis building can have an impact on the relation of science and society as much as social sciences have done throughout the decades of their existence. Thus we have to be aware that it is impossible to lean back and behave like an unconcerned observer, but much more develop awareness what our models about science and society might have as impact on the way these interactions develop. Being aware of the danger of normative engagement, what we might want to offer is 'reflexivity for power'.[12]

The second dilemma is between remaining a relatively marginal, small and less strongly institutionalised territory in academia and trying to become an obligatory passage point at a time when science/society interactions seem to be at stake in virtually all domains. Indeed this is a question that accompanied the institutional establishment of STS in particular within teaching science curricula at universities. Would it be good to have STS courses prescribed to all our students (in particular in the sciences as is often claimed) and should not also policy makers be taught science studies courses in order to understand the field s/he is moving in? This is a debate I have been often confronted with during the last few years. Indeed also here we have a clear parallel to what happened in the PUS field. What is at stake, I would argue, is far less the reflection of the own field but how its structures and the knowledge produced are interacting with society. What is hoped for is much more to develop the capacity to handle the complex and messy situations which the growing questioning and intrusion of society into science seems to cause. In the terms of Thomas Gieryn's boundary work (Gieryn 1995, 1999), we are nothing but observing a process of protection of scientific boundaries: Keeping society/politics near but out. One could even take it further: it is also about getting STS near while keeping it out. In that sense a high degree of awareness will be needed from our side and probably the understanding that we are in the process of continual negotiation of authority, without any possibility to see a definite settling of the 'claim' we are working on. Thus the question is not about the fact there is boundary work going on, but always about the when, how, by whom and to what end there is engagement in this work and ultimately, where these boundaries are (provisionally) drawn, about what we mean by human in a society with 'science as culture'.

The third dilemma is between the wish for internal coherence, between the development of a clearer paradigm within science studies on the one hand and a rather flowering openness on the other hand. This dilemma is closely linked to my first remark about the difficulty to think about ones own history. STS is aware of the meaning of border drawing and of shaping of a research field. It means above all exclusion, need for formalisation of a certain number of internal procedures and the development of criteria through which the borders are protected.

However it seems to me that in the case of this dichotomy of choices another way to go has been overlooked. I would like to address it using the concept of a boundary object, as used by Star and Griesemer (1989). A boundary object is according to their definition something which has a clear enough structure, with a certain robustness and is thus recognisable to the outside. At the same time it is however informally malleable in the sense that it allows for those actors who meet it to attribute a range of meanings. In that sense I would argue that what is needed is to develop innovative institutional settings and network structures, allowing the development of a visible structure without having to discipline what was meant to be transdisciplinary.

This brings me to my last point, namely the challenge for STS to reconsider its own publics. Much of our consideration has gone so far to the relations with scientists. In particular the 'science wars' debates have shown rather clearly the difficulty to communicate across the boundaries. In that sense it seems more important than ever to think about how we have drawn our boundaries. The question is much about whether or not and up to what degree we build relationships to our 'subjects' of studies, i.e. the scientists and science policy makers alike.

But we also have to think about the boundaries we hold with science policy makers, industry, social groups, which are active in domains linked to technoscientific developments at large, but also about wider publics. Thus we find ourselves in the midst of the problems of public understanding of science studies and we definitely also need more understanding for our own publics or at least more clearness about whom we want to address, and how. Let's hope for our field that we get further than in the public understanding of science debate.

University of Vienna, Austria

NOTES

1 I would like to thank Brian Wynne for his most valuable comments on the draft version of the paper.

2 'STS' stands as an acronyme for both denominations used, namely 'Science, Technology, Society' and 'Science and Technology Studies'.

3 Besides a large number of workshops and conferences on specialized issues, there are two big regular conferences organised in the field: the annual 4S conferences (organised by the Society for the Social Studies of Science) and the bi-annual EASST conferences (organised by the European Association for the Study of Science and Technology); every fourth year they are held jointly.

4 The central journals are among others *Science, Technology and Human Values, Social Studies of Science, Science as Culture, Research Policy, Public Understanding of Science.*

5 The problem of institutionalising science critics has been discussed very early on in the field of feminist/gender studies. The difficulty of reflecting science from within needs to be given serious consideration.

6 The focus on language as a key to popularisation remained alive in particular in guide-books for practitioneers. Jurdant quotes in an article the example of a 1992 book by the French chemist Pierre Laszlo. "Popularisers," Laszlo claims, "are advised to write in plain language, not using too much jargon; to construct short sentences, if possible without too many subordinate clauses; to provide illustrations and preferably drawings; to start with very short summaries; and so on. The average sentence length should not exceed 22 words, and the average length of the words should be less than six letters" (Jurdant 1993, 365–6).

7 The notion is used in a sense parallel to the one developed in Latour's first laboratory study. In that sense it is an 'apparatus which provides some kind of written output' (answers to a questionnaire) and "the inscriptions are regarded as having a direct relationship to the original substance (what people think and know about science)" (Latour and Woolgar 1986, 51).

8 This corpus of literature would for example embrace work such as Dorothy Nelkin's on controversies (e.g., Nelkin 1979), but also the early work by Brian Wynne (e.g., Wynne 1980, 1982) on risk and participation, to give but two examples.

9 The title is taken from a project on public risk construction: Ph. Chavot, U. Felt, A. Masseran, F. Hausser: L'aliment entre 'mise en sense' et 'mise en scène': Construction er réappropriation du risque alimentaire (www://irist.u-strasbg.fr/nrecherche.htm#aliment.

10 It is difficult to draw a unambiguous border-line and define who is regarded as belonging to the STS community and who not. Here I also considered criteria such as self-definition or association to the field by outside members (e.g., policy makers).

11 In fact when looking at the programmes of recent STS conferences the different sub-communities in the PUS area simply regroup in different sessions without entering debate.

12 This notion was developed in discussion with Brian Wynne.

REFERENCES

Ashworth, John (ed.) (1997), *Science, Policy and Risk,* London: The Royal Society.

Beck, Ulrich (1986), *Risikogesellschaft: Auf dem Weg in eine andere Moderne,* Frankfurt a.M.: Suhrkamp.

Bucchi, Massimiano (1998), Science and the Media. Alternative routes in scientific communication. London: Routledg, 1998.

Cloître, Michel and Terry Shinn (1986), 'Enclavement et diffusion du savoir', *Information sur les Sciences Sociales* 25(1): 161–87.

Collins, Harry M. (1987), 'Certainty and the public understanding of science: Science on television', *Social Studies of Science* 17(4): 689–713.

Collins, Harry M. and Trevor Pinch (1993), *The Golem – What Everybody Should Know About Science,* Cambridge, UK: Cambridge University Press.

Dierkes, Meinolf and Claudia von Grothe (2000), *Between Understanding and Trust – The Public, Science and Technology,* Sidney, Australia: Harwood Academic Publishers.

Durant, John R., Geoffrey A. Evans and Geoffrey P. Thomas (1989), 'The public understanding of science', *Nature* 340 (6 July 1989): 11–14.

Durant, John R., Geoffrey A. Evans and Geoffrey P. Thomas (1992), 'Public understanding of science in Britain: The role of medicine in the popular representation of science', *Public Understanding of Science* 1(2): 161–82.

Felt, Ulrike (1997), Wissenschaft auf der Bühne der Öffentlichkeit. Die alltägliche Popularisierung der Wissenschaften in Wien, 1900–1938, Habilitationsschrift, University of Vienna.

Felt, Ulrike (1999), Optimizing Public Understanding of Science, Network in the 5th Framework Programme/Raising Public Awareness of Science and Technology (2000-2003), Manuscript (see http://www.univie.ac.at/wissenschaftstheorie/OPUS).

Felt, Ulrike (2000), 'Why should the public 'understand" science? Some aspects of public understanding of science from a historical perspective', in M. Dierkes und C. von Grote (eds.), *Between Understanding and Trust: The Public, Science and Technology*. Berkshire: Harwood Academic Publishers, pp. 7–38.

Felt, Ulrike (2001), Evaluierung der *Science Week @ Austria 2001*: Analyse eines Experimentes der Wissenschaftskommunikation im österreichischen Kontext, Research documantation prepared in collaboration with Annina Müller und Sophie Schober (62 pages).

Fleck, Ludwik (1980 [1935]), *Entstehung und Entwicklung einer wissenschaftlichen Tatsache*, Frankfurt a.M.: Suhrkamp.

Fuller, Steve (1997), *Science*, Minneapolis, MN: University of Minnesota Press.

Gieryn, Thomas F. (1995), 'Boundaries of science', in S. Jasanoff, G.E. Markle, J.C. Petersen and T. Pinch (eds.), *Handbook of Science and Techology Studies*, Thousand Oaks, CA, London and New Delhi: SAGE Publications, pp. 393–443.

Gieryn Thomas (1999), *Cultural Boundaries of Science: Credibility on the Line*, Chicago: University of Chicago Press.

Giddens, Anthony (1991), *Modernity and Self-Identity: Self and Society in the Late Modern Age*, Cambridge, UK: Polity.

Hargreave, Ian and Galit Ferguson (2001), *Who's Misunderstanding Whom?*, London: ESRC.

Heidenreich, Martin (2002), *Merkmale der Wissensgesellschaft*, Manuscript, http://www.uni-bamberg.de/sowi/europastudies/dokumente/blk.pdf.

Hilgartner, Stephen (1990), 'The dominant view of popularisation: Conceptual problems, political issues', *Social Studies of Science* 20: 519–39.

Hilgartner, Stephen (2000), *Science on Stage – Expert Advice as Public Drama*, Stanford,CA: Stanford University Press.

INRA & Report International (1993), *Europeans, Science and Technology – Public Understanding and Attitudes*, Brussels: Commission of European Communities.

Irwin, Alan and Brian Wynne (eds.) (1996), *Misunderstanding Science? The Public Reconstruction of Science and Technology*, Cambridge, UK: Cambridge University Press.

Jacobi, Daniel and Bernard Schiele (eds.) (1988), *Vulgariser la science – Le procès de l'ignorance*, Seyssel, France: Champs Vallon.

Jasanoff, Sheila, Gerald E. Markle, James C. Petersen and Trevor Pinch (eds.), *Handbook of Science and Technology Studies*, Thousand Oaks, CA, London and New Delhi: SAGE Publications.

Jurdant, Baudouin (1993), 'Popularisation of science as the autobiography of science', *Public Understanding of Science* 2: 365–73.

Latour, Bruno and Steve Woolgar (1979/1986), *Laboratory Life: The Construction of Scientific Facts*, Princeton, NJ: Princeton University Press.

Lewenstein, Bruce (1995), 'Science and the media', in S. Jasanoff, G.E. Markle, J.C. Petersen and T. Pinch (eds.), *Handbook of Science and Technology Studies*, Thousand Oaks, CA, London and New Delhi: SAGE Publications, pp. 343–60.

Michael, Mike (1992), 'Lay discourse of science: Science-in general, science-in-particular, and the self', *Science Technoloogy and Human Values* 17(3): 313–33.

Miller, John D. (1983), *The American People and Science Policy: The Role of Public Attitudes in the Policy Process*, New York: Pergamon Press.

National Science Board (1989), *Science and Engineering Indicators: 1991*, Washington, DC: Government Printing Office.

Nelkin, Dorothy (ed.) (1979). *Controversy: Politics of Technical Decisions*, Beverly Hills, CA: Sage.

Nowotny, Helga, Michael Gibbons and Peter Scott (2001), *Re-Thinking Science. Knowledge and the Public in an Age of Uncertainty*, Cambridge, UK: Polity Press.

OECD (1997), *Promoting Public Understanding of Science and Technology*, Paris: OECD.

Paine, Robert (1992), 'Chernobyl reaches Norway: The accident, science, and the threat to cultural knowledge', *Public Understanding of Science* 1: 261–80.
Royal Society (1985), *The Public Understanding of Science*, Report of the adhoc group, London: Royal Society.
Science and Society (2000), *Science and Society: Report by the Select Committee appointed to consider Science and Technology*, http://www.publications.parliament.uk/pa/ld199900/ldsctech/38/3801.htm.
Shinn, Terry and Richard Whitley (1985), *Expository Science: Forms and Functions of Popularization*, Sociology of the Sciences Yearbook, Dordrecht, The Netherlands: Reidl.
Star, Susan L. and James R. Griesemer (1989), 'Institutional ecology, 'translations" and boundary objects: Amateurs and professionals in Berkeley's Museum of Vertebrate Zoology, 1907-1939', *Social Studies of Science* 19: 387–420.
Stifterverband (1999), Memorandum zum *Dialog Wissenschaft und Gesellschaft*, http://www.stifterverband.org.
Weingart, Peter (2001), *Die Stunde der Wahrheit? Zum Verhältnis der Wissenschaft zu Politik, Wirtschaft und Medien in der Wissensgesellschaft*, Weilerswist: Velbrück Wissenschaft.
Whitley, Richard (1985), 'Knowledge Producers and Knowledge acquirers', in T. Shinn Terry and R. Whitley (eds.), *Expository Science: Forms and Functions of Popularization*, Sociology of the Sciences Yearbook, Dordrecht, The Netherlands: Reidl, pp. 3–28.
White Paper dti (2001), *Excellence and Opporunity – A Science and Innovation Policy for the 21st Century*, Department of Trade and Industry, UK, http://www.dti.gov.uk/ost/aboutost/dtiwhite/.
White Paper on Governance (2001), *Democratising Expertise And Establishing Scientific Reference Systems*, Brussels: Commission of the European Communities, http://www.europa.eu.int/comm/governance/areas/group2/report_en.pdf.
Wynne, Brian (1980), 'Technology, risk, and participation: The social treatment of uncertainty', in J. Conrad (ed.), *Society, Technology and Risk*, London: Academic Press, pp. 83–107.
Wynne, Brian (1982), *Rationality and Ritual: The Windscale Inquiry and Nuclear Decisions in Britain,* Chalfont St. Giles, UK: British Society for the History of Science.
Wynne, Brian (1992), 'Misunderstood misunderstandings: Social Identities and the public uptake of science', *Public Understanding of Science* 1: 281–304.
Wynne, Brian (1995), 'Public understanding of science', in S. Jasanoff, G.E. Markle, J.C. Petersen and T. Pinch (eds.), *Handbook of Science and Technology Studies*, Thousand Oaks, CAm London and New Delhi: SAGE Publications, pp. 361–91.

CHAPTER 3

EMILIE GOMART AND MAARTEN HAJER

IS *THAT* POLITICS?

For an Inquiry Into Forms in Contemporary Politics

MAKING POLITICS INTO AN EMPIRICAL QUESTION

What is the relevance of science studies for political science? We want to argue here that science studies' tactic of turning transcendences into empirical questions can benefit the study of politics. By describing the empirical turn science studies have made in normative analyses, for example of "good experiments", we will seek to extrapolate a new orientation for the study of contemporary politics. The limits of this extrapolation will also be addressed.

Students of politics are repeatedly baffled by recent developments and are seeking new, more adapted terms with which to describe the shapes politics take today. Analysts of politics and public policy can be seen to define a politics polycentric, trans-national and dispersed (Rosenau 1995, Beck 1999, Dryzek 2000). Many have emphasized the difficulty of employing the familiar conceptual apparatus and methodological techniques we used to employ in the past to grasp today's political phenomena. The confrontations in Genoa in the summer of 2001 were a bitter illustration of the urgency these questions have acquired. Inside a fortress, the 'informal' and highly exclusive meeting of the elected leaders of the economically most powerful nation states (plus Russia) addressed key issues in global politics. Outside, a battle raged on between unclear parties (who were behind the masks?). Demonstrations were held, calling not just for a different set of policies to address key problems in the world, such as poverty, labour conditions and the environment. If it was difficult to disentangle the protesters' criticisms on the content of world policies, the debaters on both sides were very effective and articulate (even in their non-linguistic expression) in questioning the G8's *ways of doing* world politics. Tony Blair, for one, argued that democratically elected leaders could not be prohibited from meeting under the pretext that "people go down into the streets, devastate the city and vandalize the shops and even the hospitals and wound policemen" (*Le Monde,* 24 July 2001). Alternatively, one is tempted to argue that it was this very legitimacy of democratically elected leaders, which was at issue. Students of politics then who remark on the obsolescence of the conceptual tools to describe contemporary politics often articulate this argument with the claim that we can no longer rely on the established insti-

B. Joerges and H. Nowotny (eds.), Social Studies of Science and Technology: Looking Back Ahead, 33–61.

tutions of politics, that these must be reinvented. Another way of putting it is that it is the *form* of politics that, Genoa and other baffling events have shown, cannot be taken for granted.

There are no universally valid, fixed practices of politics (forms). We are familiar with the argument of the displacement of politics (Beck 1992, 1999), that argues that we look for politics in the wrong places. By studying the 'subpolitical' processes that take place outside the domain of formal politics we would regain a handle on political dynamics and on the agenda for institutional renewal. Another way of putting it has been that political action now often takes place *next to* or *across* institutionalised political orders, thus challenging or even cancelling out the established contracts between the respective participants. Hajer described this as 'politics in an institutional void' (Hajer 2002): the institutional void comes out in the fact that actors not only deliberate about the solutions for the problems on their agenda, but *while doing so,* also negotiate new ways of doing things, resulting in new political practices.

Of course, it is not that state-institutions have suddenly vanished. What we retain from these arguments here is that the viability of our present 'classical-modernist'[1] political institutions come with a specific idea of what democracy is, epitomized in territorially-based representative forms of government. It is this particular notion of politics, and indeed of 'good politics', that is now being problematized, both in theory and practice. We suggest a stronger focus on the fact that the ways of doing politics are being renegotiated alongside various forms of policy deliberation. They design new concrete conditions, and search for new shapes of legitimate political intervention. Drawing on Bruce Ackerman one might see this as a historical period of 'constitutional politics' (Ackermann 1992, see also Van Gunsteren 1998). This play with the forms of politics seems an important challenge to political science and policy analysis.

Classical-modernist political institutions, such as elections and centralized decision making on issues of collective interest via councils of elected representatives, are under pressure everywhere. Similarly, policy making based on expert-advice has come under siege and the classical-modernist operationalization of politics as a system of checks and balances dividing the powers between the judicial, the executive and the legislative has also come under critique. Hardly surprising, then, that political theory now reflects on alternative patterns of organising democracy. In this, the political theorist Bernard Manin presents a useful reminder that the institutions we now tend to take to epitomize democratic government began as contested forms and were only progressively historically constructed as democratic techniques of representation (Manin 1997). It is a useful reminder to refrain from thinking that we can know *a priori* what (democratic) politics look like. If we accept the idea that what we see as democratic government is perhaps better seen as a particular, contingent set of classical-modernist practices, then we should also open up our idea of democracy. Here it is the issue of form in politics that helps redefine the agenda. If politics can also be seen as a constant experimentation on a range of forms, we should refrain from thinking we know *a priori* the forms democratic politics takes.

There is another argument however to open up politics to empirical inquiry. Not simply that contemporary forms are baffling to minds conditioned by classical-modernist concepts, but that there is something inherently transient and ephemeral in democratic forms which requires from analysts just this empirical approach. Philosophers Jacques Rancière and Claude Lefort argue along these lines that the state of flux of political forms, undying contestation, urgent doubts on legitimacy is not the effect of globalization and other phenomena of the new millennium, but rather can be taken to *characterize* democratic politics *per se*. Is democracy not just this search for evolving forms of political representation, the recognition of its own incompleteness and failures, the setting up of the virtuous cycle of critique and mobilization in the invention of new forms more appropriate for the time? These two series of arguments urge us to acknowledge that *we do not know* a priori the legitimate political forms of democracy.

In this article, then, we draw on different strands of science studies and policy analysis, after explicating these unlikely roots of our take on politics, we sketch an approach to rethinking politics as well as democratic governance. In particular we will suggest how science studies' recent philosophy of (the forms of) good experiments might be articulated with a tradition of interpretive policy analysis to open up the question of 'what is (contemporary) politics' to empirical investigation. A second question can then also be formulated: under these conditions, how can we, along with the actors in the field, help to re-invent a democratic politics? The article thus hopes to shed new light on the unexpected variations and surprises of contemporary politics seeing them not necessarily as the sign of political failure, but as hopeful indicators of the existence of an actual *field of experimentation* where actors and analysts might work together to elaborate and try out new political forms (see Dryzek 2000).

THE ISSUE OF POLITICS IN SCIENCE AND TECHNOLOGY STUDIES (STS)

How can STS be employed in a research project of studying intriguing contemporary forms of politics? In this section, we give a tentative answer that does not presume to exhaust all possible associations between the two fields of political science and science studies. Rather our answer is an attempt to put into relation a small number of key arguments and skills of these two respective fields with the aim of yielding a maximum return, to use an industrial expression, for both. We start with the account of good experiments influenced by actor-network theory (ANT) and subsequently seek to put this into relation with accounts of a form-aware search for deliberative democracy in political science and policy analysis. STS and policy analysis might at first sight be unlikely bedfellows, and yet we will suggest that much is to be gained in their comparison. In particular, with this contrast, the habitual ways in which the question of politics has been approached in the two fields loses their self-evidence. This is the first step to turning politics into an empirical question, to asking in all earnestness 'Is *that* politics?'

In the past two decades, STS authors have been preoccupied with the problem of rendering their works politically relevant. Annemarie Mol has made a distinction between two ways in which this has been done which we would like to build on.[2] First she suggests, for several authors, the political question was 'Who participates in designs, experiments and scientific policy?' A handful of scientists and technocrats or all those affected by these practices and policies? Politics then for these authors had to do with *who* acts, decides, chooses, participates. It is among these authors that we find suggestions for 'democratising' science. A second way of phrasing the 'political question', Mol argues, has been to focus on the relations and norms, the politics, inscribed *in* technologies, scientific and medical practices. Thus the question, in the latter approach, is no longer who acts and whether this participation follows expected democratic procedures, but *how specific capacities to act are embodied or performed in science, medicine, or technology*. How does a practice 'engage'[3] persons? How, for example, do different medical practices transform his/her experience of her/his body and affect his/her trajectory? How do these various performances coexist? Are these practices *good* for the subject? Here then, for authors such as Annemarie Mol and John Law, the question of politics becomes tied to question of morality and of how different 'goods' get performed. We would like to review both of these approaches to underline the arguments useful to policy analysis as well as to discern the specificity (and thus the limits) of politics for both these approaches. In the end, our aim is to propose a radicalisation of both these approaches of politics and attempt to pry open the question "What is politics?" to empirical inquiry a little further.

Let us turn then to the first version of politics in STS: *who* acts or participates? For certain authors, this has meant a commitment to asking who takes part in decisions that count.[4] A democratic approach to science and technology has been developed by those who criticize scientific experts' reluctance to include local knowledges and stakeholders in scientific debates. They make a plea for the participation of citizens, users or people concerned by the consequences of a scientific or technological innovation.

Rein de Wilde has criticized this work writing that if STS has been radically innovating about the manufacture of knowledge, if STS succeeded in problematizing what had previously been self-evident in *science* for 'moderns', the *politics* these same ("who") authors develop is uncritically modernist. Many are content with simply repeating the call for the extension of local and direct democracy. In a curious inversion of the Enlightenment dream that technology was a moral machine through which 'man merged with man', continues de Wilde, STS authors allow that *democracy* is the moral machine through which the bonds of man to man and to nature are strengthened and improved (De Wilde 1997, 11). Targeting the works in particular of Wiebe Bijker, Langdon Winner and Richard Sclove, De Wilde argues that these authors adopt democracy wholesale without problematizing it. De Wilde thus criticizes these authors for their a priori preference for one specific form of (direct) democracy as the 'real' form of democracy. Its processes of aggregation and representation of 'We, the People' are taken to be unproblematic, unequivocal and efficient mechanisms for ensuring the inclusion of all and for guaranteeing the legitimacy of an engineer's activity. "Democracy [will be realized] ...if citizens have an opportunity to participate directly in debates on affairs relevant to them." De Wilde scathingly concludes that, in a time marked by the displacement of politics, the crisis of legitimation of party-based and national politics, to assume that the simple participation of the user in design is sufficient for democratic design is to indulge in facile politicizations drawn from older political forms is an anachronism and a sign that these STS authors have not done their fieldwork.

A further critique might be made. These works rest on an assumption that individuals are political actors if they are from the start capable of a specific sort of action (taking part in rational negotiations and debates, deciding, choosing). The striking resemblance between this political actor and the 'autonomous individual' of liberal political philosophy is not problematized by these authors, in contrast with much of contemporary social scientific work which has sought to show that such capacities as 'reasonable' action or discussion are not inherent properties of individuals but granted from the setting in which actors move[5] or laboriously acquired through years of discipline[6].

A second strand of research in science studies would emphasize the importance of forms of participation that do not just allow access but also fabricate the very capacities of participants. Political scientists have emphasized that the techniques of political participation (votes, polls, deliberation forums) are not simple explorations and aggregations of 'public opinion' but processes through which 'the public' is constructed and transformed. In contrast, authors who focus on 'who participates?' seem to be reluctant to allow that this 'who' might be constructed. These authors seem to take participation to be a moment of purity. Even as these STS authors (including de Wilde) accept that *science* is not a pure space of representation without transformation, *politics* becomes, in their arguments, that space where the user, stakeholder, lay citizen can tolerate no reduction, assimilation, deformation.[7]

While preserving the concern of these authors 'who acts?', we would like to allow that this 'who' might benefit from political forms that construct it as a competent participant. The second strand of STS (the politics of 'what') then becomes useful, not in opposition but as an amendment to this first approach.

This second strand of 'political' science studies focuses on the little shifts practices perform in the trajectories of things and persons. Attention then is keenly turned to the concrete *settings* or *dispositifs,* which allow them to emerge endowed with specific qualities. Particularly developed in studies on the mediated body, this analysis of settings is the point of departure of our reflection on political forms.

According to this second approach, the setting of a medical intervention or of a scientific experiment is a mediator through which the object investigated gains capacities it did not have before. We begin then with a constructivist take on action: choice, reasonable participation and other traditionally defined political capacities are achievements of the setting rather than unproblematic properties of persons. As we shall see below, in policy analysis too studies focus on a search for settings that might construct the possibility of certain forms of political action.

Our approach to politics and its roots in the philosophy of scientific experiments can be further clarified in the following example where a setting performs capacities: the cages used in psychological experiments on rat sexuality in the 20th Century. According to Glickman's fascinating history of rat experiments, until the mid 1950s, the male rats were habituated to the small enclosure of the experimental cage while the females were retained in a small compartment in the ceiling of the apparatus. When the experiment began, the females were literally dropped from above. These experiments described the male sexual role as 'active', 'aggressive'. The male rats 'mounted', 'ran after' and 'cornered', 'palpated the flanks' of their mate. In contrast, female behaviour was described as the automatic turning upwards of her haunches after 'proper stimulation' by palpation of her flanks, the 'lordosis reflex'. The sexual role of the female rats was 'essentially passive' (Frank Beach cited in Glickman 1996, 21).

In later experiments, different cages were used. This transformed the phenomenon of female sexuality in a way unexpected to the experimenters. With a different cage, the female rat did after all seem to be sexually active. One experimenter – who was a feminist – made the cage 'semi-natural' and large enough for the rats to move. Sexual behaviour, in that situation, became a "[dyadic] negotiated activity". McClintock, a feminist ethologist, writes,

> The female rat signals her readiness to mate by placing herself in proximity to the male and then controls the pace of mating with appropriate mixtures of approach, avoidance, solicitation and rejection (McClintock cited in Glickman 1996, 21).

Suddenly, in this new cage, the female rat 'approaches', 'grooms' the male rat, presents her hindquarters if the male seems disinterested, and finally "turns and runs directly away from the male either with a dart-hop gait, or the stiff legged run of an estrous female" (McClintock cited in Glickman 1996, 21).

Hence, as the experimental setting (the cage) is transformed, so is the phenomenon of female rat sexuality. This surprising event can be articulated to two very different propositions of what it takes to make a good experiment.

A first analysis is the one put forth by the experimental-psychologist Glickman, and consistent with a traditional philosophy of science. For him, these examples show that material elements of the experimental apparatus determined the outcome of experiments on female rat sexuality. Elements of the first cage, he suggests, made, forced the female rat into a behavior, which was not 'naturally' hers. More specifically, the cage had been so small that it held no corners large enough for the female rat to hide if she tried to resist the male rat (Glickman 1996, 25). The devices of the experiment had limited the sexual behavioural repertoire of the female rat: her only sexual action was a reflex. For Glickman, the problem was not just that the technical apparatus had interfered with results. The experimental apparatus, the small rat cages, had materialized the biases of the scientist. The conservative, male experimenter had unwittingly implemented his own cultural expectations in the experimental devices and had let these expectations determine the results of his experiment. When experimenters abandoned the small cage, the weight of the imposed biases became lighter and the phenomenon became less artifactual. He concludes (citing McClintock, the feminist ethologist): "Studying female [rat] [sexual] behaviour in a small cage is akin to studying swimming behaviour in a bath tub...: it cannot be done" (McClintock cited in Glickman 1996, 23).

In another example, Glickman tells that experimenters introduced an electronic transducing device that allowed the deciphering in 1972 of a male post-ejaculatory song that has impact upon subsequent mating behaviour. It was not before 1991 however that experimenters recognized that females were 'singing' as well. The frequencies of the female vocalization had been so high that their detection required sophisticated electronics that only recently were introduced to the laboratory. Glickman argues again that results had been determined by our own biological bias, here "limitations [imposed] by our own sensory equipment", the human ear (Glickman 1996, 23).

In a manner reminiscent of a traditional sociological criticism of science, Glickman writes that the trouble with these psychological experiments is that during the thirty years of experiments with small cages and unsophisticated hearing devices the scientific setting had implemented, effectively carried out, the biases of the experimenter. The results of experiments had been corrupted by these biases. Glickman assumes that the ideal experiment is the one in which the experimenter has a pure access to the phenomenon and can observe without letting his theoretical and material apparatus interfere. The good experiment, to use other terms, accurately and precisely represents the phenomenon. It is a transparent looking glass, effacing itself to allow only the phenomenon to be seen. The early experiment with small cages had told us more about the scientist (and his biases) than about the phenomenon of rat sexuality itself.

We want to develop a second take on the role of the setting, also inspired by the event of the changing behaviour of the female rats as the experimental cage got larger. Prolonging the works of philosopher Isabelle Stengers in particular (see e.g. Stengers 1993, or Latour, in preparation; Gomart and Hennion 1999; Gomart in press). Our starting point is very close to Glickman's. We too want to begin with a criticism of experiments where the experimenter's expectations are not challenged by his results, where the results mirror the biases of the experimenter. However Glickman's criticism rests on the assumption that biases *deformed* the phenomenon, made the experimenter incapable of seeing the true nature of female rat sexuality. By mobilizing a different definition than the one Glickman uses, we will try to open the possibility of another argument. Glickman's definition betrays a nostalgia for a pure science, unmarred by social concerns, and a direct, unmediated access to the phenomenon. Recent approaches in science studies, in contrast, describe the inevitability of the entanglement of science and society. More radically, several authors have argued that science acquires its autonomy through a process of construction (Latour 1999): the biases materialized in the cage then do not just deform the phenomenon, they are crucial in fabricating the phenomenon's autonomy. As a result, for those within science studies who attempt to define a 'good experiment', the goal can no longer be to purify science from deformations caused by social biases or 'exterior' constraints.

An alternative goal is instead to criticize bad experiments where biases (the small cage, the human ear, bathtubs) *do not allow the object of the experiment to subvert* the expectations of the experimenter. The problem with the female rats experiments was that the experimenter had not been able to define "the questions which [the rats] had forced them to reflect upon" (Stengers 1993, 76). Or to use other terms, the small cage had not "given [the rat] a chance" to surprise the experimenter and transform his initial hypotheses (Latour 2000). When the transducing device was introduced, or when the cage was enlarged, the rat became suddenly capable of surprising the experimenter. We will argue that it is not when biases were eliminated but when specific other artifices were introduced to bias and constrain differently that the female rat became capable of intervening in the experiment and surprising the experimenter. The rat acquired this new capacity when biases and artifacts were varied and added (Latour 2000). The good experimental setting then, and perhaps, by extrapolation, the political form, is *not one which is neutral but one which deforms, constrains and enables* in interesting ways.

Drawing on these recent contributions to the philosophy of science, then, it is clear that the problem of the early rat cages is not that they were inscribed with biases. The problem is not that certain elements of the experimental setting refused to be sufficiently silent, transparent, passive, absent (*in*determinate enough). Impurity is not the problem. Rather, a good experiment is on the contrary the one where the setting is *present*, active *enough*. When there are enough biases and deformations, and when these are of a specific quality. In a similar manner, Stengers defined a good experiment as an occasion for sudden reversals, unexpected results. The aim is not to reduce the influence of the setting, but to construct an influential setting, which deforms the phenomenon in an *interesting* way, giving a novel spin to the ordinary word 'interesting'.

> *Interesting someone to something* means above all to construct an arrangement in which a thing [s'arranger pour qu'une chose] – apparatus, hypothesis, reasoning, in the case of scientists – might become capable of concerning him, intervene in his life, eventually transforming him (Stengers 1993, 108).

'Interesting' here is not a mere synonym of 'good'. The interesting setting is one where the person or creature or thing is not left alone, authentic, but transformed by what occurs, and transformed in ways which induce its interference with the project. What is at stake, Stengers concludes, is to "constitute the phenomenon into a reliable witness who participates in the construction of the questions that count".

Stengers, by preferring the verb 'to interest to'[8] (an intransitive verb which turns the attention to the setting to which the rat must be interested), urges us to focus on the material *forms*. Her original take on the role of the setting evokes a specific understanding of the term participation, which has little to do with ideals of direct democracy put forth by the first strand of 'political' STS authors above. What is at stake, according to these studies of good experiments, is not an accurate representation of an entity out there. Rather, experiments are analysed as mediators, active forms, which transform and fabricate new capacities in those who pass through them. They associate these rats to other elements in ways that modify what they are and what they are capable of doing. The intervention of the setting is what is being evaluated now. Further, this intervention is taken to be successful when it occurs in a peculiar manner: the setting should not simply determine, force or oblige the rat to behave in a certain way. A circular movement is described where the rat is transformed at the moment that it is 'interested to' the experiment. For Stengers, the action of the object of an experiment (in our case the rat) is characterized by the fact that it surprises the experimenter. If it had been expected, this action might be obedience. Good experiments are those where surprise occurs.

The argument is not simply that surprise is the criterion for good experiments. Instead, we argue that there are no transcendent criteria for good experiments: surprise does not attempt to bypass this uncertainty but instead insists that criteria are inherently *immanent* and cannot be picked a priori to guarantee outcomes. If the setting is conceived as a site of emergence (that is, as a place where rats acquire entirely new competencies) it is impossible to attempt to define once and for all the universal criteria of good experiments. This does not mean that there is no 'good' but that each experiment proposes a new definition of what good might be. The definition of the 'good' in rat experiments *acquires* content at the very moment that the female rat 'sings out'. The 'good' is, literally, a sophisticated hearing device. When the rat surprises the experimenter, behaves in unexpected ways, the rat defines the quality of the experiment and distinguishes among the devices of the setting those which make it so. The incapacity to formulate transcendent criteria for a good experiment is tied to the experimenter's inability to predict which property is necessary for this entity to be interested to the experiment: no one knows beforehand what are the essences, and therefore the vulnerabilities or the resistances of the entities that pass through a setting. Attempting to fix the essence of a creature is the opposite of a good experiment where the point is to transform just what this creature is capable of. Nailing down essence (determining who or what one *is,* once and for all) is a sin we would not dare to commit.

The good experiment is then described not in terms of its accuracy of representation but in the concrete terms of the kind of intervention the setting has: what becomes of the rat, if it is treated as sexually passive or sexually active? What does the experimenter's hypothesis, materialized in the setting, *do* to this rat, what does this setting *'fait faire'* ('make one do' or 'engage'), what does it articulate the rat to as it renders it articulate? To what industries, social movements, intimate gestures ... does it tie this rat to? How does this tie transform the rat as well as those to which it is now attached? This alternative definition of good experiments is not concerned with the question of whether the female rat *is* more *like* the 'natural' female rat; instead it is concerned with what the experiment treats the female rat like (Latour 1999, 143). In the case of the female rat, *to treat her like* an active sexual agent transforms in an unprecedented and interesting way relations between feminist ethologists, their rats, their older 'biased' colleagues, and because of the parallels constantly drawn in ethological debates between rats and humans, this proposition tentatively transforms relations between male and female humans.

This philosophy of scientific experiments then urges us to develop another definition of politics: what does a setting (practice, form) *do* to those who are engaged in it? One danger however with this formulation is that it risks to miss the specificity of contemporary politics. Indeed works on 'good experiments' are constructivist alternatives to *moral* arguments for realism or pure 'sociologism' and we might then be led to assimilate too fast the moral and the political. This might not be an interesting bias, in a Stengerian sense, for an empirical investigation of contemporary forms of politics. But before exploring the limits of the empirical tactic in STS, let us turn to analyse one example of policy making which benefits from it.

POLICY MAKING AND THE RE-INVENTION OF POLITICS

The debate within STS on the research of 'who' versus 'what' helps to rethink politics. In particular the work in STS on the notion of the 'good experiment' raises pertinent questions for political-science research as it brings the issue of form back in. There is of course a political-science scholarship on the analysis of forms and settings. Most notable is the work on political symbolism. Yet this has never been more than a minority-interest in political science. As Murray Edelman wrote in 1964, "Although every act takes place in a setting, we ordinarily take scenes for granted, focusing our attention on actions." (Edelman 1964, 95). Edelman, however, analysed politics as drama, and the setting was a key element of his analysis of politics. Once politics is conceived of in dramaturgical terms, the *mise-en-scene* is no longer regarded as pre-given, it becomes a topic for political analysis: "In the drama, the opera, the ballet, in the display of paintings and in the performance of music setting is plotted and manipulated, just as often as in the staging of governmental acts." (ibid, 96) Edelman drew on the work of Kenneth Burke to try to fill this hole in political analysis. Burke worked on what he called the 'scene-act ratio' calling attention to the fact that, as he put it, the 'scene contains the act' or the 'scene contains the agents' (1969, orig. 1945, p. 3). The argument he made was that scenes have to be regarded as 'fit containers' for certain acts, that acts cannot be understood without understanding the context within which the acting takes place. The analysis of political symbolism never really took off however within the political science discipline. It is telling that the next key contribution in the analysis of symbols in politics came from Joseph Gusfield, a sociologist of social problems, analysing the policy making on drinking-driving as a symbolic order (1981). A conceptual combination of this work with the STS-scholarship on 'good experiments', and the appreciation of the *dispositif* in particular, to study the new developments in policy making is, it seems to us, a timely new take on political analysis.

The second reason why a further collaboration with STS is promising, relates to the way in which the concept of 'bias' is employed within STS. 'Bias' is a key concept for political science. Indeed, it is fair to say that political science is preoccupied with tracing bias. Yet bias is characteristically seen as a way in which some interests are favoured over others. Political analysis aims to spot the 'mobilization of bias' in organizational structures and see how some people get more than others and why. In the famous words of E.E. Schattschneider (1960) politics is defined as *mobilization of bias* as 'some issues are organized into politics while others are organized out'. For Schattschneider this was an argument to try and cancel this bias (especially business interests) by building up countervailing powers: "The object of the game is not to destroy business power but to match it with governmental power." (1960, 127). Later Steven Lukes put forward the mobilisation of bias as the 'second dimension' of power in his influential book on the three dimensions of power (1974). We take this to epitomize the stand of the discipline: any notion of good or fair politics takes as its starting point that mobilization of bias has to be *limited and constrained.*[9] Good politics is a politics in which people are allowed to express their own ideas without being constrained by the exercise of power. The STS take on bias, in which a 'good experimental setting' is not seen as one that is neutral but as one which constrains, deforms in 'interesting' ways, as argued above, opens up new insights into the ways in which politics is played out today and may, ultimately, help define new ideas of 'good' politics.

One field is particularly interesting for this empirical search for experiments in politics. This is the debate on public involvement in politics, a key concern within the political sciences. For that reason we will dwell on it at some length, before returning to the argument on 'who' and 'what'. The debate on public involvement is marked by a widely shared concern about the position of key political institutions (elections, political parties, parliamentary decision making) and a lack of 'civic engagement'. The work of Robert Putnam on civic engagement in the United States (2000) and Manin's analysis of the emerging 'audience democracy' are key markers in this debate. In the more applied subfields of political analysis, such as policy analysis and planning (that both constitute intellectual traditions in themselves, originating in the work of authors like Dewey and Merriam, or Karl Mannheim respectively) there is a growing dissatisfaction with the ways in which present forms of public participation function. The recent polemical contribution to a meeting of practitioners by its leading theorists Judith Innes in an article with practitioner David Booher is telling:

> It is time to face the facts we all know, but prefer to ignore. The traditional methods of public participation in government decision making simply do not work. They do not achieve genuine participation in planning or decisions; they do not provide significant information to public officials that makes a difference to their actions; they do not satisfy members of the public that they are being heard; they do not improve the decisions that agencies and public officials make; and they don't represent a broad spectrum of the public. Worse yet, they often antagonize the members of the public who do try to work through these methods. Moreover, they pit members of the public against each other as they each feel compelled to speak of the issues in polarizing terms to get their points across – making it even more difficult for decision makers to sort through what they hear, much less to make a choice. Most often

> these methods discourage busy and thoughtful individuals from wasting their time in going through what appear to be nothing more than rituals designed to satisfy legal requirements (Innes and Booher 2000, 1).

The default response to such challenges has not been the detailed investigation of these practices. Rather, the discipline has been captured by a macro-sociological determinism in which such analyses of 'crisis' of politics are combined with a macro-sociological account of societal change.[10] Hence waning civic engagement and loss of community are understood in terms of rising levels of emancipation, individualisation and de-secularisation; while audience democracy is related to the growing role of mass media in society. In Putnam the analysis of diminishing engagement gets combined with a nostalgia for non-political forms of civic engagement that used to support traditional political institutions. Likewise, Manin's analysis of an audience democracy comes with a suggestion of a 'staged' conflict between parties rooted in their competition for media-attention, that fails to relate to 'real' social cleavages in society.

There are at least two problems with this type of analysis. First of all the macro-sociological determinism suggesting that the course of events is the 'inevitable' outcome of broader processes of societal change. Secondly, there is a lack of appreciation of the new emergent forms of politics that coincided with the demise of the classical-modernist political institutions. There is no need to reiterate here the work of Ulrich Beck on the 'displacement' of politics. Yet one could add that much is happening much closer to the state as well. Elsewhere Hajer has shown that policy-making processes should be examined in a multi-dimensional way (Hajer 1996, 2002). Far from being 'merely' about finding solutions for problems, policy making now functions as a 'public domain' in which actors discuss a wide diversity of issues.

People do not necessarily seek representation via political parties or elections, but want to be heard on particular issues they pick for themselves. Conceptually this leads to a reversal of the political order: whereas one would traditionally conceive of 'policy' as the outcome of 'political deliberation' (i.e. the political debate in councils of representative democracy), we now see how political deliberation results from policy making initiatives. There are many reasons for this shift (Hajer 2002). Yet what stands out is that party politics and decision making of elected representatives alone cannot uphold the classical-modernist order of politics. Citizens seek multiple forms of representation, and thus the ordering of politics that allocates the authority of legitimate decision making to elected representatives is challenged.

Governments have responded to this challenge. As a consequence, all through the West we can see how traditional strategies of imposing policy programs on society now are in competition with experiments where societal actors are involved in 'making policy'. This opening up has been addressed under an array of conceptual labels, such as the move from 'government' to 'governance', the emergence of 'co-production' of policy, 'collaborative dialogue' or 'interactive policy making' (see Pierre 2000).

This is also where the issue of 'form' comes back in. We argue that these experiments in policy making constitute a key site for political analysis. Indeed, as we will show below, the new emergent practices include politics in settings that are often not recognized as political but do lead to collective deliberation on public problems and, as such, seem politically important. What emerged as a response to political failure now constitutes an actual field of experimentation where actors and analysts do indeed work together to elaborate and try out new political forms. This is, we would argue, a new empirical turn. Such experiments do not simply measure the characteristics of the public at large to understand macro-phenomena like the level of 'civic engagement', they investigate the qualities of particular practices. Yet in order to do this productively, political science must rethink its methods. It is here that the link to STS and the work on 'good experiments' opens intriguing possibilities.

The new experiments with forms of public involvement raise the question of what the settings (practices, forms) do to those engaged in them in concrete cases of public participation. The idea that it is the productivity of bias that needs to be studied potentially leads to a new division among political scientists who study policy-making practices. This public policy strand in political science has been characterized as being about 'the struggle to know the world and change it at the same time' (Nelson 1996:558). It is a long-standing orientation within political science that is associated with such scholars as Charles Merriam, Harold Lasswell, Herbert Simon, Charles Lindblom and Aaron Wildavsky. For a long time public policy analysis was dominated by 'scientist' approaches (for a discussion, see Fischer 1998 as well as Nelson 1996) dedicated more 'to the development of methods and practices designed to *settle* rather than stimulate debates' (Fischer 2000, forthcoming). Yet starting from the early 1970s this 'positivist' orientation in policy analysis received bitter criticism from what became known as the 'post-positivist' orientation in policy analysis. The post-positivist literature should be credited for systematically spelling out the limitations of mainstream positivist policy analysis (Tribe, 1972, Frank Fischer 1980, 1998; Bobrow and Dryzek 1987; Deborah Stone 1988). In their analysis the issue of form played a role. In his bitter critique of mainstream policy analysis in *Philosophy and Public Affairs,* Laurence Tribe for instance picked out the 'antiseptic terminology' of policy makers as a form that steered a particular play of politics (Tribe 1972). Later Doug Torgerson spelled out the influence of the usual judicial form of inquiries. In his work on the Berger Inquiry Torgerson pitted the traditional inquiry against the experimental open 'fact finding mission' of Chief-Justice Berger in the Alaska Pipeline Inquiry (Torgerson 1997). However, although this post-positivist strand in policy analysis paid attention to issues of form, its main strength was its programmatic critique of the positivist mainstream in policy analysis. Knowing this argumentative 'constrast space' (Garfinkel 1981) of the post-positivists helps understand their reiterated call for a more 'participatory' policy analysis. It is the critique of positivist-inspired expert-led governmental programs (Fischer 1990) that explains the importance of citizens participation, and later of 'local knowledge' (Yanow 2000). Citizen-participation and local knowledge are to correct the 'bias', the blinders, of rationalist policy analysis.

Yet it seems fair to say that detailed empirical research of the ways in which participation contributed to the 'intelligence of democracy' was never a key concern. Much of the work on participatory policy analysis and planning is based on the idea that excluded groups need a better access to policy deliberation. We think this research could take a next step bringing in what Mol called the question of 'what' (see above), focusing on how participants can be understood as 'products' of the engagement in a particular setting.

In political science bias is mostly seen as the institutional force that precludes people from taking part in the 'policy conversation' and speak in their own voice and with influence. In this context 'deliberation' is then used to define those qualities of a discussion that allow for this free exchange. Publicity, reciprocity, accountability, are among the criteria that are often used to define notion of what is called 'deliberative democracy' (see Gutman and Thompson 1996; Bohman 1996). To be sure, this is an important development in itself. If deliberation is key to good political practice, then democracy should not be reduced to the democracy of 'majority rule' in representative councils. Within political science, this is by no means self-evident. For instance, in the latest edition of the 'Oxford Concise Dictionary of Politics', edited by the Oxford political scientist Iain McLean, the entry of democracy simply states: "Greek, 'rule by the people'. Since the people are rarely unanimous, democracy as a descriptive term may be regarded as synonymous with majority rule ..." (1996, 129). Only to continue with a lengthy exposition of the many ways in which political science has sought to make sense of the political dynamics of majority rule. Here deliberative democracy has a different orientation suggesting to shift the research emphasis.

Yet there is an important distinction between those that see deliberation in terms of 'rules' for the proper exchange of arguments and those who would argue that deliberation cannot be understood without taking the role of 'practice' into account. The established Habermasian tradition would focus on the development of a 'discourse ethics' as a set of rules of conduct for a fair and intelligent deliberation. The latter notion of a deliberative policy analysis would argue that listing abstract criteria for deliberation misses the point. New research should investigate how actors are constructed, transformed or empowered into participation. In that perspective, bias might be perceived as a purely negative force standing in the way of good deliberation. Instead it might gain a more positive connotation if its power to produce engagement is insisted upon. The quality of a deliberation would then be investigated in terms of the variation of practices, acknowledging that the criteria of a deliberative democracy will change and vary and cannot be fixed a priori.

This scholarship thus does not put the search for a more deliberative practice over board, but seeks to bring in the issue of form, employing the concept of 'practice' to investigate what happens in particular settings. The empirical issue here would first of all be to see what is *made* political in these cases and how this happens; the understanding of 'democratic' politics would then be postponed to a later phase of research. Whereas traditional policy science focused on "problems" and "decisions" this new scholarship on 'deliberative policy analysis' sees knowing and doing as intimately related. It emphasizes the active, searching quality of knowledge acquisition and the intrinsically situated character of knowledge (both of which converge in Dewey's concept of 'inquiry', see Dewey 1938; for a longer treatment, see Wagenaar and Cook, forthcoming). So while there are several attempts to think 'beyond' certain traditional conceptions of democratic forms, the difficulty seems to restrict oneself to proposing *retractable* hypotheses, working tools useful in attempting to capture the unique historical specificity of contemporary political forms. Below we illustrate how we might go about this. Just as our analysis of rat experiments refrained from proposing transcendent criteria of good experiments, here too the trick is to find the criteria immanent in each political form.

SEARCHING FOR 'INTERESTING' POLITICAL PRACTICES

In order to examine how form or setting influences politics we examined one case from regional planning. The case focuses on the Hoeksche Waard, an island south of Rotterdam. The Hoeksche Waard also happened to be an area for which plans tumbled over one another over the last couple of years. The land use on the island is mainly agricultural. Being located between the industrial and harbor complexes of Rotterdam to the North, and the Belgian harbor city of Antwerp to the South it stands under tremendous pressure from developers, both from within government and the private sector. To many protagonists, the Hoeksche Waard is the 'obvious' location for various new forms of urbanization and development. Some years ago it was seen as a potential location for a second national airport. As we write the island is confronted with the construction of the line of the High Speed Rail link to Brussels and Paris, which will chip off a slice of the island. The extension of the A4 motorway from the north to the south is still under discussion. It would result in a major highway intersection in the center of the Hoeksche Waard. The Hoeksche Waard is also of great interest to environmentalists. In particular the southern side to the island allows for large-scale creation of new wetlands, good for birds and a obvious way to act on the European Habitat Directive. This southern part of the island is also of potential value for water management, a prominent issue in times of frequent flood threats: the Hoeksche Waard is often looked upon as a location for a 'water retention area'.

Being an island, the Hoeksche Waard is more than the mere product of an arbitrary 'zoning' exercise. It has some symbolic meaning as a territorial unit yet it does not in itself constitute a formal level of government: it comprises six municipalities and is part of the much larger province of South Holland.

The Hoeksche Waard constitutes an interesting case as it was the subject of discussion and analysis of several different forms of planning politics over the last five years, ranging from a formal participation phase in the drafting of a statutory regional plan to a cultural manifestation that focused on the future of the area. Question is if form played an important role in the politics of planning of how these five practices. Is there reason to believe that the setting influenced the politics of the future of the area? To get to an answer the following practices will be examined:

- a so called 'area-oriented sketch' (*Gebiedsuitwerking*): as part of the preparation of a national white paper on land use planning (*Actualisering Vinex*);
- the formally required public participation on a draft of the new regional plan to be published by the Province Zuid-Holland;
- *AIR Southbound,* an 'interdisciplinary architectural manifestation' with as its theme the city and the country, organized by Foundation AIR from the city of Rotterdam (Architecture International Rotterdam);
- *AIR-plus:* a subsequent debate of the plans from the AIR manifestation in the Hoeksche Waard itself;
- the making of the manifesto *De Hoeksche Waard weet wat zij wil!* ("The Hoeksche Waard knows what it wants!"), published via the region Zuid-Holland Zuid.

We analysed the interaction between form and planning discourse examining the practice in which new ideas and new coalitions emerged (for an explication of 'argumentative discourse-analysis, see Hajer 1995, Hajer and Wagenaar, forthcoming). Our search was not for the mechanisms that allowed for, or inhibited the representation of the 'authentic' Hoeksche Waard in these practices. It was rather targeted at seeing how each practice constituted politics in its own way, and examining that *form* of politics, examining the particular sort of engagement it enabled or delimited, each creating their own Hoeksche Waard not another one. Following Kenneth Burke (1945/1969) we examined the scene-act ratio, exploring the fit between scene (or setting) on the one hand and the act on the other hand.

Form 1. The Development of an 'Area-Oriented Sketch'

The idea of the Area-oriented sketch[11] was to develop four markedly different models for spatial development of the island. This was intended to facilitate national strategic planning in the years to come. Planner-experts were asked to show different 'possible futures': one emphasizing the quality of openness, the other showing a model of urbanization, the third emphasizing a stronger industrialization for the area, the fourth emphasizing the exploitation of the Hoeksche Waard for recreational purposes. The four scenarios were represented in four maps with added text. Subsequently policy makers and organizations from civil society discussed their preferred models. The input of individual citizens was very low. In the end the provincial report to the Minister of Spatial Planning spoke out strongly in favor of the scenario of the 'Hoeksche Stad'. In this scenario the northern part of the island would be used for a densely built extension to the Rotterdam conurbation, in fact amounting to a continuation of the existing land use planning strategy. Although not an unusual planning practice, the Area-oriented sketch was experienced as a 'creative chaos and order' of mutually contradictory plans (in the words of J. Wolf, responsible politician for spatial planning at the time). The setting of the area-oriented sketch was marked by expert-deliberation, and made the Hoeksche Waard very much an object for future development. It was not meant to generate a broad-scale debate, as it was meant to be policy-input, and written and staged primarily with expert participants and readers in mind.

Form 2. Participating in the Determination of the Draft-Regional Plan

Provinces are legally required to periodically update their regional development plans (*streekplan*). In this document the province is supposed to outline its planning strategy and legitimize its intentions in terms of the framework outlined by the national government. The official draft of the new updated plan came out in 1998. What was new was its attempt to contextualize the spatial developments in the Hoeksche Waard, relating it to 'the general socio-economic developments in Northwest-Europe'. In this light, it seemed inevitable to address new issues, such as water management, nature conservation policy and infrastructure development. In terms of concrete policy proposals the regional plan put forward the intention to develop a large enterprise zone in the middle of the island, at a new motorway junction. The remarkable connectivity of this enterprise zone was to result from an new extension of the A4-motorway from the north to the south. Apart from this, it suggested to create a new 265 ha big location for glasshouses for intensive horticulture.

In this setting the Hoeksche Waard was seen as a 'search space' for various plans of the provincial and national governments. This was legitimated by quantitative research showing an assessment of future 'spatial claims' of various sectors of trade and industry. It legitimized its intervention with the claim that provincial planners had as their task to cater for 'bovenregionale claims', claims surpassing the interests of particular regions. It actually drew on the four models from the 'arera-oriented sketch' and other techniques, and made planning into an expert-led matter of finding the optimal location for various spatial claims.

The draft-regional plan was primarily the product of politico-administrative deliberation, coordinated by experts from the province of South Holland (the governmental level legally responsible for bringing out regional plans). Initially, the plans generated a broad political and administrative support. The various politico-administrative bodies published a 'Declaration of Intent' on 7 April 1999, arguing for a "qualitative positioning of the Hoeksche Waard" in its larger urbanized context.[12] It argued for a need for enhanced co-ordination and announced a yearly program of implementation. The 'ambition' was summarized as follows:

> ... conservation, development and strengthening of the qualities of the area, that have been marked as carriers of identity; recognition of the *internal impulses* for change and evaluation thereof in terms of potential value to exploit or strengthen the qualities of the Hoeksche Waard; recognition of the *external impulses* for change (spatial claims) and evaluation thereof in terms of chances to exploit or strengthen the qualities of the Hoeksche Waard (emph. added).[13]

It thus emphasized the needs of both local and provincial governors. As a planning document the text held together a broad politico-administrative coalition. The local authorities were pleased with the recognition of 'internal impulses', the province claiming success because of the recognition of the right to use the Hoeksche Waard as search space ('external impulses'). A 'steering committee' with representatives from the various agencies was announced to oversee the implementation. It was announced that actors from civil society (*maatschappelijk middenveld*) would be involved later on, and not at the level of strategic planning but in the context of individual projects. In many regards the setting was one of traditional elitist policy making, with policy makers and key players from business and politics agreeing on a plan, without aiming to include other voices. Perhaps because of this form, the public contributed a very significant way, once it was allowed to do so.

Following the acceptance of the concept-regional plan by the Provincial parliament on March 26, 1999, a round of public consultation started.[14] People could read the draft-plan and comment on it, by writing their comments to the province. People from the area seized this – first formal – opportunity. As the time for participation closed on June 28, 1999 more than 6.000 letters of protest had been handed in. By far the largest part thereof focused on the intended enterprise zone in the middle of the island. Nearly all of the 6000 letters were a pre-printed letter of protest, handed out by an organization that fought the further development of the Hoeksche Waard.

The form of political participation effectively produced a political deadlock. In the face of 6000 letters of protest parliament, which has the statutory duty to approve or refute all regional plans, one could not possibly accept the plan unchanged. On April 25, 2000 the draft-plan was withdrawn, as it became clear that the national parliament would overrule the province in case it would insist on it going forward.

Form 3. Air Southbound

Against the background of the societal turmoil over the draft-regional plan a cultural manifestation on the relationship between the country and the city started in Rotterdam. It was by no means intended to be the 'next' step in the debate on the area. Far from it: it was meant to be a manifestation on the ways in which the city of Rotterdam might think about its surroundings as for this the future of the Hoeksche Waard seemed a nice case study. The cultural manifestation, called *AIR-Southbound – where the landscape begins,* was initiated by the Rotterdam-based cultural Foundation AIR and started on April 4, 1998. The Foundation AIR had gained a reputation for successfully causing cognitive shifts among the minds of politicians and citizens on a variety of issues in urban design and planning in the years before. Each manifestation of AIR was meant to "*confront* and *stimulate* policy makers, politicians and the world of urban renewal in Rotterdam via international ideas on local problems of urban planning and architecture, by posing relevant questions for research, hand out commissions for design and raise attention to developments in urban planning and design."[15] So far its debates had been restricted to problems within the city of Rotterdam. For AIR Southbound the Hoeksche Waard was chosen because the initiators were convinced this was the area most likely to be mobilized for the next big city extensions. Hence it seemed the best object to get a debate going about the city-country relationships.

If anything, AIR Southbound was targeted as an intervention in the Rotterdam debate and, at best, the national planning debate. In the national debate it aimed to criticize the then standing 'Vinex-strategy' which, it then became clear, created the very same extension of Dutch cities all over the country.[16] According to the initiators, the Vinex-landscape was the result of the dominant way of conducting politics: as negotiation among politicians and administrators. Here all tried to have it their way, but this to the detriment of the landscape. In the circles around AIR this was a big theme. In his capacity as Chairman of the Association of Dutch Urban Designers, Joost Schrijnen had argued for a new role for 'design' in the country, both in the city and outside it (BNS 1995). As they had not been heard through a manifesto, they searched for new sorts of interventions. A few years later the *New Map of the Netherlands*, was presented. A map showing all the plans that had already been accepted, but had not yet been built. This visualization of the future caused a major upset in politics and the media.

AIR was connected to this line of thinking. Here too the idea prevailed that specific (forms of) practices of politics produced bad planning and ruined the countryside. In this regard AIR was also a practice that saw 'good design' as the way out. In all its previous manifestations AIR aimed to show that merely saying 'No!' was not a good strategy. It wanted to put the value of good design and professionalism across, not by playing in the context of the existing political practices, but by *staging* concrete design interventions. Their hypothesis was that politicians and policy makers paid far too little attention to the ways in which cognitive shifts occur, how people are suddenly captured by new insights, indeed, how settings can produce a particular politics and a 'quality' landscape.

In this regard AIR aimed to promote a notion called 'research by design', thus blurring the typical division of labor in planning processes. The manifestation was an attempt to try and trace the particular characteristics of the island, its needs and potentials. More concretely, AIR aimed to generate two types of research[17]: research to generate *images*; and research to find *"real [actually] existing meanings and adherences" (reëel bestaande betekenissen)* in the area. In this the practices of AIR were based on a particular analysis of the shortcomings of the prevailing planning practice. Whereas the provincial planners employed the 'general socio-economic developments in Northwest-Europe' as a starting point of their intervention, it was the conviction of the program committee that the best way to avoid that the Hoeksche Waard would be overrun by the "inevitabilities of the new 'network society'" was to create a 'story line' for the Hoeksche Waard (D. Sijmons). Only if there would be a widely shared idea of the qualities of the Hoeksche Waard, would it be possible to get out of a run-of-the-mill politics.

Hence AIR was a step out of line with the regular practices of policy making. And, as it was successful in achieving some of its aims, it is worth dwelling upon. AIR comprised four phases: (1) exploratory research; (2) an international conference on the city and the country; (3) the commissioning of six designs for the area; and (4) the exhibition and discussion of the designs. In the first phase visual artists, photographers, designers, anthropologists and writers were sent out to try and 'capture' the qualities of the Hoeksche Waard, all under the banner of 'The discovery of the Hoeksche Waard'. Key was the intention to break through the then dominant deductive ideas of 'urban landscape' and 'carpet metropolis' by investigating the concrete case of the Hoeksche Waard. This was published in a magazine, called 'The Discovery of the Hoeksche Waard'. The idea to speak of a 'discovery' is indicative of the fact that the organizers really had the urban audience in mind. From that point of view the title was apt: what might have appeared to them as an island without identity, was now brought to life in diverse and original ways. It was a 'discovery' in that sense.

Yet to the people living in the Hoeksche Waard the title of 'discovery' had a different meaning. For them, the idea of 'being discovered' was offensive and the reactions were very negative: "We are not some rare tribe from the rainforest".[18] Locally the term 'discovery' was interpreted in terms of the frame of reference in which the city of Rotterdam (with its ever expanding harbor) was primarily seen as a 'well-known oppressive force'. This reading of the meaning of 'discovery' also had repercussions for the local attitude towards the manifestation AIR. It was one more 'threat': 'We are prepared for the changes to come'.[19] Hence in the beginning AIR was seen by some as the 'arty vanguard' of the coming Rotterdam-industrialists and the mayors of the six municipalities knew what to do. AIR was positioned as 'yet another plan' 'for us, but not by us', and a row emerged over the fact that the Foundation had not set aside enough tickets for local people to attend its meetings.

Hence AIR, which was intended to be a critique of the prevailing way of conducting the politics of planning, was locally understood *in the very terms of* the practices it aimed to criticize. This was not only due to the title of the first magazine. Apart from that AIR had sent busloads of students of architecture (not easy to overlook because of their dress codes) to the island, had politicians looking at the area from the top of a mobile crane, had others traveling over the island with a hot air balloon, all as part of its attempt to document the 'meanings' on the island. So next to the undoubtedly important discursive framing there was a host of interventions in which artists entered the area in a way that put off the tightly-knit local communities (the area is known as the Dutch 'bible belt'). So its very success in being original in fact-finding in order to be appreciative of many 'structures of adherence' in the area had a flip-side. The staging of AIR had the effect that it appeared as a variation by outsiders with clever ideas for the Hoeksche Waard

A new phase within the manifestation AIR started when the six designer teams presented their plans. Some designs gave people the idea that the future of the Hoeksche Waard could be discussed in several different ways. Without going into the details, in this phase AIR became a stage at which the Hoeksche Waard was discussed in different terms. This phase was illuminating in a different regard: not only were the plans different than the ones from previous occasions, *the setting* was different too. Provincial administrators and politicians, used to having a central role in planning meetings, saw designers and landscape architects center stage. The designers that had been commissioned to elaborate a plan discussed these plans with a wide variety of people in the audience. Different designs created different audiences. Depending on the way in which the designs were presented, the debate took a particular turn. Most notable was the opening up that occurred in the discussion of a plan by H+N+S landscape architects. Whereas most presentations triggered questions and queries from colleagues, their presentation drew in many others. Their plan and presentation allowed the constitution of prolixe and intrigued others who were suddenly interested in and motivated to participate in the debate.

Form 4. Air Plus

A fourth practice in which the future of the area got discussed was AIR plus. Five individuals, all coming from the Hoeksche Waard, felt that the manifestation was too much a matter of talking *about* the Hoeksche Waard and not enough about talking *with* the people from the Hoeksche Waard itself. They had participated in the manifestation and felt that the type of reasoning and visioning might help to bring the political debate on the Hoeksche Waard back to life. They started the 'Hoeksche Waards Initiatief' or *Initiative Hoeksche Waard.* As a first step they brought the exhibition of the AIR designs to the island. The exhibition 'Where the Landscape begins' was put on show in the hall of the headquarters of the water management board in the town of Klaaswaal. It was the first element of what became *'AIR-Hoeksche Waard: what is next?'*, a series of regional debates on the future of the Hoeksche Waard. The exhibition, guided tours and debates resulted in a relatively broadly shared vision on the future of the island. It was laid down in *Het Hoeksche Waards Manifest.* Here the plans the provincial government had presented earlier were criticized. An alternative idea for the development of the area was promoted, based on respect for key qualities such as 'emptiness' and 'greenness', and acknowledging the fact that the area represents different qualities to different groups.

AIR PLUS was an initiative of five individual citizens – a dike warden, a businessman, an architect, an architectural historian and a journalist. Without too much preparation and without a professional publicity policy they managed to intrigue more than

1000 people with the exhibition and the debates, including farmers, teenagers, 'urban import' and politicians. In the resulting *Hoeksche Waards Manifest* key ideas from AIR were reiterated.

Form 5. 'Spatial Organization Hoeksche Waard'

The fifth and final practice was the initiative *Ruimtelijke Inrichting Hoeksche Waard (RIHW).* The RIHW was essentially a practice of administrators and politicians that had emerged to cater for the need to coordinate policies and interventions. Yet the RIHW got a new role through the protest against the draft regional plan and the discussions following AIR. RIHW was a practice that connected these debates back to the circles of government. It published its own manifesto "*The Hoeksche Waard knows what it wants!" (De Hoeksche Waard weet wat zij wil!)* in 2000. This was for the first time that the politicians from the six municipalities came up with a vision on the island could rejoice in widespread support from the residents on the island.

The RIHW has many of the features of what in public administration is called 'network management'. The key is not to have decision-making powers but to organize coalitions that cut through the diverse governmental agencies and to create visions with which people can work. On 30 March 2000 the RIHW staged a debate with approx. 120 councilors and representatives from the water management boards to discuss the future of the Hoeksche Waard. On 11 Oktober 2000 it handed its manifesto to the national Minister for Spatial Planning, Jan Pronk.

The case of the Hoeksche Waard illustrates not only the limits to the form of traditional post hoc participatory practice (the 6,000 letters of protest) but also indicates the many practices through which in this case, a public emerged that shared a particular idea of the Hoeksche Waard and its future. The debate wandered, taking several – mostly unplanned – turns thus constantly changing the discourse on the Hoeksche Waard. In actual fact the players in this debate often comprised the same core of politicians, planners and administrators. Yet the emerging ideas about the Hoeksche Waard were not merely a matter of 'learning' through reiterated deliberation. In this sense the five practices cannot be seen as five 'stages' in a process. Of course, participants had their intentions but in some cases these intensions were markedly different from the outcome that was produced by the scene-act-ratios of the various practices. So for instance, the provincial authorities co-financed the Manifestation AIR Southbound thinking that it might help to prepare the minds for the 'inevitable developments'. The manifestation did indeed help to break through local inhabitants' initial 'don't change anything here!' attitude, still, afterwards they felt 'cheated' by the organizers as the manifestation only made things more difficult for them because it brought about preferences that ran directly against provincial plans. The practice of AIR Southbound transformed how the rapid changes in society would or indeed should impact on the Hoeksche Waard.

Moreover, AIR Southbound and the subsequent AIR-plus! produced a shared identity through an interaction of designers, mediators and inhabitants. And it was an uttered identity which constructed them as political actors who could not be ignored in future plans for the area.

In more general terms the case of AIR and AIR-Plus! underlines the importance of the issue of form in understanding public involvement. Although it had reduced the room for maneuver of the provincial politicians they accepted it to be an exceptional case. Politician J. Wolf spoke of AIR as 'sjeu in de besluitvorming'[20] (the most exciting moment in the decision making) and former regional manager J. Ilsink saw AIR as 'a gift from heaven' ('Een geschenk uit de hemel')[21]: "through the drafting of the agreement HOP we had set free substantial societal forces that subsequently lent their support to the AIR-project. A sort of gift from heaven, AIR, because it took a very different point of entry as HOP which had been inevitably administrative: it was artistic and socio-cultural, which helped to involve more people in thinking about the future of the Hoeksche Waard than I could bring in."

The idea of AIR as the 'sjeu' underlines the importance of staging in understanding the significance of AIR in the political process. Politicians and administrators were well aware that AIR was out of the ordinary in its form. Although we would obviously see AIR as a highly, indeed purely political practice, to politicians and administrators it *appeared* to be a 'safe space' (Forester 1999), i.e a 'quasi non-political' practice. It is not difficult to see why: it started with an 'exploration' (as if it was an expedition), it then continued with a 'conference' (which is a form from academia), to be followed by 'presentations' of the designs, (a form borrowed from the world of architecture and design) and it was concluded with an 'exhibition' in a national museum (a form from the art scene). Yet like in the rats experiment the cage could never be benign, but was always intrusive, AIR had this effect too. It was never the theatre for the 'staging of inevitable things to come', it was a risky form, that transformed the politics of planning in the area. Point here is that it served as an 'interesting' practice. Indeed, it appeared as a good experimental setting of a politics of political involvement as it clearly was a form that was not neutral but which constrained and deformed politics in interesting ways.

What gave AIR this interesting quality? Compared to the two predecessors it was the first practice that broke with the 'policy talk', the expert-dominated conceptual way of doing planning that was evident in the previous practices. It created new entry points into a complex policy conversation, not always pleasing its audience(s) but constantly refracting the debate. In this AIR worked as an 'investigatory design': it was based on a blurring of the discourses and practices of research, narrative, art, design and planning. The enabling effect of this form on public involvement had something to with the wide employment of slides, video, photos, maps. This use of visuals is one we are currently investigating in our joint research on democracy and design.

Next to the actual form of the public presentations AIR introduced various 'staged' acts: excursions, openings, presentations that constantly helped to come back to the issue of what to do with the Hoeksche Waard. If something emerged, it always emerged in the process, never was there a moment of a *fait accompli* against which everybody would revolt. In actual fact one might say "AIR" itself only emerged in the course of its own activities: it was not until at least halfway into the manifestation that people started to realize that this was something special, that everybody wanted to be part of it, and that engaging in its practices became a key moment in the politics of planning.

The quality of AIR in this case study was the fact that the scene-act ratio allowed for a discursive transformation, and one that was not imposed on people but that emerged in the interaction. The quality of a moment like AIR would be that it produced a new sort of discourse, one that created a new and persuasive idea of how people wanted to live.

Both in science and politics there is transformation in the construction of political collective/scientific objects. We have been fascinated by these practices where the issue has not been to limit the influence on the collective/objects during this construction, but to tentatively describe how such influence might occasion qualitatively interesting effects. Thus, the traditional emphasis on (un-coerced) participation is shifted to an interest in the forms of influence (indeed as a form of coercion), which might facilitate participation rather than annul it. The fact that AIR allowed a variety of people and voices into a conversation, without giving up the important role for designers in this ('I am holding the pencil!') produced its particular quality.

The case of the Hoeksche Waard shows not only that forms are productive, that they participate in the emergence of a group of 'Hoeksewaarders' that cannot be overlooked in later policy-making, but the case also draws our attention to a group of actors who are explicitly attentive to just this phenomenon, the setting up of forms which impact decision and participation. These actors, landscape architects and designers, have been increasingly mobilized in the Dutch policy process to improve the quality of spatial planning. Our analysis could take two turns if we were to do classic STS or policy analysis: (1) we might emphasize (like the authors criticized by De Wilde) whether the experts (architects) efface themselves to allow the laymen to take part in the debate; (2) we might emphasize (like Annemarie Mol or John Law) the norms or modes of life implicitly inscribed in the practices of the architects and how these are transformed in their encounter with practices of policy making. We prefer however to (3) investigate when the architects' plan becomes a 'robust political plan'; what are the traditional architectural techniques used and how are they modified to tentatively construct this plan as 'political'. How does a design drawing circulate and what work does it do as it suddenly takes on a political role? What is the concrete difference between an *architectural* design and a design engaged in *political* action? Following the drawing then is one way of making politics into an empirical question. This third approach of politics, we suggest, is one we way we hope to articulate STS with policy analysis' project to describe new and as of yet unacknowledged political forms.

CONCLUSION: RETHINKING POLITICS

This paper is an argument for employing the empirical tactic of STS to investigate techniques (of drawing, speaking, writing, building, showing, arguing etc.) of contemporary politics. It is an argument aiming to strengthen the development of a new policy analysis and thus open up prevailing academic practice in political science. In the end, it might also offer a glimpse of forms of new democratic governance.

Inspired by literature in both STS and political science, we are committed to the description of a wide range of contemporary forms of political practice. We tentatively described the orientation of policy analysis and the potential contributions of STS to that field. Such an alliance between STS and policy analysis would concern the investigation of forms and settings: an attentiveness to the *concrete conditions* of political practice and a hypothesis about the *positive* effects these biases and constraints might have on political action.

Behind this search for the forms of politics, however, is a concern about democracy and democratic politics. Much of the policy analysis work that derived its tone and orientation from a fight against positivism or for more democratic state-institutions (such as in the call for participatory policy making in the previous decades). Yet we are less sure about what the practices of a democratic politics in our age will and should look like. We doubt that there are institutional arrangements that are unproblematically democratic, no matter what is the task at hand. We argue we should open up our idea of politics, before turning to the redefinition of democracy. This position is markedly different from the idea that this is the time to introduce a 'new' set of arrangements (e.g. "*Deliberative* Democracy – Washes twice as democratic!"). It is much more the realization that we have to rethink politics, and *from there* address the question of democratic politics. Put differently, just like STS has a long tradition of warning for 'technological fixes' we would like to warn politics for new 'institutional fixes'. No institution can *guarantee* a democratic process. There is a tendency among those who recognize a lack of democracy to grab for the 'new institutions' (see Haas et al. 1996; Held 1997). Instead, we attempt to open up the definition of politics, calling for an appreciation of the *meandering,* searching, quality of the contemporary political processes. Each form we describe proposes a retractable, temporary hypothesis about what it is and how it is to be done. This particular take on politics we would like to call 'a symmetrical anthropology of politics,'[22] to analyze with the same methods what we take to be evidently democratic and non-democratic practices of constituting collectives.

Despite our professed desire to 'minimalize' our definitions of the political for the moment, we do not suppose that our opening up is a simple reduction or suspension of biases and assumptions. Instead, we try to devise another set of biases, another way of phrasing the question ('what is politics?') which might allow us to be *surprised* by contemporary 'politics'.

As we stated above, surprise is not meant as the new, funky transcendent criterion for good anthropology or good politics. Rather it is one legible sign that might suggest that the results of our study are not a simple reification of our prior biases but something slightly different. A good experiment is one where the setting did not force the rat to obey and repeat the experimenter's first hypothesis.[23] It is one where the rat was given a chance to deviate the experimenter's expectations.[24] Thus in the case of the rats, one crucial question is: How did the larger cage constitute the occasion for the experimenter to put herself and her hypotheses at risk? The setting did not become risky simply because the cage was large. Rather it became risky when the cage of the rats was enlarged. More precisely, the variation in the sizes of the cages make room for differences to appear between the phenomena. And as the contrast appears, the rat sings surprisingly and the experimenter becomes capable of hearing songs she had never heard. The different cage sizes when juxtaposed allow an unexpected event to appear. The *changes* in the cage is the set up which drags the experimenter out of her day to day dullness and into a 'fine awareness' (Nussbaum), a capacity to be shocked, to perceive difference and novelty. This renewed capacity to perceive difference is *not* the elimination of bias and impurities in an otherwise neutral setting, it is *"not a pulling back of a curtain, breaking a veil that impedes knowledge"*. Instead it is through the uncertain variation of the setting, through the *oblique* mediation of an unstable experimental set up, the addition of another bias (the large cage) that the experimenter "*comes into concussion*"[25] with a surprising rat. Extrapolating to the Hoeksche Waard, we might say that it is the meandering succession of political forms, which gives this case its specificity. The slight shifts and differences in political stagings, presentations, in the drawings and plans circulated were as many occasions for the actors to react differently and perhaps interestingly. Of course, in politics there is a moment of 'formal' decision making and the Hoeksche Waard is no exception to that. Yet it would be bizarre to reduce the political analysis to that moment, just as it would be wrong not to see it as a staged moment in a string of stagings.

If politics is less clearly defined and demarcated, there are no fixed criteria that might allow us to guarantee the quality of our choices when building an experimental set-up or, by extrapolation, designing a political form. The absence of guarantee is daunting. No one can know in advance whether a particular shift in the setting will make this a 'good' experiment or 'good politics'. Further, no one can define a priori what is 'politics'. Both in the case of the Hoeksche Waard and of the rat cages, the risk is built in to every variation of the *dispositif*[26] as well as on the thin chance that one of these variations will make a new planning perspective emerge, or make the rat "sing out". Both in politics and psychology, the experimenter then must relentlessly transform his/her setting, knowing that each shift can have potentially dramatic consequences. This definition reveals a stuttering and sleepless (scientific and political) experimenter who as soon as s/he settles upon a form pertinent to a particular representation, has to throw it out for a new one.

Nussbaum draws the portrait of this experimenter (though for her, it is an artist). For Nussbaum, our highest task is to make ourselves people 'on whom nothing is lost' (Nussbaum quoting James, in *The princess Casamassima*). Artists, according to her, have a finer awareness of the particularities, the details of the world. Through art, the reading of novels in particular, others can learn from the artists' attentiveness to detail and "strain to be people on whom their subtleties are not lost". Risky scientific and political forms then are those which tentatively make the designers of these settings into "someone on whom nothing is lost". This fine awareness is precisely what makes these actors aware of the total uncertainty of the experimental/political situation. This awareness is not a cognitive attribute or moral value of the subject itself, it is constructed by the setting, the form. The *setting* is what can "*grind [the experimenter] to a halt, suspend him in time, slacken the countdown of seconds*"[27]. The chaos produced by a setting (when capacities are swapped and new entities emerge and mediate) is what *artificially* slows it all down, breaks the anaesthetized rhythm of routines and allows one to be affected by what had not been visible.

Our will to empiricize the question of politics explains our desire to be surprised. Only if we are surprised have we succeeded (temporarily) in actually opening up the question "what is politics?" to the extent that the answer might have been unexpected at the start. Such an approach finds echo and support from one classic author in democratic theory. In the work of John Dewey, 'politics' can be understood as the specific work of evoking and constructing collective actors (like 'society', 'state' and 'the people') which would not exist otherwise. The specific task of the political is to construct these collectives and their interrelation. What characterizes the range of Western democratic politics for Dewey is not a specific fixed institutionalized form, like free elections or a parliament of representatives. Democracy, we read in Dewey, is precisely the constant flux and experimentation with different political forms that are spurred by the contestation existing ones elicit and which construct different collectives and articulations among them.

These collectives, it is noteworthy, are not pre-existing entities they emerge through political action. 'The public' for Dewey is not a population of a country (numbers fixed, identities flattened out in surveys). It refers, he explains, to the negative consequences of the development of scientific, economic activities at a given time, expressed as the "need [of] and harm [to]" a group.

> The public consists of all those who are affected by the indirect consequences of transactions to such an extent that it is deemed necessary to have those consequence systematically cared for. Officials are those who care for the interests thus affected. Since those who are indirectly affected are not direct participants in the transactions in question it is necessary to [have officials to] see to it that their interest be conserved and protected. The buildings, property, etc. involved in the exercise of this office are res publica, the common wealth. The public as far as organized by means of officials and material agencies to care for the extensive and enduring indirect consequences of transactions between persons is the Populus (Dewey 1956, 16).

'The public', then, is a category emergent through scientific, technological and political practices. The 'state' too is an emergent category referring to the actions necessary to defend and protect this harmed 'public'. Politics then is the simultaneous construction of 'state' and the 'public', individuals and collective, represented and representative.

What is key for us, who appreciate surprise, is that the state, its forms, the actors and practices it refers to cannot be fixed in advance. They (must) change incessantly to ensure that the 'public' is defended by 'the state'. If new practices emerge and create new needs, existing political forms are by definition insufficient. Political forms should vary, Dewey writes, not only because every attempt to define and defend a public is imperfect and none deserves to be made eternal, but because the 'public' changes. Just as we are trying to do, Dewey then makes a plea, not for a certain kind of political form, but for their incessant variation.

> By its nature, a state is ever something to be scrutinized, investigated, searched for. Almost as soon as its form is established, it needs to be re-made (Dewey 1956, 31).

As soon as a political form is established, prepare for a new one! Democracy, the good politics we are in the end preoccupied with, is not the guaranteeing of process through the fixing of certain institutions. Rather it is this openness, this preparation for unexpected forms and their surprising positive effects.

Amsterdam School for Social Science Research, Amsterdam, The Netherlands

NOTES

1 This terms is intentionally paradoxical: referring to the present political institutions in this way we aim to highlight the fact that they are spatial-temporal constructions, the defense of which might be based in the social and political theories that were the product of the same *discourse* that gave rise to the institutions in the first place. See Beck 1994, Hajer 2000, but also Wagner 2001.

2 Annemarie Mol, *The body multiple (mimeo)*, chapter 6.

3 See Anemarie Mol, *The Body Multiple*, under review at Duke University Press.

4 See for example Sclove, R. 1995 and Pinch, T.& Bijker, W. 1989.

5 (Goody; Hutchins; Norman; Lynch; Berg and Akrich: Gomart and Hennion 1999)

6 (Foucault, Elias).

7 De Wilde also criticizes these authors for being constructivists for science and purists for politics. Yet he too worries about construction (the transformations and deformations that occur as citizen jury members for example are trained in preparation for their work). He can only be concerned about the social engineering which this construction might imply and is incapable it seems of conceiving of a defomration that might benefit the political process and the participants. He does not see the benefiot of bias, a concept we develop below.

8 Another word she uses is pertinence, showing that to interest then has something to do with a concern for the specificity of the phenomenon. This is not a regression to a search for the authentic essence of the rat. Specificity is constructed in the same experiment which cares for it.

9 To the extent that the value and importance of the *dispositif* or *bias* for a notion of 'good politics' is being recognized this is the case in the tradition of postmodern political theory (e.g. Shapiro 1992) or in work on aesthetics in politics (Ankersmit 1997).

10 To be sure, the aforementioned authors Innes and Booher do not fall into this trap.

11 Officially called the "Gebiedsuitwerking Hoeksche Waard 2010-2030 / Verkenning Inrichting Lange Termijn Hoeksche Waard (VILT-HW).

12 In Dutch: 'Kwalitatieve positionering van de Hoeksche Waard in het spanningsveld van het stedelijk gebied van Rotterdam en de Drechtsteden en het landelijk gebied van de Rijn-Schelde Delta.' The agencies involved were the province South-Holland, the watermanagement board (Waterschap) De Groote Waard as well as the six municipalities from the Hoeksche Waard: Binnenmaas, Cromstrijen, 's-Gravendeel, Korendijk, Oud-Beijerland. en Strijen.

13 http://www.pzh.nl/index.html?/actueel/streekplan/szh-4-3.htm

14 Please note that this differentiation between policy making and participation was always seen to be a positive feature of the Dutch planning system. The idea being that the state should not bother its citizens with plans that it was not in complete agreement on. As a consequence, the plans are only announced once the various governmental players are in agreement (and in this case some key players from trade and industry as well). For an explanation of the peculiarities of the Dutch Planning model, see, Scientific Council for Government Policy 1998, Hajer & Zonneveld 2000.

15 Notitie Programmawerkgroep AIR-Rotterdam Zuidwaarts, april 1997.

16 Minutes of the third meeting of the program committee, 4 November 1996. 'Vinex' was the acronym of the then binding national white paper on urban planning.

17 Minutes of the Sixth meeting of the program committee, 5 March 1997.

18 Mayor B. Kolbach in *Kompas,* 10 September 1998.

19 Mayor B. Kolbach in *Kompas,* 10 September 1998.

20 *Algemeen Dagblad 19 juni 1998.*

21 J. Ilsink, voormalig streekmanager Zuid-Holland Zuid, *RIHW Nieuwsbrief,* Jrg.1, nr.2, p.4.

22 Symmetry here refers to the methodological aim constitutive of STS not to presuppose that science and non science is a difference which can be explained by the inherent qualities of science: objectivity, neuitrality etc. It suggests that you allow that both science and non-science can be treated with the same social scientific tools because both are social, political thru and thru. Though symmetry has recently been contested in STS, it can be useful again to constituted a field of anthropological study of politics. The symmetry here is between what we had assumed to be politics, good politics, democracy and what we had not. That we are prepared to explain with the same methods what we might have previously distinguished on normative grounds.

23 Diverse works in the sociology of technology have emphasized this point: an entity cannot be defined apart from the practices that engage (others with) it:The bridge is "low" in Winner's example, only in relation to the public busses that are blocked by it. The gun, in Latour's example, is light and wieldy only in the angry amateur's hand. The patient is prolixe, in Mol's case, in the medical practices that give him cause to speak...The entity's very specificity, its authentic qualities, its *essence* acquires meaning only in relation to

the others it is connected to. Essence is a *construct*, the *result* of these practices, rather than the raw material, the given with which these practices must begin with (See for example, Latour 1999). The exact nature of the construction of this essence has been the topic of much controversy in science studies: If the object cannot be defined independently of the setting can we say that the setting determines and fixes what the object is and does? ANT authors have preferred to describe "mediators", objects that are not entirely determined by the setting and whose actions "exceed" any predictions. The descriptions of the delegations to objects and the unexpected outcomes, re-distributions of tasks, competences, responsibilities incurred, all serve to illustrate that objects "mediate", "resist", "prolong and deviate", "faire faire".

24 The unexpectedness of action is the characteristic of mediation as opposed to determinism. Objects in the sociology of technology of ANT are characterized by the unpredictability of their action. They do more than what was expected. Here this "excess" is taken to stand for what characterizes the action of non-objects in experimental settings. Because animals and human beings are much more influencable and vulnerable to suggestion than objects, Stengers for one argues, to say that a human mediates becomes a normative criteria, the sign that something went well as the human was constituted as an entity just as strong, just as capable of resisting as objects in physics experiments (which can explode, break, remain still with much more stubbornness than humans). Latour illustrates the importance of biases for interesting transformation: "*Speaking of her new study, [another ethologist working not w rats but w sheep] stated one of her "biases" in the following way: "I tried to give my sheep the opportunity to behave like chimps, not that I believe that they would be like chimps, but because I am sure that if you take sheep for boring sheep by opposition to intelligent chimps they would not have a chance". A whole new philosophy of scientific practice resides in this extraordinary statement: "to give the opportunity to behave" is not the same thing as "imposing a bias onto" animals that cannot say a thing. ... By importing the notion of intelligent behavior from a "charismatic animal" [the monkey] ... she might modify, subvert, elicit, in the understanding of sheep behavior features that were until then invisible... It is* because *she artificially and willingly imposes on sheep another resource coming from elsewhere that "they could have a chance" to behave intelligently* (Latour 2000, 10).

25 The italics here are expressions borrowed from Nussbaum, who used these in a different argument.

26 For the notion of *dispositif*, see Michel Foucault (1975) and also the debate, partially summarized in Latour (1999), on the roile of the experimental setting in the construction of the scientific object.

27 James has described a moment of pure experience, as a moment when one is aware of one's respiration. This has been beautifully developed by Desprest, op cit. xx.

REFERENCES

Ackerman, Bruce (1992), *We, the People. Foundations*, Cambridge, MA: Harvard University Press.

Ankersmit, Frank (1997), *Aesthetic Politics: Political Philospohy Beyond Fact and Value*, Stanford, CA: Stanford University Press.

Beck, Ulrich (1992), *Risk Society – Towards a New Modernity*, Cambridge, UK: Polity.

Beck, Ulrich (1994), 'The reinvention of politics: Towards a theory of reflexive modernization', in U. Beck, A. Giddens and S. Lash (eds.), *Reflexive Modernization. Politics, Tradition and Aesthetics in the Modern Social Order*, Oxford: Polity Press, pp. 1–55.

Beck, Ulrich (1999), *World Risk Society*, Cambridge, UK: Polity.

Berg, Marc and Madeleine Akrich (eds.) (in press), *Bodies on Trial* (Special Issue of *Body and Society*).

Bobrow David and John Dryzek (1987), *Policy Analysis by Design*, Pittsburg, PA: The University of Pittsburgh Press.

Bohman, James (1996), *Public Deliberation: Pluralism, Complexity and Democracy*, Cambridge MA: MIT Press.

Burke, Kenneth (1969 [1945]), *A Grammar of Motives*, Berkeley, CA: University of California Press.

Dewey, John (1999 [1927]), *The Public and Its Problems*, 2nd ed. 1954, Athens, Ohio: Swallow Press.

Dryzek, John (2000), *Deliberative Democracy and Beyond*, Oxford: Oxford University Press.

Edelman, Murray (1964), *The Symbolic Uses of Politics*, Chicago: University of Illinois Press.

Elias, Norbert (1994), *The Civilizing Process* (original German version 1939), Cambridge, UK: Blackwell Publishers.

Fischer, Frank (1980), Politics, Values, and Public Policy – The Problem of Methodology, Boulder, CO: Westview Press.
Fischer, Frank (1990), Technocracy and the Politics of Expertise, London: Sage.
Fischer, Frank (2000), Citizens, Experts, and the Environment: The Politics of Local Knowledge, Durham, NC: Duke University Press.
Fischer, Frank (1998), 'Beyond Empiricism: Policy inquiry in postpositivist perspective', *Policy Studies Journal* 26(1): 129–46.
Forester, John (1999), *The Deliberative Practitioner,* Cambridge, MA: MIT Press.
Foucault, Michel (1975), *Surveiller et punir: Naissance de la prison,* Paris: Gallimard.
Garfinkel, Allan (1981), *Forms of Explanation. Rethinking the Questions in Social Theory,* New Haven, CT: Yale University Press.
Glickman, Stephen (1996), 'Disciplines, subdisciplines and cultural influence: A view from comparative psychology', paper presented at the International Symposium of the Wenner-Greeen Foundation for anthropological research: *Changing Images of Primate Societies: The Role Of Theory, Method, Gender,*" Rio de Janero.
Gomart, Emilie (in press), 'Eloge of Methadone', in M. Berg and M. Akrich (eds.), *Bodies on Trial* (Special Issue, *Body and Society*, under review). (Dutch version appeared in *Krisis,* 2000, vol.4).
Gomart, Emilie and Antoine Hennion (1999), 'A sociology of Attachment: Music amateurs and drug users', in J. Hussard and J. Law (eds.), *ANT and After*, Oxford: Blackwell, pp. 222–47.
Goody, Jack (1977), *The Domestication of the Savage Mind*, Cambridge, UK: Cambridge University Press.
Gunsteren van, Herman (1998), A Theory of Citizenship. Organising Plurality in Contemporary Democracies, Boulder, COL: Westview Press.
Gusfield, Joseph R. (1981), The Culture of Public Problems: Drinking-Driving and the Symbolic Order, Chicago: Chicago University Press.
Gutman, Amy and Dennis Thompson (1996), *Democracy and Disagreement*, Cambridge, MA: The Belknap Press.
Hajer, Maarten A. (1995), *The Politics of Environmental Discourse. Ecological Modernization and the Policy Process*, Oxford, Oxford University Press.
Hajer, Maarten A. (1996), 'Ecological modernisation as cultural politics', in S. Lash, B. Szerszynski and B. Wynne (eds.), *Risk, Environment & Modernity. Towards a New Ecology*, London: Sage, pp. 246–68.
Hajer, Maarten A. (2000), *Politiek als Vormgeving* (inaugural Address), Amsterdam: Vossius Press.
Hajer, Maarten A. (2003, forthcoming), 'A frame in the fields: Policy making and the reinvention of politics', in M. Hajer and H. Wagenaar (eds.), *Deliberative Policy Analysis: Understanding Governance in the Network Society*, Cambridge, UK: Cambridge University Press.
Hajer, Maarten A. and Hendrik Wagenaar (forthcoming), 'Introduction', in M.A. Hajer and H. Wagenaar (eds.), *Deliberative Policy Analysis: Understanding Governance in the Network Society,* Cambridge University Press.
Held, David (1997), *Models of Democracy,* 2nd ed., Cambridge, UK: Polity Press.
Hutchins, Edwin (1995), *Cognition in the Wild*, Cambridge, MA: MIT Press.
Innes, Judith (1996), 'Planning through consensus building: A new view of the comprehensive planning ideal', *Journal of the American Planning Association* 62(4): 460–72.
Innes, Judith E. and David E. Booher (2000), *Public Participation in Planning. New Strategies for the 21st Century,* Berkely, CA: University of California, Institute of Urban and Regional Development.
Lasswell, Harold D. (1971), *A Pre-view of Policy Sciences,* New York: American Elsevier.
Latour, Bruno (1999), *Pandora's Hope: Essays on the Reality of Science Studies,* Cambridge, MA: Harvard University Press.
Latour, Bruno (2000), 'A well articulated primatology: Reflections of a fellow traveller', in S. Strum and L. Fedigan (eds.), *Primate Encounters: Models of Science, Gender and Society,* Chicago: University of Chicago Press, pp. 358–81.
Latour, Bruno (fortcoming), 'Being affected', in M. Berg and M. Akrich (eds.), *Bodies on Trial* (Special Issue, *Body and Society*, under review).
Latour, Bruno (in preparation), "Si l'on parlait un peu politique?".
Lefort, Claude (1981), *L'invention democratique,* Paris: Fayard.
Lukes, Steven (1974), *Power: A Radical View,* London: MacMillan.
Lynch, Michael (1991), 'Laboratory space and the technological complex: An investigation of topical contextures', *Science in Context* 4: 51–78.

Lynch, Michael and Steve Woolgar (eds.) (1988), *Representation and Scientific Practice*, Cambridge, MA: MIT Press.
Manin, Bernard (1997), *The Principles of Representative Government*, Cambridge, UK: Cambridge University Press.
Mol, Annemarie (2002), *The Body Multiple*, Durham, NC: Duke University Press.
Nelson, Barbara (1996), 'Public policy and administration: An overview', in R.E. Goodin and H.-D. Klingemann (eds.), *A New Handbook of Political Science*, Oxford: Oxford University Press, pp. 551–92.
Norman, Don (1987), *The Psychology of Everyday Things*, New York: Basic Books.
Nussbaum, Martha (1990), *Love's Knowledge*, Oxford: Oxford University Press.
Pierre, Jon (ed.) (2000), *Debating Governance: Authority, Steering, and Democracy*, Oxford: Oxford University Press.
Pinch, Trevor and Wiebe Bijker (1989), 'The social construction of facts and artifacts', in W. Bijker, T. Hughes and T. Pinch. (eds.), *The Social Construction of Technological Systems*, Cambridge, MA: MIT Press, pp. 18–50.
Putnam, Robert (2000), *Bowling Alone*, New York: Simon & Schuster.
Rancière, Jacques (1998), *Aux bords de la politique*, Paris: La fabrique.
Rosenau, James N. (ed.) (1995), *Governance Without Government. Order and Change in World Politics*, Cambridge, UK: Cambridge University Press.
Schattschneider, Elmer E. (1960), *The Semi-Sovereign People*, New York: Holt, Rinehart & Winston.
Sclove, Richard (1995), *Democracy and Technology*, London: The Guilford Press.
Shapiro, Michael J. (1992), *Reading the Postmodern Polity: Political Theory as Textual Practice*, Minneapolis, MN: Minnesota University Press.
Stengers, Isabelle (1993), *L'invention des sciences modernes*, Paris: La Decouverte.
Stone, Deborah (1988), *Policy Paradox*, Durham, NC: Duke University Press.
Torgerson, Douglas (1997), 'Policy professionalism and the voices of dissent: The case of environmentalism', *Polity* 29: 345–74.
Tribe, Laurence (1997), 'Policy science: Analysis or ideology?', *Philosophy & Public Affairs* 2: 66–110.
Wagenaar, Hendrik and Scott N. Cook (2003, forthcoming), 'A frame in the fields: Policy making and the reinvention of politics', in M. Hajer and H. Wagenaar (eds.), *Deliberative Policy Analysis: Understanding Governance in the Network Society*, Cambridge, UK: Cambridge University Press.
Wagner, Peter (2001), Theorizing Modernity – Inescapability and Attainability in Social Theory, London: Sage.
Wilde, Rein de (1997, January), Ironies of the Eighth Day. Reflections on the Modern Faith in Compatibility of Community, Democracy and Technology, paper presented at the MTV Conference on Technology and Democracy, Oslo.
Yanow, Dvora (2000), 'Seeing organization learning: A cultural view', *Organization* 7(2): 247–68.

CHAPTER 4

YARON EZRAHI

SCIENCE AND THE POSTMODERN SHIFT IN CONTEMPORARY DEMOCRACIES

The rise of the modern conception of politics, unguided and unsecured by divine benevolence, the emergence of politics as the exercise of human freedom with its potential defects no less than its promise, rendered the issue of restraint, of the limits on arbitrary political power, perhaps the most acute dilemma in the modern polity. Much of modern political theory is preoccupied with the issue of how can free political agents be restrained and what are the limits of the political.

My purpose in this discussion is first to show that from a historical perspective, science was often conceived as a principal source of apolitical authority with the capacity to discipline political action, criticize the decisions and programs of governments, and generally set limits to politics in the secular state. I shall also discuss the possibility that in some cases the very uses of science and technology to rationalize and depoliticize the exercise of power in fact has allowed power to be exercised uninterrupted in disguise. I would like to suggest that in the context of politics, scientifically certified 'reality', or the 'necessary laws of nature' could gradually replace the will of god and the injunctions of religion as usually latent constraint on the free and, therefore, potentially arbitrary will of political agents. I shall continue to argue further against this background that the shift from modernity to post-modernity in the sphere of politics, has involved a significant erosion of the ability of scientific knowledge and institutions to continue and constitute such a principal source of apolitical authority with the capacity to appear as setting limits and imposing some discipline on political power. I shall suggest further that due to deep changes in the structures and dynamics of contemporary politics the issue of limiting arbitrary power has been recast, focusing attention on alternative strategies of limitation and protection which depend much less than before on any substantive or cultural aspects of science.

The perceived authority of science to set limits to legitimate public actions and discipline the exercise of political power in modern democracies depended on the evolution of an instrumental concept of politics in terms of which the exercise of political power is at least partly defined and controlled by experts who apply technical standards of action (Ezrahi 1990). The authority of science and of science-based-technology to instrumentalize or de-politicize the exercise of political power in key spheres of public action was

B. Joerges and H. Nowotny (eds.), Social Studies of Science and Technology: Looking Back Ahead, 63–75.

facilitated by, three, partly interrelated, key conditions. The first was the belief that science provides a non-arbitrary, impersonal and therefore rarely contestable account of the realities or the facts of the world, which are relevant to effective actions. The neutrality of scientific knowledge is here secured by the assumed uniform lawfulness of reality as a whole and its status as a unified and independent object of scientific intelligence unaffected by diverse social, political, and cultural contexts. The second condition was the existence of hegemonic inclusive normative perspective on the social and political order and the place of 'nature' in relation to it. The presence of such foundational-normative system or state perspective on politics, society and nature implicitly facilitated the delineation of a sphere of dogmas, or conventional truths, which could be perceived therefore as apolitical, and facilitate the social framing of science and technology as indifferent to politics and therefore as having the capacity to depoliticize attitudes and actions based on, or even merely compatible with, expert knowledge. James Scott had shown how the synoptic gaze of the state which empowered many monumental, often disastrous, state projects like planned cities and scientific agriculture, consisted of such convergence of scientific outlook with a collective normative state perspective (Scott 1998). As I indicated above, however, depoliticization understood as a check on politics could function in these cases also as a rationale for increasing the actual scope of state interventions. The accommodation of science and technology to 'work' within the confines of apolitical mandates of action was in any case a prerequisite for scientific authority in order to appear distinct from political authority. In many respects this condition relates to the ethos and history of the autonomy of science, the freedom of research, and the institutional separation of academic or scientific and political institutions. The third condition for the emergence of science and actually, even more so, technology, as a means of limiting political authority in the modern (usually, but not exclusively, democratic) state, was the belief that scientific truths about the world not only give rise to universally valid knowledge of physical (and later social and economic and even political) reality, but also to a universally valid, politically neutral, best ways to act upon the world in the pursuit of human goals. This belief has rested on the assumption that technology combines universally valid scientific knowledge of the structure of the world with universal human goals and can therefore escape the sphere of partisan ideological or political contests (condition b). These three conditions are, of course, not entirely separable. In a society where enlightenment culture is hegemonic, scientific accounts of the world are more likely to be perceived as neutral or objective than in a traditional, religious, or normatively divided society. In many respects a hegemonic secular value perspective is not only congenial for the admissibility of scientific accounts of the world (condition a) but also for the willingness to privilege science based technologies as neutral modes of action (condition c). The power of science and technology to impose limits on political claims and actions has derived largely from the capacity of the polity to extricate areas of tacit or explicit consensus from the sphere of regular political contests. This is not entirely different from the process through which issues, which might have been in the past the subject of political contests, are recast as issues regulated by the law. Also here a basic norm which is elevated above a multiplicity of competing values and interests facilitates the separation between law and politics. So in the final analysis conditions a and c depend to some extent on condition b. But condition b can be congenial for the integration of scientific knowledge and authority and politics only when conditions a and c are also present.

The first factor, the authority of scientific accounts of reality, has derived largely from the religious metaphysical origins of modern science and particularly from the claims of scientists to have inherited the God's eye synoptic view of nature and society (see, for instance, the dedication in Nicolas Machiavelli's *The Prince*). Early theological notions of the world as a coherent whole, a lawful clock-like object created by the divine engineer, were later replaced, of course, by metaphysical assumptions about nature, and later still even by leaner methodological postulates (Della Mirandola 1942; Burtt 1954; Manuel 1974; Funkenstein 1986). But even as an intellectual requirement, or merely an aesthetic ideal, the notion that the truths of science are universal and that scientific knowledge, regardless of its particular agents or spokesmen, is not just a local knowledge, was essential to enable science to represent as uncontroversial, a slice of experience called, among other things, 'nature', and fix it as a referent of the apolitical aspect of public actions, a limit on the potentially arbitrary human will of political agents.

It was here that the 'inexorable laws of nature', 'objective reality' or 'hard facts', could enter the contexts of political discourse and action as limits imposed by necessities beyond human control, a role formerly played by the divine will. This scientifically certified conception of reality, to be sure, has never fully conquered the images of reality in the common experience. Religion, magic, superstition, myth, conventions and fantasy continued to mediate large parts of ordinary life (see, for instance, Ezrahi 1986). Nevertheless, the introduction of the modern concept of lawful and intelligible universe, reinforced a deeper appreciation of the dualism between 'fantasy' and 'reality'. It opened the way for modern political theorists like Machiavelli and Hobbes to limit the impact of the religious imagination on the exercise of power, or help secularize politics by making distinctions between the ways peoples and societies 'really' are, and the ways an 'unfounded' imagination leads us to imagine or desire them to be. Thus, for Machiavelli, the prince must limit and discipline his actions in light of what comparative political and military history instructs are the probable causal links between different kinds of actions and their effects.

Because the 'reasons of state', the arguments for action which are based on the supreme goals of maintaining order and securing the survival of the state which are supposedly grounded in a cool, amoral, and a-religious conception of reality were still hidden from the sight of the larger public, and often subversive and repugnant to its sensibilities, the prince is advised to live and act in a dual universe. In matters of state he must act according to knowledge of the probable causes of the stability or collapse of states. But in order to elicit the approval of his subjects (which is also an aspect of 'real politics'), he must manage his public appearances and representations such as to repudiate or disempower their moral criticisms. In order to do that he needs to appear as a 'natural' actor in what in fact is an imaginary political world created by the public's belief that politics and morality are inseparable. A good political actor, according to Machiavelli, must know both what works to preserve the stability of the state and what overt theatrical postures are necessary to legitimate his actions before the public. The Machiavellian statesman must be a master of both statecraft and stagecraft, of 'real politics' and its necessary or useful instrumental misrepresentations, as two parts of the same act. He must be an actor in both senses of the word. In the Machiavellian political universe, knowledge of reality as a means of limiting and disciplining action, includes knowledge of the distinctions between fantasies which can support and those which can subvert informed instrumental action. The science of politics includes, therefore, also the knowledge of how to use political theatrics in order to cover up and protect the sphere of coldly instrumental political actions. This is necessary, according to Machiavelli, because of the tendency of the subjects to perceive the conduct of the prince in moral terms. It is, of course, remarkable how persistently relevant this Machiavellian instruction has been regarded by modern political leaders, and how much this dualism between necessary actions and overt gestures continues to be a part of political experience. The focus of contemporary politics on image building and image management, is in fact but a contemporary variation on the same theme.

With Thomas Hobbes the logic of political reality is, of course, much more widely transparent. Hobbes worked within a political universe in which the subjects where expected to play a more active role in the construction and sustenance of political authority, a world in which laws define more inclusive areas of freedom and constraint for political agents who formerly operated within the more concealed universe of courtly politics. It is, of course, very instructive that Hobbes uses 'nature' and the ideas of 'laws of nature' to limit and shape the political universe, and that the structure of his argument is so deeply influenced by the contemporary scientific imagination and by mathematics and geometry as models of knowledge and discourse. In different forms, the idea of knowable social, political, and economic reality continues to reign in the works of leading modern political and social theorists like Montesquieu, Condorcet, Marx, Benthem, J. S. Mill, Durkheim, Schmitt and modern academic political scientists. All enlisted science to criticize and discredit what they held where sheer human fantasies and rationalize the exercise of political power in a world conceived as consisting of 'real' and discernable causes and effects. Of course, also among the above thinkers, often, what was a political fantasy or baseless imagination for some, were hard facts of political reality for others. What is important for our purposes is, however, the persistent dualism they sustained between scientific and imaginative notions of politics, regardless of where the demarcation lines are drawn. In addition, of course, also the physical and the biological sciences played an increasingly important role in defining and representing nature as a factor in the shaping of public actions in spheres like public health, transportation, war, the environment and the like.

As I already indicated above, ironically the scientification of the structure of matter, the nature of epidemic, or the properties of chemicals, and the limits nature-through-science imposed on the religious, magical, or moral rationales of conduct, were also the very conditions which enhanced the powers of the modern state. Science, by advancing secular amoral and apolitical frames of state action both constrained and empowered the state. The perception of scientific and technological reasons for action as apolitical, the sense that they impose limits on the arbitrary uses of political power could facilitate an actual extension of political power not just its limitation. The fact that such processes as deploying systems of transportation and communications, the provision of medical services and electrification involved various kinds of visible hardware only helped the state to represent its interventions in these areas as scientifically informed applications of techniques to the physical world thus concealing the political 'software' of technical choices. What is important for our discussion is that at first this empowerment appeared consistent with the perception of politics as a field of action constrained by extrapolitical facts. In this respect the continuity between the role of science and religion as sources of extrapolitical authority was crucial for the evolution of the modern conceptions of legitimate state powers. Political and social facts, to be sure, could not be represented with the same confidence and decisiveness as physical facts. The drive to geometrize or mathematicize social and political discourse and action could not go nearly as far as it did in relation to the discourse on nature and physical actions, but it went far enough (Porter 1986, 1995). Much of the attraction of statistics and economics has derived from their capacity to stir up such hopes that social and political facts would become progressively quantifiable, that they will provide the tools to better distinguish facts from fictions in the contexts of human affairs and furnish the resources for checking arbitrary powers and authorities. With the proliferation of technologies the success of the physical sciences in 'capturing' physical nature encouraged the aspirations to use social technologies to similarly capture and direct the Leviathan – the state and its powers. Despite their obvious limitations, by sustaining these aspirations, the social sciences, following the natural sciences, participated in framing the modern collective political imagination such, that state and society were constructed as systems of, at least partly, predictable regularities, whose knowledge could make the uses of power less arbitrary. As a projection of the rational-secular version of the physical world as a whole, the idea of society – or the state – contained the theological imprint of the idea of God's eye view from which the whole can be seen in its entirety .This vision furnished the basis for claims of knowledge which fit and explain particular elements as parts of a coherent whole. Again, secularized version of the God's-eye-view lent science an authority by virtue of which it could claim to be above the contests of partisan opinions and human interests. The scientific outlook on political reality could thus generate a duel strategy for checking arbitrary power .By defining political reality as distinct from political fiction it could cast arbitrary power as capricious power guided by fantasy or illusion unaccountable to facts. By evolving a conception of political reality as a series of relations between causes and effects, it could provide the grounds for a notion such as 'reasons of state' which could be invoked by the leadership for the purpose of rationalizing the selection of courses of action which, although often controversial, are most likely to produce the most desirable effects from the perspective of the state. Without such a scientific, or rather semi-scientific, view of reality, the concept of 'reasons of state' could not have the power to 'objectify' a policy or an action in systemic instrumental terms. Machiavelli the father of the 'policy sciences', could tell the prince what actions to take or not to take in order to save the state from collapsing, claiming to have the scientific ability to see politics from a synoptic perspective that includes both the, not necessarily compatible, perspectives of the prince and of the people. In fact, however, Machiavelli had to write two separate books (*The Prince* and *The Discourses*) in order to do a scholarly justice, to both perspectives. In the context of the modern state where the perspective of the state claimed to be more integrated with that of the public, for such a scientific authority to appear apolitical, it had to be perceived at least not subversive to the perspective of the state. Only within such a frame which is sufficiently self-evident, sufficiently embedded yet concealed in the foundations of the socio-political order to be taken for granted and remain almost unnoticeable, could science, as well as other cultural enterprises, appear national, autonomous or, in other ways, public rather than politically partisan. Thus the second condition for the special status of science as a source of apolitical authority, namely a privileged synoptic normative perspective to which it could be attached, is, in fact, entirely independent of science. In many respects scientific authority was here but a surrogate extension of a first-order political authority, the authority that constitutes and upholds this synoptic perspective and which sets in turn the boundaries of second-order politics, of what is the recognized sphere of the political, of the contest over material and symbolic resources. This is a way in which politics divides into two or more levels in order to limit itself. It is important to recognize here that the comprehensiveness and the stability of such a normative perspective, regardless of whether it is hierarchical or egalitarian, are in many ways much more important for the apolitical status of science within a particular socio-political order, than its specific content. Insofar as the status of being apolitical or neutral is always relative, and indeed possible only in relation to a particular definition of politics and political conflicts, science needs first to be integrated into a particular stable system of foundational values in order for it to appear objective, or apolitical in relation to the specific perspectives and actions taken within that framework. The strictly intellectual strength of the Copernican account of the solar system was relatively indifferent to the normative biases, which mediated its favorable or antagonistic receptions in hierarchical or egalitarian societies, and the same can be said about Keynsian economic theories or Einsteins's physics. The willingness to regard any one of those or other scientific theories as apolitical account of a slice of phenomena has depended in each society upon the status of science in relation to the hegemonic normative system and, more specifically, the interpretation of the contextual import of each of these theories. The fact that scientific theories can be enlisted sometimes by opposing political opinions does not in itself diminish their internal validity. The point is rather that the very external or contextual import of science, which politicizes its status and authority when it is subject to competing political uses, in fact depoliticizes the status and authority of science when it is controlled instead by a hegemonic normative bias. In selective areas such depoliticization was especially evident in the course of the twentieth century in societies at war. During world war II, for instance, western democracies largely suspended their domestic political rivalries and significantly expanded the sphere of the apolitical in which science and technology could be enlisted to a variety of causes connected to the war effort. Years after the war, scientists and technologists recalled with nostalgia the authority, freedoms, and resources they enjoyed during the war. But under normal, conditions which do not warrant temporary suspension of politics, scientists must continually negotiate their claims to represent neutral and objective knowledge by means of a variety of intellectual, institutional and rhetorical strategies (see, for instance, Lindblom 1990).

Comprehensive, stable, relatively coherent systemic normative frames which become hegemonic during war time, are, of course, always short lived. No political regime can have sufficient control of competing and subversive normative perspectives to be able to completely repress adversary knowledge, art, and politics. The very history of early modern science as illustrated for example in the case of England, was actually largely the history of knowledge whose social status changed as its agents gradually moved from the periphery integrating it into the normative perspective of the center. But the difference between a vulnerable normative systemic bias and its absence altogether, is still very significant and consequential. As I shall show below, it is precisely the failure to sustain such a systemic state perspective, which has so profoundly altered the condition of science and its authority in many, although by no means all, contemporary (especially democratic) states.

In order to appreciate this change, it is useful to examine the belief widely held, up to approximately the seventh decade of the twentieth century, that democratization goes hand in hand with modernization, which includes the spread of scientific literacy and technological development, in other words, the 'theory' that all good things go together. Underlying that vision combining freedom and progress was, of course, the enlightenment idea and the various concomitant versions of the notion of history as a movement, occasionally interrupted by regressive forces, but generally progressing, towards a more perfect human society. It is this convergence of beliefs about politics, science, and technology, which largely empowered modern democratic governments to pursue vigorous programs of development at home and abroad. While the harmony between democracy and activist state interventions, partly framed by science and technology, was occasionally tainted by clashes between workers and industrialists, poor and rich, or economic crises, the attitudes upholding the apolitical-instrumental status of science and technology remained on the whole firm enough. At times it was even reinforced by the nationalism promoted by the cold war.

This state of affairs contributed to the third condition of the depoliticization of science in the modern state, the belief that scientifically informed state policies constitute apolitical grounds for public technologies. Instrumental technological actions are based, according to this view, upon the same politically and normatively neutral grounds as scientific explanations whose universal validity is secured by the unity and uniformity of the world as the object of science and the neutral context-free soundness of scientific methods. Scientification and technologicalization, were strategies which modern, particularly democratic, states often evolved to legitimate state power by appearing – and sometimes even succeeding – to control its potential arbitrariness (see Ezrahi 1990).

If the second condition for the depoliticization of science, the presence of all-encompassing hegemonic normative state perspective, depended upon socio-political and cultural conditions largely external to science, the third condition was in addition based on wide-spread public misunderstanding of the inherent discontinuities between scientific knowledge and technology. It was a misunderstanding particularly of the fact that knowledge not only constrains human action and shapes it but also enlarges the range of possible scientifically informed actions and, the necessity of choices. This in turn enlarges the scope of contesting values which must be taken in consideration in justifying or legitimating actions actually taken. Under this condition the need to legitimate public actions encouraged governments to unwarrantedly stretch the claims of scientific certainty to the field of action, thus trying to avoid explicating the more controversial nonscientific and non-technological ethical and political preferences implicit in their choices. Such tendency often converged with the propensity of governments to stretch the argument of national defense in order to protect themselves from political criticisms even in fields of action actually unrelated to the security needs of the nation. Combining the incontestable value of security with claims of scientific certainty served many governments as a powerful strategy of avoiding the pains of political criticism and accountability. The wide current understanding that science and technology do not eliminate the needs for hard value choices but in fact enlarge them, developed only over the last decades due to a remarkable process of collective social and political learning. This process of learning was promoted by some monumental technological disasters and controversies which exposed the much more complex relations between science, technology, ethics and politics (Jasanoff 1986, 1994). These developments gave birth to such enterprises as technology assessment agencies, and new paradigms of science for policy.

I am arguing then that the roles of science and technology as sources of apolitical authority in the modern nation state, an authority which can facilitate the checking of arbitrary political power, has been resting upon a cluster of normative perspectives on science, technology and politics. To reiterate then, the first sees scientific knowledge as a mirror of the given structure of the universe as a lawful whole accessible to human intelligence regardless of the particulars of socio-cultural context. The second is the more specific intra-state condition namely the existence of a hegemonic or shared framework which fixes demarcation lines between what is regarded as political and what is regarded as beyond politics. The third is a socio-cultural premise that science is indeed the 'conscience' of technology, what guards the technical, rational, and apolitical integrity of technological action. This condition requires that the public, or at least a large part of the public, will be blind to the internal normative, often political, choices inherent in both technological hardware and more abstract technologies of action, as well as to the political effects of their applications.

What then accounts for the contemporary impoverishment of science and technology as sources of apolitical authority, the very resources with which so much of the legitimation of state actions and intervention was obtained in the modern state? I would like to suggest that the single most important factor explaining this development is the partial disintegration of hegemonic normative system, or state, perspective on the collective goals and public policies of many contemporary democratic states (see, for example, Eisenstadt 1999). Some observers believe that the decline of the synoptic state perspective (the collapse of condition b) is partly caused, partly complemented, by the evolution of an international normative frame which is connected with the process of Globalization. The most obvious illustrations include the international spread of legal commitments to human rights, the wide international condemnation of terror, and the increasing readiness to preserve the environment and treat the glob as a single physical system requiring collective human responsibility. Do such developments suggest the possibility that a hegemonic international normative frame will replace the particular national ones? Dose the erosion of socio-cultural conditions for the depoliticalization of science and technology within the modern nation state is occurring as the status of the authority of science at the international level is being reinforced? Do we witness a process of convergence between a universal normative frame and scientific outlook on human problems consistently with some of the utopian visions of the enlightenment? The decline of the authority of science and technology at the state level does not, of course, imply that the international alternative is around the corner. The latest consolidation of some international norms is also insufficient to warrant such expectations. The foreseeable future is likely instead to bring about some sort of amalgam of national and international norms .The normative environment of science and technology is likely to be characterized by continual instability which inhibits long term depoliticalization of scientific authority as well as limits large scale projects which require stable political mandates and scientific or technological rationales. Still science and technology are likely to help consolidate some international practices which win wide support .

Another factor contributing to the partial erosion of science and technology as sources of apolitical authority at the state level is the deterioration of what I called the third condition, public beliefs or public perceptions which upheld the trust that technological or technical actions in public affairs extend scientific certainties into the field of human actions thus depoliticizing government policies and programs. Even without touching the faith in the universal validity of science as an intellectual enterprise (the first condition) or the neutrality of technology as its natural offspring (the third condition), the erosion of institutionalized *systemic* normative bias (the second condition) would have been sufficient, to be sure, to reduce the privileged status of science and technology in the context of contemporary democracies. As a matter of fact, that development has underlay a massive decline of authorities, which is often seen as a postmodern syndrome, across diverse spheres. Beyond the obvious deterioration of authorities in other spheres including public institutions, politics, and fields of expertise such as engineering and medicine, the shrinkage in the distinct social authority of science and technology in the context of public affairs has been only a part of a larger erosion of authorities in such cultural fields as history, art criticism, education and the like (Brint 1994). The relativization and the politicalization of the social status of science and technology did not require, then, more than the breakdown of the systemic normative bias which created a sphere of apolitical authorities in the modern state. Put yet another way, we are dealing with another phase of the decentralization of political and cultural powers in contemporary states. *The principal development I am referring to here has been leading to the general impoverishment or deterioration of the splendid isolation of scientific reason and knowledge as sources of distinct standards of discourse and action. This enlightenment notion has been replaced by new configurations in which knowledge and power, private and public values converge and interact unrestrained by the fences which effectively protected the discreteness of former normative and institutional domains.*

Contemporary political configurations no longer seem to rest as firmly as they did in modern nation and democratic states on the idea of the polity as a relatively coherent whole or of the individual as a stable discernable elementary unit. In the context of deep differences and fluid group boundaries (Geertz 2000) the expectation that science would provide an incontestably neutral perspective on issues of public policy could no longer be sustained. Stripped off of its former aura, science has lost much of its former power as a political resource in the justification of state policies. The activism of the modern state (Scott 1998) has been replaced by much less interventionist strategies and regulatory regimes in which expert knowledge is less distinctly salient as an ingredient of public choices and actions, and more tentatively and elastically integrated into a complex, and continually shifting, webs of legal, economic, organizational, ethical, and political frames (Lindblom 1990). Similarly the contemporary individual is much more elusive and shifty than his, or her, modern ancestor. Among other things, the insights of contemporary psychology, made richer and more daring by literature and other humanistic disciplines provided a fresh basis for criticizing strictly economic and political models of the individual as a rational decision maker whose choices are based on a single, complete, transitive, preference ordering (Steedman and Krause 1987; Albright 1994). The individual, represented by a final ordering has been widely replaced by a multi-faceted individual having many such, often disharmonious, orderings. "Agents may terminate deliberation and take decisions without having resolved the moral, political, economic and aesthetic conflicts relevant to their predicaments" (Levy 1982). Once it has been recognized that humans are so deeply diverse and complex it becomes more evident why the notion that individual behaviors can be aggregated for purposes of predictions and comprehensive planning turned out to be so problematic.

Considering the multiple factors and considerations involved in the making and implementing public policies it is not surprising that as Björn Wittrock aptly observed "Professional social science will in most cases just be one of many inputs, and unless it somewhat supports or is supported by other forms of beliefs, convictions, and views, it is unlikely to have a major impact on policy making" (Wittrock 1991).

Moreover, to the extent that modern public life is mediated by mass electronic communications, the non-rational affective elements making the public sphere, become more pronounced (Boltansky 1999). In such a universe, science and technology still count. But with the decline of Enlightenment models of public discourse, policy making and citizenship science and technology lost much, although not all, their powers as privileged sources of authoritative standards for guiding and assessing public choices and actions.

Considering such deep processes of change and their impact on the status of science in contemporary states it appears futile to regard people such as Michel Foucault, Jacques Derrida, Bruno Latour, David Bloor or Stevan Shapin as the slayers of cherished certitudes, coherences, and objectivities. The sociological and philosophical critique of foundationalism is more a source of illumination of the macro-cultural and socio-political processes, which have been shaking established authorities and structures, than even their remote cause. It is much more promising and fruitful, I think, to search for the forces which have been eroding modern coherences and blurring former boundaries in the accumulative effects of the evolution of modern individualism, the mass media culture of our time or other aspects of Globalism than in articles published in journals like *Social Text.* One cannot exaggerate the enormous influence of the electronic mass media on the transformation of contemporary modes of persuasion and structures of authority. The effects of the mass media revolution include among other things the increase in the relative weight of pictures compared to words and arguments in mediating the political process. This development, and its relation to the shrinking of public deliberation in contemporary democracies, have been affected also by the acceleration of the pace of communications due, among other things, to the commercialization of the media and the subordination of programming, even of news, to the interests and the schedules of advertisers .Other contributing factors include the proliferation of radio and television channels and the omnipresence of entertainment in contemporary communications culture .All such developments have been radically altering the sociocultural universe in which the authority of scientists as well as standards of truth and objectivity could be validated in the past. Leading contemporary policy scientists and other experts or scholars who follow closely trends and changes in the relations between science and politics seem to differ only about the causes or the indicators of these changes not about the fact of their occurrence. Prominent among the accounts given of the changing sociopolitical environment of science and technology, especially in the context of their application, is the erosion of formerly wide agreements on the validity hegemonic ideas such as the welfare state which have supported for decades vigorous government interventions in the pursuit of equality and social justice. Scholars such as Martin Rein and Donald Schon pointed out, for instance, that in place of such former areas of consensus, experts, who are enlisted as advisers in the formation of public policies and their implementation, are facing a multiplicity of often incommensurable ideas of the desirable social order which give rise to competing frames of policy problems and the measures required to cope with them. What they regard as rapid 'frame shifts' and the proliferation of frame conflicts raises "an increasing doubt about the neutrality of science as an agent in debates over the framing of policy issues" (Rein and Schon 1991). The record since this observation was made in the early 1990s, can only reinforce the diagnosis that these developments have not been just temporary or easily correctable but, as I have suggested above, are the result of deeper structural changes. By now it may be easier to recognize why the belief that the attrition in the authority or status of science lies with a group of social scientists or scholars from the humanities who are spreading the culture of relativism and are undermining public trust in scientific standards of truth, objectivity, or the existence of 'real facts', has been so grossly misplaced. Scholarly works, to be sure, can influence social and political processes, but the spread of the forces which lead to the postmodern condition cannot be stopped or slowed down by attacking Foucault, Lyotard, or Latour any more than the process of secularization in the 17th century could be stopped or slowed down by the attacks on the publications of Descarte, Hobbes, or Galileo. Those concerned with the declining authority of science in contemporary society will do much better to examine the changes which are blurring the boundaries between the public and the private spheres, or between political and economic activities and try to figure out how scientific standards of truth can best be represented and contextualized under such conditions? They could also benefit by following works of scholars who examine the relations between the conditions which have been undermining the sphere of the apolitical at the macrosocial level and have enabled the relatively apolitical or neutral status of the expert to be recast or reconstructed in relation to local configurations of values and interests.

In the final analysis, the humbler authority of science in the context of public affairs does not diminish the relevance of knowledge to the understanding of phenomena nor of technology and technical skills to the advancement of human goals. No serious observer can underestimate the importance of scientific expertise in areas such as genetics, energy, food etc. But the absence, in many areas of public concern, of a stable synoptic normative perspective backed up by state authority and power, especially in contemporary democracies, no longer allows the parameters of scientific diagnoses and prescriptions, or of technological interventions, to be fixed beneath the surface of public perceptions as given, precast, pre-political or apolitical guidelines. In contemporary democracies each public decision or program is likely to require open and elaborate political processes of negotiation and compromise in order to produce a relatively stable, if only rarely coherent, political mandates for scientifically or instrumentally sound actions. If in modern states science and technology could often join latently, although not necessarily undeliberately, as the extensions of politics disguised as apolitical reason, in contemporary democracies any exercise of political power is more openly opposed by adversary political powers as each of the sides can advance a scientific position which is perfectly valid in relation to its own goals and interests. In the much more pluralistic reflexive adversarial context of contemporary politics, one can find scientists and other experts working in good conscience for all sides. Hence also 'nature,' 'science,' 'technology,' and 'hard facts' as rhetorical assets enlisted to the external justifications of decisions and actions, are more equally distributed among competing political agents and, therefore, are also radically devalued. Such multiple and contradictory uses of scientific and expert authority obviously constitute a major reason for their weakening in the socio-political context. This shift reflects a profound change in the respective standards of political and scientific authorities in contemporary states and, therefore, an equally significant shift in the nature of their relations. In the modern state, science entered a niche formerly taken by religion as a source of authority based on the presumption of certainty, unequivocal answers, and solutions, which transcend the ambiguities and conflicts of ordinary human life. In many respects the twentieth century was a century of monumental politics and monumental scientific and technological projects. Since the closing decades of the last century both politics and science lost their foundations in the certainties and convictions which appeared to sustain the hope that human decisions and actions can be grounded in authorities that transcend inherently fragile human judgments. In our time political power and authority and the fear of arbitrariness seem to be checked by the fragmentation and the parochialization of the political rather than by the apolitical. The radical shrinking of the sphere of the apolitical, represents in fact another more far-reaching stage in the de-theologization of politics and culture, the erosion of the remaining secularized deposits of religion and the adaptation of both power and knowledge to humbler human goals.

Hebrew University of Jerusalem, Israel

REFERENCES

Albright, Daniel (1994), 'Literary and psychological modes of the self', in U. Neisser and R. Fivush (eds.), *The Remembering Self, Construction and Accuracy in The Self-Narrative*, Cambridge: Cambridge University Press, pp. 18–41.

Boltansky, Luc (1999), *Distant Suffering: Morality, Media and Politics*, Cambridge: Cambridge University Press.

Brint, Steven (1994), *The Age Of Experts. The Changing Role Of Professionals in Public Life*, Princeton, NJ: Princeton University Press.

Burtt, Edwin A. (1954), *The Metaphysical Foundations of Modern Science,* Garden City, KS: Doublday.

Della Mirandola, Pico (1942), 'Of the dignity of man' (E.L. Forbes (transl.), *Journal of The History Of Ideas* 3: 347–54.

Eisenstadt, Shmuel N. (1999), *Paradoxes of Democracy, Fragility, Continuity, and Change*, Washington: The Woodrow Wilson Center Press.

Ezrahi, Yaron (1986), 'Science and the civic spirit of liberal-democracy', in L.S. Rouner (ed.), *Civil Religion and Political Theology*, Notre Dame, in University of Notre Dame Press, pp. 59–75.

Ezrahi, Yaron (1990), *The Descent of Icarus, Science and Contemporary Democracy*, Cambridge, MA: Harvard University Press.

Funkenstein, Amos (1986), *Theology and the Scientific Imagination*, Princeton, NJ: Princeton University Press.

Geertz, Clifford (2000), *Available Light. Anthropological Reflections on Philosophical Topics*, Princeton, NJ: Princeton University Press.

Jasanoff, Sheila (1986), *Risk Management and Political Culture*, New York: Russell Sage Foundation.

Jasanoff, Sheila (1994), *Learning from Disaster, Risk Management After Bophal*, Philadelphia, PN: The University of Pennsylvania Press.
Levy, I. (1982), 'Conflict and social agency', *Journal of Philosophy* 79: 246.
Lindblom, Charles E. (1990), *Inquiry and Change, The Troubled Attempt to Understand and Shape Society*, New Haven, CON, Yale University Press.
Manuel, Frank E. (1974), *The Religion of Issac Newton*, Oxford: Clarendon Press.
Porter, Theodore M. (1986), *The Rise of Statistical Thinking, 1800-1900*, Princeton, NJ: Princeton University Press.
Porter, Theodore M. (1995), *Trust in Numbers*, Princeton, NJ: Princeton University Press.
Rein, Martin and Donald Schon (1991), 'Frame reflective discourse', in P. Wagner, C.H. Weiss, B. Wittrock and H. Wollmann (eds.), *Social Sciences and Modern States*, Cambridge: Cambridge University Press, p. 283.
Scott, James C. (1998), *Seeing Like a State, How Certain Schemes to Improve the Human Condition Have Failed*, New Haven, CT: Yale University Press.
Social Text, A quarterly published by Duke University Press, sponsored by the Center for the Critical Analysis of Contemporary Culture at Rutgers University.
Steedman, Ian and Ulrich Krause (1987), 'Goethe's *Foust,* Arrow Possibility Theorem and the individual decision-taker', in J. Elster (ed.), *The Multiple Self,* Cambridge: Cambridge University Press, pp. 197–231.
Wittrock, Björn (1991), 'Social knowledge and public policy: Eight models', in P. Wagner, C.H. Weiss, B. Wittrock and H. Wollmann (eds.), *Social Sciences and Modern States*, Cambridge: Cambridge University Press, p. 347.

PART II

STS AND THE SOCIAL SCIENCES

CHAPTER 5

BJÖRN WITTROCK

HISTORY OF SOCIAL SCIENCE: UNDERSTANDING MODERNITY AND RETHINKING SOCIAL STUDIES OF SCIENCE

The formation of modernity in the late eighteenth and early nineteenth century is often described in terms of a fundamental transformation of political and economic order. It also involved a profound change in the nature of discursive practices. These discursive transformations were not mere 'ideological' epiphenomena. Rather, deep-seated epistemic and ontological ruptures and redefinitions opened up new horizons of expectations and hopes. These shifts made possible or, more literally, conceivable, if only as projects and imaginations, new institutional practices.

The conjunction of new epistemic and ontological presuppositions and the range of institutional proposals put forth on the basis of these presuppositions constitute – what, in another context (Wittrock 2000), I have called – 'promissory notes'. Such promissory notes form focal points for long-term processes of contestation and interpretation. In the course of such processes, they may travel in time and space far beyond the imaginations and expectations of any of the progenitors of the original shift. They entail a range of possible societal states of affairs that were previously – and literally – not 'conceivable' but that can thereafter not be made, as it were, 'unthought'.[1] Discursive and cosmological shifts are not 'ideological' reflections; they are necessary, but not sufficient, for the constitution of new institutional practices.

An understanding of the transformations, which jointly constitute the formation of modernity, is only possible if the social and agential nature not only of political and economic practices but also of technological and discursive practices is recognised. In this sense all studies of science and technology depend on social and historical theory and scholarship. In the sequel, the history of social science at three crucial junctures during modernity will be highlighted. These three periods are, firstly, the time of the formation of distinctly modern societal practices in the late eighteenth and early nineteenth centuries, secondly, the high time of organised modernity and of industrialism and imperialism at the turn of the nineteenth century, and finally, the period of intensified global interactions and North American pre-eminence in the second half of the twentieth century.

B. Joerges and H. Nowotny (eds.), Social Studies of Science and Technology: Looking Back Ahead, 79–101.

For each of these periods, an effort will be made to indicate, if only and by necessity in broad outlines, ways in which social sciences attempted to achieve a measure of institutional, epistemic but also professional distinctiveness and autonomy. An effort will also be made to indicate both that a social theory in terms of a functional-evolutionary account of these events as just a reflection of a uniformly increasing societal differentiation is seriously misleading. However, also the inverted form of such an analysis in terms of what Shinn and Joerges (2002) have called the new orthodoxy of knowledge will be criticised, an orthodoxy that denies a differentiation between scientific practices and other human practices and that often operates with assumptions of a given set of agential underpinnings, essentially in terms of power-wielding and strategic rhetoric.

MODERNITY AND ORIGINS OF DISCIPLINARY CLOSURE

In the late eighteenth century natural philosophy gradually gave way to a range of natural science disciplines. Moral philosophy was slowly transformed into a range of separate discourses. In the course of the nineteenth century these discourses came to distinguish themselves both from natural science and from literary discourse and came to label themselves social sciences. In this process of an epistemic sea change, historical reasoning, which had formed a central component of moral philosophy, came to form a discursive realm of its own, separate from philosophy and separate from the social sciences.[2]

Simultaneously, the type of critical historical reflection on the European political experience relative to that of other parts of the world, an in particular that of East Asia and China and which was most closely associated with Voltaire, but echoed also in Kant's critique of European imposition and in his appreciating comments on Chinese society, disappeared. In its stead came first a distinctly Europe-centred conception of world history with Hegel's lectures on the philosophy of history as an emblematic expression. Later history emerged as an academic discipline with its focus on the European experience of the formation of a range of nation states.[3] In its formative stage in the early nineteenth century, in the foundational works of Ranke, the main theme was the shaping of Europe through the confluence of the cultural traditions of Latin and Germanic peoples.[4] Later most European historians would narrow their focus further and write narratives about the trajectories of individual nations. Sometimes these studies, as in majestic works at turn of the nineteenth century by Meinecke and Hintze, had a strong comparative perspective, sometimes their orientation was limited to the achievements of one particular nation.

In the late nineteenth century, at a period in time when European global pre-eminence was at its peak, history largely came to be a scholarly exercise that served as a discursive parallel to the formation or reform of European nation states. The new investigations of social conditions and the back side of processes of industrialisation, urbanisation and modernisation were to become institutionally embedded, if in a slow and uneven process, in the research-orientated universities and other new higher education institutions.[5] They formed an analogous parallel to the efforts of those nation states to cope with 'the social question'. Thus the relationship of the new social sciences to an historical conception, other than that, which took the life of a given nation state as its starting point, could not be but a tenuous one.[6] The notion that the European experience should not be taken us the self-evident yardstick for the achievements of a civilisation was even more remote. To some extent this is true also of the intellectual giants of early social science, including Max Weber. Thus even if Weber's most seminal works, such as his collected essays on the world religions, are masterpieces of global history, they stand in a complex and never quite resolved tension to other parts of his works. In the interpretation of his legacy for future generations of social scientists, as handed down most prominently perhaps by Talcott Parsons, it were these latter parts, emphasising the unique nature of Western modernity and its historical trajectory, that became the most visible and most frequently cited ones.[7]

It was not in Europe, however, that a full-blown professional separation occurred of what had previously been a broadly conceived movement of social science into a range of individual social science disciplines. This took place in the United States already at the turn of the nineteenth century. One of its consequences was a more secure professional position for the practitioners of these forms of scholarship. Another one, however, was the emergence of a widening chasm between on the one hand history and on the other hand different social science disciplines such as sociology and political science. A corresponding development did not take place in most European countries until well after the Second World War. There can be little doubt that in the wake of this development an interest in history, and particularly an interest in global history, came to occupy a less prominent part within the newly differentiated social science disciplines than had previously been the case. In a number of European countries, political and sociological studies had from the mid-nineteenth century onwards remained intimately linked to historical research. However the focus had, as argued above, been on the development of a national polity rather than on global developments.

MODERNITY AND HISTORY: THE REDISCOVERY AND REJECTION OF GLOBAL HISTORY

It was only the disaster of the First World War, which came to shake the conviction of a historically assured pre-eminence of Europe in particular and a more vaguely defined Western world in general. In the wake of the war, a first wave of efforts appeared to write the history not of civilisation but of the rise and decline of different civilisations, accounts in which the achievements and predominance of Europe were cast in serious doubt. Often enough, such accounts were written from the perspective of a generalised conservative cultural pessimism. In a more conventional form this is true of Spengler, in an abstracted and radicalised form this is characteristic of Heidegger's programmatic writings from the late 1920s and early 1930s. In other cases, authors tried to reflect upon the cultural foundations of different political and societal orders from the vantage point of an historical scholarship characterised by the highest degrees of critical reflexivity. This is to some extent the case in the philosophical writings of Jaspers. It is even more tangible in the historical writings of such diverse authors as Marc Bloch, Franz Borkenau, and Arnold Toynbee. In their writings one may discern the outlines and the scholarly promise of a social science that brings in a comparative and critical account of world history.

However, these potentials were not the ones that became predominant or were realised when social science finally became institutionalised across the board in the 1950s and 1960s. Rather this occurred in the non-historical form of social science that had long been predominant in the United States and that became transposed abroad after the Second World War, linked to the notion of the so-called behavioural revolution. These efforts were often promoted within the framework of the new international social sciences associations that had been established with links to UNESCO and other forms of international scientific collaboration, shaped by the United States.[8] Thus social science not only became less historical than it had tended to be in most parts of Europe at the turn of the century. It also became shaped by the fact that social science disciplines, and most notably so perhaps political science and sociology, in their theoretical core came to reflect the pre-eminent position of the United States in the post-Second World War world.

Thus even if social science, in its own long-standing self-conception, remained a discourse of modernity, these presuppositions tended to entail a social science that was reticent to theorise either world history at large and even those upheavals that came to constitute the particular Western trajectory. The particular Western trajectory to modernity tended to be assumed rather than examined. Furthermore, the relationship of a European trajectory to global historical developments tended to be ignored. These types of questions, so prominent in earlier and overtaken forms of philosophy of history, were simply irrelevant to the behavioural sciences of modern industrial societies and their increasingly urbanised and differentiated forms of organised social life.

Within the discipline of history, a gradual process of disciplinary demarcation led to analogous results. Thus scholars, to whom questions of world history had been a central concern, tended to lose their standing as exemplars and be regarded as falling outside of the bounds of the discipline and rather occupying a role as civilisational critics. As such they might be interesting perhaps but ultimately they were seen as failing to conform to proper standards of modern historical science. Arnold Toynbee is an obvious case in point, and despite his strenuous efforts to argue for the empirical and scholarly basis of his writing of history he was often depicted as a speculative writer. In other cases, such as that of Franz Borkenau, their works have simply fallen into relative oblivion. These scholars appeared as hopelessly overtaken in methodological terms long before the era of the behavioural revolution of the 1950s and 1960s, an event which further contributed to their neglect, as did in some cases their different political allegiances.

GLOBAL HISTORY AND GLOBALISATION: RETHINKING SOCIAL THEORY

It is really only towards the end of the twentieth century that global interactions have become so prominent and immediately visible as to make obvious the existence of distinctly modern, yet clearly different, societies also across the globe.[9] It is in this context that there are renewed efforts to understand the different civilisational legacies and to explore various modes of interactions over long periods of time. Within social science this renewed interest often has come under the label of studies of globalisation. Paradoxically, globalisation studies often seem premised on assumptions close to those of earlier forms of theorising about convergence and modernisation. They describe the global and all but inevitable diffusion and impact of market interactions and capitalist forms of production. Thus present day globalisation studies and theories about global networks are in conceptual terms strangely reminiscent of modernisation theory.[10] Notions of structures may be replaced by those of networks, and Eurocentrism by globalism. However the core assumptions of earlier modernisation theory in the form of a functional evolutionary account of history and a functional and non-agential account of society is remarkably familiar.

In historical research the renewed interest has taken the form of an interest in what is often termed global history. What holds this wide area together is essentially an insistence, from a variety of different perspectives, on the legitimacy and scholarly viability of asking questions about long-term developments that transcend the borders of any given polity, or indeed geographical region. This, indeed, is a process that is now going on among historians on a worldwide scale. In this process, it is clearly the case that historians seek for ways of engaging with questions of global history that will not commit them to a representation of the world in terms of the interaction of a number of states and polities, nor of cohesive civilisational blocs. Rather there is a search for more sensitive modes of representation that highlight cultural and institutional legacies that are shared across such boundaries. Terms such as 'connections', 'encounters', and 'entangled histories' recur frequently. It is difficult, for scholarly and maybe also for normative reasons, not to feel sympathy for these efforts. They seem to hold every promise to yield important insights. However, they do not relieve us of the need to go beyond a mere amassing of interesting insights in the hope that we might eventually be able to discern the contours of global historical developments. This is the point where social theory must confront global history.

THE FORMATION OF MODERNITY AND THE RISE OF SOCIAL SCIENCE

It is possible to depict the formation of modernity in Europe as the result of a series of basically continuous processes where political, economic and intellectual transformations mutually reinforced and conditioned each other. There is indeed a specific path of development that originated in those parts of the European continent that bordered on the Atlantic seaboard and that had precisely the plurality of intellectual and political fora as a key characteristic. It would be possible to trace the diffusion of analogous forms of societal organisation in space and time. It would involve an analysis of Western Christendom but also of parliamentary assemblies, urban self-government and of universities as sites of learning across the vast regions of Central and East-Central Europe. It would trace the impact of the Renaissance, the Reformation, the Scientific Revolution but also the establishment of a system of mutually balancing territorially delimited states, and the Enlightenment but also a period of great revolutions ushering in the formation of new types of political order. It may seem reasonable not to impose artificial notions of different epochs or ages in the continuous flow of loosely structured events in historical time.

Yet such an account would run a risk of just reproducing the inevitable complexities of historical changes without contributing to an understanding of them. In particular it would underestimate the rupture that occurred in both institutional and intellectual terms in Europe in the late eighteenth and early nineteenth centuries. The events involved did not emerge *ex nihilo*. However, jointly they involved the appearance of a radically new configuration of phenomena, the formation of distinctly modern societies.

Analogously, one may point to developments over long periods of time in discourses on social and political phenomena. There are innumerable examples throughout history of efforts to give guidance to rulers. Advice of this type has also been written down in books or manuals. Economists and educational researchers may also point to the perennial nature of the aspects of human life that constitute their scholarly domains. In the case of legal scholarship a tradition, if only in a weak sense of the word, may be said to extend back to Roman Law and to have been institutionalised at least from the twelfth and thirteenth centuries. The development is in fact coterminous with the creation of European universities (Kelley 1990).

In the case of the study of politics, it is equally clear that the form of polities that emerged in Europe in the course of the fifteenth and sixteenth centuries as national and composite monarchies gave rise to new types of education both of princes and of diplomats. In the seventeenth century parallel developments in the direction of the establishment of university-based training in politics and rhetoric for diplomats and high administrators occurred in different parts of Europe, including France, Germany, Italy, the Low Countries, and Sweden. In the same period, interest in matters of education grew dramatically, particularly in Protestant countries in Europe. A universalist scholar such as Comenius may be seen as an early pioneer in educational studies.

During the eighteenth century the state and cameral sciences became an important instrument of the rulers of European states, often of a more or less absolutist nature, in their efforts to survey and govern their domains (see Raeff 1983; Tribe 1988). During this period professorships of economics were created at various universities. During the second half of the century statistical reasoning was also undergoing important developments, often with clear links to administrative concerns but similarly related to developments in arithmetic and geometry (Brian 1994). Also during the eighteenth century, the experience of European expansion stimulated of what might be termed the emergence of anthropology as a scholarly subject (Fox, Porter and Wokler 1995). All these types of examples are interesting in their own right. However in none of these cases is there an unbroken scholarly tradition that links these early efforts to present-day activities in university settings.

The key institutional features, which emerged at the turn of the eighteenth century, involved dramatic ruptures with traditional forms of political and social order. This involved not only the emergence of a general sense of openness and contingency concerning the structural conditions of the political body. The institutional transformations also depended on processes inherent in the deep-seated cultural and cognitive shift that involved, as one manifestation, a passage from political and moral philosophy to a social science. Ever since scholars have called themselves social scientists, with the first recorded uses occurring in France in the 1790s, they have also seen these new forms of knowledge as characterised by an effort to understand the world of modernity. They have tended to describe key features of this new world in terms of processes of industrialisation, urbanisation, and political upheaval, originating at the north-western edge of the Eurasian landmass but eventually having repercussions on a world-wide scale.

In the self-understanding of the social sciences there is a long-standing and predominant view about the formation of modernity that highlights transformations in the late eighteenth and early nineteenth centuries. It sees these transformations in terms of a conjunction of a technological and a political transformation – the industrial and the democratic revolutions, respectively. This traditional interpretation, as already argued, radically underestimates the deep-seated epistemic transformation that occurred at the turn of the eighteenth and nineteenth centuries.

In fact, in this period there occurs a fundamental transition from earlier forms of moral and political philosophy into social science. This transition is also linked an institutional restructuring not only in forms of political order but also in the forms in which human knowledge is brought forth and in which claims to validity are ascertained. One feature of this institutional transition is the emergence of a public sphere that gradually comes to replace arenas of a more closed nature such as aristocratic literary salons. Another one is the rise of new or reformed public higher education and research institutions that come to replace both the laboratories of wealthy amateurs and the academies under Royal patronage and partial control.

Thus social science emerged in a period of deep-seated transformations of political and cultural institutions. Social science was, from its very inception, characterised by an inherent duality (see Wagner 1994 and 1998). On the one hand it was a form of knowledge that was premised on a previously unknown openness and contingency of the world. It is a world in which human agency and interventions may crucially contribute to a transcendence of the legitimacy of tradition and the familiarity of local experiences and inherited bonds and communal obligations. In this sense the new social science also serves to supply concepts that came to be constitutive of a set of new institutional projects of a specific nature (Wittrock 2000). Thus, there emerged institutions that were not just new but came to be vehicles for the enhancement of a continuous process of innovation. On the other hand, these institutional frameworks and the social sciences themselves were to be endowed with a certain degree of stability. One way to achieve this was to demonstrate that the new forms of knowledge, in contrast to the cameral concerns of an older absolutist state, were premised on universalistic rather than on particularistic assumptions about human beings, human agency and human societies.

The rise of the social science disciplines must then be cast in terms of the fundamental transformation of European societies that the formation of modernity entailed. One shift in intellectual and cultural transformation in this period pertains precisely to the concepts of society and history and to the new awareness of the structural and constraining nature of societal life. Pierre Manent has put forward the notion that society is a 'postrevolutionary discovery' (Manent 1998, 51, but see also Manent 1994, 80–3). True enough, and as is convincingly demonstrated by Keith Baker, the term society undergoes a long conceptual development in the French context in the course of the seventeenth and eighteenth centuries – with a dramatic increase in the utilisation of the term in the mid-eighteenth century. It is also true that, in his critique of Louis Dumont's analysis of Western individualism[11] and holism, Marcel Gauchet argued that (this is Baker's elegant summary):

> Individualism was not simply a symptom of the dissolution of the primacy of the social whole, as that had been understood in traditional religious terms. It was also a necessary condition for what he once again called (following Karl Polanyi) the 'discovery of society' – its discovery in strictly sociological terms, disengaged from the religious representations in which it had hitherto expressed its existence. Not until the ideological primacy of individual interests was postulated, he argued, could constraints upon these interests be discovered in the operation of an autonomous social order subject to its own laws (Baker 1994, 112).

Johan Heilbron has pursued an inquiry into the constitution of individual interests (Heilbron 1995, 1998, 77–106). In the course of the seventeenth and eighteenth centuries, such interests were conceived as amenable to the constraints of various notions of sociability. In particular, given a human condition short of true religious virtue, was there a prospect for a human existence beyond the borders of a Leviathan-like imposition of absolute order that would involve socially acceptable outcomes of the pursuit of the self-interests of human beings? Such inquiries were pursued in various ways in the different parts of Europe throughout the late seventeenth and the eighteenth centuries. However, Heilbron and many others today agree that, even if there is a long process of gestation of the modern concept of society, the unique event of revolutionary upheaval requires that discursive controversy and political practice become joined in the formation of a distinctly modern era. Pierre Manent has elaborated a similar argument: "The Revolution offered the original spectacle of a political change of unheard-of scope, yet having no stable political effects, of a political upheaval impossible to settle, of an interminable and indeterminate event" (Manent 1994, 82).

This description of the Revolution as an irreversible and interminable process of fundamental change was formulated perhaps most clearly by one of the most well known thinkers of the nineteenth century, Alexis de Tocqueville. In his memoirs, Souvenirs, written in the summer of 1850, he describes the revolution as one long upheaval "that our fathers have seen the beginning of and which, in all likelihood, we shall not see the end of. Everything that remained of the old regime was destroyed forever" (Tocqueville 1964, 30).

In fact, Reinhart Koselleck's conception in his early work *Critique and Crisis* is quite similar. He also links the temporal duration of the process of upheaval to its spatial, and indeed world-wide, extension, as well as to its increasing intensity in terms of modernity as a process that affects all human beings, not just, say, those in central political institutions or certain major cities:

> The eighteenth century witnessed the unfolding of bourgeois society, which saw itself as the new world, laying intellectual claim to the whole world and simultaneously denying the old. It grew out of the territories of the European states and, in dissolving this link, developed a progressive philosophy in line with the process. The subject of that philosophy was all mankind, to be unified from its European centre and led peacefully towards a better future (Koselleck 1987, 5f).

In this process horizons of expectations, to use one of Koselleck's key notions, opened up that were previously unknown. It is also this sense of openness and contingency that serves as a forceful impetus to an examination of the structural conditions of the political body and entails a passage from political and moral philosophy to a social science.

The transition meant that five key problematiques – which today are more acutely open to reinterpretation than they have been for decades if not for a century – are being formulated or at least fundamentally reformulated and are entering into the new social science discourse.

First, the whole role of historical inquiry becomes a crucial one. On the one hand, historical reasoning becomes an integral part of the intellectual transition, and even abstract reason itself becomes historicised in early nineteenth-century philosophy. However, on the other hand, the moral and political sciences break up into a variety of new discourses that in the course of the nineteenth century coalesce and are reduced to a number of disciplines. This means that the stage is set for the divergence between a professionalised historical discipline and the other social and human sciences, a divergence that we still today experience as a major intellectual divide.

Secondly, interest in language and linguistic analysis enters into all domains of the human and social sciences as a key problematique. One outflow of this is the constitution of textual and hermeneutic modes of analysis. A second one – familiar from contemporary debates on linguistic analysis and poststructuralism – is that of the relationship between text, interpretation, and consciousness. A third one is the effort to historicise language and linguistic development itself. Thereby a crucial link was provided to various collective entities such as the historic construction of notions of different peoples.

This leads to a third problematique, namely that of constituting new collective identities. Membership in a collectivity could no longer be taken for granted in terms of the life experiences of the inhabitants of a certain village or region. Nor could a relationship of obligation and loyalty between the princely ruler and his subjects continue to constitute an unquestionable core of the body politic. That, however, meant that even the most basic categories of societal existence were open to doubt.

Fourthly the whole problematique of the relationship between notions of polity, society, and civil society was succinctly and acutely reformulated in this period of transition. The fact that once again these notions are probed and fundamentally re-examined should not conceal the fact that they were indeed in many ways not just reformulated in this period but rather discovered or even invented.

Fifthly, assumptions about what prompts human beings to act and how to interpret their actions within a broader framework are at the very core of any scholarly program in the social and human sciences. At the turn of the eighteenth and nineteenth centuries the fundamental categories that we still by and large draw upon were elaborated and proposed.

Three or four such fundamental categorical conceptualisations were propounded. Each of them corresponded to a conceptualisation of what 'society' was constituted by. These categories might be described as follows:

- *Economic-rationalistic;* with a corresponding view of society as a form of *compositional collective.*
- *Statistical-inductive*; with a view of society as a *systemic aggregate.*
- *Structural-constraining;* with a view of society in terms of an *organic totality.*
- *Linguistic-interpretative;* with a conceptualisation of society as an *emergent totality.*

The transition from a discourse on moral and political philosophy to a social science – analysed by, for example, by Robert Wokler (1987) – in rudimentary form has already taken place in the mid and late 1790s in France after the Revolution. It entails a decisive shift from an agential – some would say voluntaristic – view of society to one that emphasises structural conditions.

To some extent, a similar shift occurs in economic reasoning away from a broad concern about moral and political agency. In the course of the nineteenth century, 'average economic man' instead becomes cast in a web of structural properties and dynamic regularities rather than in a moral universe of individual action.

The shift in epistemic and institutional regimes that occurred at the turn of the eighteenth and nineteenth centuries did not immediately usher in the set of disciplinary configurations in the social and human sciences that we now all too often take for granted. This occurred only in the late nineteenth and early twentieth centuries – and then only in an uneven and partial process that did not become a universal pattern of ordering until well after World War II. However, it did entail, in a more or less rudimentary way, both the institutional form for intellectual activities and the epistemic forms that became constitutive of the discourses on society in the age of modernity.

THE SHAPING OF THE SOCIAL SCIENCE DISCIPLINES: THE SOCIAL QUESTION, THE RESEARCH UNIVERSITY, AND THE NEW NATION-STATES

Already relatively early on in the nineteenth century economics, or rather political economy, had become differentiated from moral philosophy. Furthermore it is also in this period that history emerges, as a scholarly field with its own canon of rules but the full disciplinary formation of history is a highly extended process. However, as emphasised already in the introductory section, social science as an institutionalised scholarly activity performed within a series of disciplines was by and large a phenomenon of the late nineteenth and early twentieth centuries. It is a process that directly and indirectly reflects concerns about the wide-ranging effects of the new industrial and urban civilisation that was rapidly changing living and working conditions for ever-larger parts of the population in many European nations during the nineteenth century.

These changes, which were often collectively referred to as 'the social question' (*die soziale Frage*), were gradually forcing themselves upon the agendas of parliamentary bodies, governmental commissions, and private reform-minded and scholarly societies. Often the impetus for the search for new knowledge came from modernising political and social groupings that favoured industrialisation but that advocated more or less far-reaching social reforms. These groupings gradually came to embrace the notion that political action to alleviate 'the social question' should be based on extensive, systematic, empirical analysis of the underlying social problems. In country after country, the political agenda of the nineteenth century was being formed by two macro-projects: the search for a solution to 'the social question' and to the question of national identity and nationhood within new or reformed nation-states. The main features of this stage in the emergence of social science in Europe and the United States are relatively well known, at least in their broad outlines (Wagner, Wittrock and Whitley 1991; Rueschemeyer and Skocpol 1996).

Between 1870 and the turn of the century, all signs seemed to indicate that new social science approaches would gain access to the scientific institutions. The thinking of their proponents was widely diffused among the intellectual and political elites. This occurred at a time when traditional liberal economics was undergoing a crisis. Thus traditional political economy was seen to be unable to deal with the 'social question' and the widespread deterioration of living and working conditions due to urbanisation and industrialisation.

However, another important, and indeed competing, field of study and training with overlapping concerns to those of the new social sciences were the legal sciences. The strength of the legal scholar's position in continental Europe arose from two factors. In the first place, the training function of the universities for the state service produced officials and judges. This meant that legal scholarship came to exert a considerable impact on the general outlook of the servants of the state. Secondly, legal scholars sought to provide a doctrine, a body of concepts that were based on elaborate technical distinctions and would enable lawyers and judges to act with promptness and precision, clarify the deliberations of the law-maker, and bring coherence and order into the legal system (Dyson 1980, 112). This doctrine was legal positivism, which was first developed for private law and then also for public law, in the latter case known as the legal theory of the state.

When the new social sciences stated their claims, legal scholarship had not yet developed an extensive interpretive framework commensurate to the new forms of political order. Instead it was heavily reproached, e.g., by Italian and French political scientists, for not being able to respond to the substantive questions which the modern state faced. Legal scholarship however remained an important alternative to social science and one that was in fact able to strengthen its position towards the end of the nineteenth century. Thus, to take but one interesting example, the so-called 'state sciences' in the German states did not really develop into anything that might be labelled 'political science'. Rather it became increasingly embedded in the legal constitutional scholarship that was expanding in Germany in the wake of German political unification.

Conversely, in the Nordic countries a similar tradition of 'state sciences' was at the turn of the nineteenth century characterised by a dual legacy of constitutional legal scholarship and, as in Britain, of studies of philosophy and of political history.

By and large most approaches to a social and political science did not succeed in institutionalising themselves in academia. In some cases they tried but failed. In other cases, they did not perceive the relevance of academic institutionalisation, but they blossomed during a passing period of a favourable political climate and decayed with the changing societal situation. The context of this unfinished transition from reform-orientated studies of the 'social question' to the establishment of social science disciplines, may be briefly outlined.

In Germany the unified historical, legal, and political thinking of the early nineteenth century was the common root for the historical approach to social science, as pursued by the *Verein für Socialpolitik*. This entailed a strong orientation towards public measures and action but not an analytical focus on the state as an institution to be empirically studied in its own right. The historical school, in the tradition of the administration-orientated policy sciences of the absolutist period, thus had a clear political dimension through its focus on social policy, but did not make political phenomena an object of its research. On the other hand, the legal theory of the state reduced all political aspects to legal aspects.

The *Verein für Socialpolitik* served as a focal point for an important grouping of reformers. A major group in its foundation was historically and empirically minded economists with Gustav Schmoller as one of the most prominent representatives. The *Verein für Socialpolitik* became the main initiator and organiser of empirical research on the 'social question', and its work was clearly of an applied or policy-orientated nature. Gradually a reorientation occurred towards a more theoretically ambitious research program. The concomitant scholarly struggles between representatives of the early and the late – most notably of course Max Weber – historical school have been extensively analysed. The particular blend of social concern and a staunchly professed non-partisan position and a scholarly style of activities differed markedly from the posture of the Fabian society (Rueschemeyer and Van Rossem 1996). It has also been underscored that the empirical stance of the association left surprisingly little lasting imprint on German scholarly life (Ringer 1969; Jarausch 1983).

In Britain, reform-minded individuals, often belonging to the establishment of Victorian England, came together in a number of private reform societies, some of which had close links to the scholarly world (Soffer 1978; Rueschemyer and Van Rossem 1996). In a sense the whole Victorian era was characterised by an intense preoccupation with the social problems of industrial society. Sometimes this was expressed in the literary form of Dickens, sometimes in the aesthetically and reform-orientated writings of Ruskin and Morris or the romantic historical work of Carlyle but also in the activities of the emerging labour movement and in connection with government activities proper. Concern for health mounted, for example, when recruitment to the army during the Boer War revealed the appalling conditions under which large sections of the British population lived. Most famous among the reform-minded scholarly groups was perhaps the Fabian Society, which came to play a key role in the establishment of what is still one of the most distinguished institutions of social research, the London School of Economics and Political Science. Even given this very name, however, it remained an open question how 'political science' was to be demarcated from the study from a broad genre of historical-philosophical studies of matters of political order in general (Collini, Winch and Burrow 1983).

In the case of sociology the term had been coined already in the early nineteenth century. However even the classics of sociology of the late nineteenth century, Weber, Durkheim and Pareto, were all broad social science generalists. Their contributions and professional allegiances were intimately related to a range of fields, including the study of politics, economics, education, history and religion, and the term sociology often referred to a broad historical-comparative study of society. However it was a type of study that saw itself as scientific and separate from reform-orientated activities of a more general nature.

In France, social research had been encouraged and pursued since the early nineteenth century by 'enlightened administrators' who had grown up with the intellectual traditions of the Revolution and the institutional innovations of the Napoleonic period. They were inclined to have an active modernisation-orientated view of society and of the State's role in bringing about reforms. By mid-century a more conservatively orientated alternative arose with the thinking of Le Play, who aimed at maintaining and restoring the traditional structures of society, but whose political views were equally based on systematic observation of society (Wagner 1990).

In Italy and Spain, early social science grew out of formative events in the history of these countries. In Italy, it was national unification; in Spain, the traumatic experience of losing imperial status in the wake of the Spanish-American War (1898) and having to redefine national identity within a purely European context. In both cases, several individuals and groupings advanced the idea that social research could serve as an underpinning for the modernisation and the reforms that were deemed both desirable and necessary to resurrect the nation.

In the United States, social science research originally had the same characteristics of associational organisation and ameliorative orientation as it did in European countries. For example, the American Social Science Association (created in 1865) played a pioneering role. This organisation embraced the notion that the social scientist was a model citizen helping to improve the life of the community, not a professional, disinterested, disciplinary researcher. During the final decades of the nineteenth century, this model of ameliorative, non-professional social science was gradually replaced by the emerging disciplinary associations such as the American Economic Association (1885), the American Political Science Association (1903) and the American Sociological Society (1905) (Furner 1975; Haskell 1977; Manicas 1987; Ross 1991).

These efforts did not have to face the kind of entrenched opposition or ideologically motivated hostility that many similar efforts in Europe met with. However, it would be misleading to portray these processes as the harmonious unfolding of disciplinary consolidation. In fact, even within the discipline, which was most successful in achieving early epistemic grounding, namely economics, the direction of this process was highly contentious. Similar struggles as those that had ensued in Europe also took place in the American setting in the 1880s between advocates of a liberal research program and a more historically and state-orientated type of economic research.

By the turn of the century, a particular pattern of differentiation and professionalisation had emerged in the American setting and proved to be compatible with an active role in government service early on. Economists and psychologists were, for instance, able to play a role in government service as early as World War I, just as demographers were able to do through the Bureau of the Census. At the same time, disciplinary and professional recognition was being achieved within the setting of American universities which at this point in time started to become ever more orientated towards research undertakings (Geiger 1986; Rothblatt and Wittrock 1993)

This process of successful disciplinary consolidation marked the beginning of a divergence between American and continental European social science. Late nineteenth century American social scientists, many of them German-trained, defined their intellectual projects in a society undergoing a process of rapid transformation: industrialisation, immigration, urbanisation, and concomitantly emerging massive social and political problems. As social scientists with a professional legitimacy, they tried to mark out their ow n scientific territory and establish their own systems of credentialing. These ambitions entailed the establishment of separate social science disciplines (Manicas 1987). In Europe, on the other hand, the professoriate often already had an established position; one writer (Ringer 1969) even uses the term 'a Mandarin class' to describe the situation of the leading German academics at the turn of the nineteenth century. The situation in some other countries, such as Sweden and Norway, was no different. In others however, mainly those in Southern and Eastern Europe, the situation did not allow for the emergence of 'a Mandarin class', nor for the formation of a set of social science disciplines along the lines that had been successful in the American case.

We may summarise the argument by stating that the social sciences emerged as forms of knowledge about societies undergoing fast and deep societal transformations. The relative success of given research programmes depended of course to a large extent on the intellectual coherence and viability of these programmes. The cases of Max Weber and Emile Durkheim are exemplary in this respect, but so is the ability of marginalist economics programmes to assert themselves even under political conditions which may have appeared as far from advantageous. The institutionalisation of the social sciences was, however and as already indicated, also clearly related to two other forms of institutional processes.

Firstly, the late nineteenth century is the period when the idea of the university as a key vehicle for research becomes a predominant one in a number of countries across the globe from California in the west to Japan in the east. This process creates the possibility for the new social sciences, if in a highly uneven way, to find a relatively stable basis for continued research and training.

Secondly, the late nineteenth century was also a period of deep institutional change of the political and administrative order in a number of countries. Administrative reforms were undertaken and new objectives assigned to the State. Furthermore demands for wider political representation meant that institutional reforms of the State, not least the demand for a culturally coherent Nation-State, created demands upon the social and historical sciences to contribute to these reform processes. The particular forms of interaction between research programmes, efforts to deal with the social question and the relationship of that question to the restructuring of universities and of the State, differed greatly across countries. However, in a number of cases, opportunities arose for the successful institutionalisation of some of the social science disciplines in ways that came to structure disciplinary developments in these countries for many decades.

CONSOLIDATION AND CONTESTATION: THE SOCIAL SCIENCES IN THE CONTEXT OF THE INTERWAR WORLD

The end of the First World War entailed the triumph of liberal democracy and the reorganisation of the political order across the European continent as well as a decisive weakening, if not the end, of European global predominance. Soon it became evident that the triumph of democracy was being replaced by its crisis and the emergence of new forms of political and social order with claims to represent the future of humankind in Europe and beyond. The social sciences had been shaped by efforts to come to terms with the social problems and the problems of state formation within the broad context of a West European and North American society with a basic respect for a range of liberal rights and freedoms. Somehow politicians and scholars in the late nineteenth and early twentieth centuries across a wide political range seem to share a sufficient respect for values of free communication to allow for both scholarly and public dialogues across those different political allegiances.

In the world of authoritarian and totalitarians regimes of the 1930s, this was no longer the case and the very value of an inquiry into social conditions could by no means be taken for granted. It was in this period, however, that pre-eminent representatives of a range of social science disciplines engaged in a self-critical reflection on the history of their own disciplines. On the basis of such historical reflection they were also able to formulate research programmes that came to serve as focal points for scholarly endeavours for decades to come.

Within just a few years in the middle of the 1930s a range of path-breaking programmatic formulations occur. Thus these are the years when economists, including Gunnar Myrdal and the other members of the so-called Stockholm school in Sweden but much more prominently so of course Keynes in England, were able both to take stock of the historical experience of their discipline and to formulate a coherent long-term research programme. In sociology, Parsons' *The Structure of Social Action* was equally historically and programmatically orientated. Its influence extended far beyond the domain of disciplinary sociology and came to affect developments in a number of other fields, including political science and social anthropology. In statistics, Fisher was able to achieve an encompassing synthesis that became a landmark. In political science and sociology, scholars at Chicago and Columbia opened up new areas of empirical research. In Europe the early work on electoral geography by André Siegfried in France was complemented by sociological and socio-psychological studies by Paul Lazarsfeld and his collaborators in Austria, and by Tingsten and Wold's minor classic on political behaviour, *The Study of Political Behaviour* (1937).

The comprehensive philosophical programmes outlined by the Continental logical empiricists and their Anglo-Saxon counterparts came, despite their relative numerical weakness, to set agendas for years to come. In this respect Alfred Jules Ayer's polemical and programmatic volume on *Language, Truth and Logic* (1934) became one of the most noted examples ad did Otto Neurath's *Empirische Soziologie* published at roughly the same time. In the whole area extending between the historical and the social sciences the French Annales School chartered a whole new terrain of research that forever changed scholarly efforts in these fields. This was also the period when the first edition of the *International Encylopedia of the Social Sciences* appeared, as a further testimony to the need for reflection, for stocktaking but also for the setting of new scholarly agendas.

Furthermore in the 1930s, there emerges research programmes that position themselves in conscious opposition to disciplinary developments in the social sciences and the humanities. One prominent example of this is the broad synthetic programme associated with the so-called Frankfurt school. A very different one is the effort of philosophical phenomenologists, most prominently Husserl and Heidegger, to elaborate a conceptualisation of human activity from a point of view at odds with that of both analytical philosophy and dominant forms of empirical social research.

Many of the efforts mentioned above proved to be of a lasting importance and stand out even today. However, in institutional terms the devastating effects of the European political landscape in these years and ravages of the Second World War entailed that it is difficult to discern clear institutional continuities in the continental European case. The developments in the interwar years had often been followed by deep ruptures that made the resurrection of the social sciences after the Second World War appear as a new phenomenon.

INTERNATIONALISATION, EXPANSION AND UTILISATION: CHANGING CONTEXTS OF SOCIAL SCIENCE

The full-blown institutionalisation of the social science disciplines on a global scale is to a large extent a phenomenon of the era after the Second World War. One manifestation of this was the establishment, originally under the auspices of *UNESCO*, of the *International Political Science Association (IPSA)* and the *International Sociological Association (ISA)*. A process of professionalisation was set in motion and came to exert a truly profound influence world-wide in the wake of the expansion of higher education systems in a range of countries in the 1960s and early 1970s.

In Western Europe and North America there also occurred sweeping processes of administrative reform, which furthermore in many countries coincided with the coming to power of new political majorities. Major new public policy programs were launched across the board in these countries. It is in this context that the social sciences came to be finally and firmly entrenched as academic disciplines in university settings. In some cases this occurred for the first time – as in Denmark – in others – as in Britain, Germany, Sweden and Italy – in a renewed and greatly expanded form. This expansion was paralleled by a growth of the social science disciplines on a global level that for the first time tended to make them truly international.

On all continents, the full array of disciplines and subdisciplines appeared. There was also an impressive expansion of research methods being utilised. Thus a previously predominant concern for institutions and processes on a national level was gradually being complemented by a stronger research orientation towards the study and management of clearly quantifiable processes. In terms of research methods, the 1960s were the period of the break-through of the behavioural revolution. To a large extent this revolution had been pending in and prepared by European scholarly efforts in the interwar period. No longer could historical, juridical and philosophical reasoning alone or in combination be considered sufficient for the analysis of social and political phenomena. Methods and techniques previously elaborated in statistics, sociology, psychology, and economics were now being tapped by social scientists on a vast scale.

This shift in research methodology coincided with the numerical expansion of the social science disciplines and in Europe was often complemented by the introduction of more formalised graduate education programmes, normally with compulsory courses in research methodology. At the same time it became possible to see the emergence of an informal 'invisible college' of younger scholars in Europe and in other continents as well, in marked contrast to the much more national orientation of scholars of older generations. In those universities and countries where this shift was most decisively pushed through, there were certainly instances where the older juridical, historical and philosophical competence was either partially lost or could at least not be developed on par with developments in these neighbouring disciplines. Thus it became gradually antiquated and lost touch with what had been the rationale for the use of these traditional methods in the first place.

Apart from external political-societal reasons for this pattern of disciplinary development, a fundamental fact of a long period after the Second World War was that it was only in the United States that the social sciences had sufficient size and scope to make widespread international emulation appear feasible. Thus, to take but one example, the American political science association had more than five thousand members in the early 1950´s, when corresponding figures for European nations were generally only on a scale of one percent or less of that number. Even in the early 1990s the combined enrolment of West European political science associations was but roughly one third of that of the American association.

Furthermore, by and large only the American version of a science of society and of politics could present anything like a clear institutional lineage dating back to the late nineteenth and early twentieth centuries. In this perspective the history of the social sciences in a number of European countries constitute exceptions to this general tendency only in a weak sense of the word. This is perhaps particularly clear in the case of political science. Thus there has been a persistent notion that political science is intimately linked up to the particular intellectual and institutional history of the United States. This can be discerned already in the address by the then-president of the American Political Science Association, Charles Merriam, in 1923. The same theme was to recur in the presidential address thirty years later by Pendelton Herring, and again in to the one delivered by Samuel Huntington another quarter of a century later. Political science is then seen to constitute an originally American-conceived science that has successfully spread to a range of nations that by virtue of their effective, if belated, acceptance of a pluralistic political system had come to qualify to serve as proper recipients of this intellectual heritage.

Even if such a perspective is understandable, it rests on a misperception. Thus it identifies political science inquiry with one particular form of such inquiry – related to the concerns of liberal individualism. In this and many similar cases, a lack of historical perspective risks to unduly narrow both the range of problems addressed and the theoretical traditions explored and advanced.

HISTORY OF SOCIAL SCIENCE AND THE RETHINKING OF SOCIAL STUDIES OF SCIENCE

In the late eighteenth and early nineteenth centuries, the transformation of discursive practices involved, as emphasised above, changes on both an institutional, epistemic and professional level. Institutionally, new public spheres and the renaissance and reform of universities in some countries entailed that new spaces where created for these discourses. The epistemic, and indeed also ontological, shifts that occurred have been described by terms such as the invention of history and the discovery of society. These shifts have had effects on conceptualisations of history, of language, of agency and of collective identities that are still echoing in the writings of social and human scientists. However, they were also of crucial importance for the conceptualisation, and indeed, constitution, of new macro-societal institutions.

Thus there was a deep-seated epistemic and institutional rupture that by no means can be accounted for within the framework of a functional theory of linear modernisation and continuous and smooth differentiation. Nor can it be captured in terms of a seamless web where changes are but the result of more or less inherent power- and prestige driven interactions. They can however be made meaningful from the point of view of a social theory that emphasises human agency and the key role of processes of agential creativity and cultural contestation in historical change.

In the second major period discussed above, it is clear that the foremost social scientists of the late nineteenth century had to demonstrate the intellectual and institutional legitimacy of their activities relative to competing forms of scholarship in law, economics and philosophy. Institutionally they had to find a more secure place for the reproduction of their discourses that those provided by various commissions and social movements. The research-orientated university turned out to be such a place in a number of countries, but one that also entailed clear boundaries for professional activities, boundaries however that differed considerably between countries. Epistemically, they had also had to demonstrate that entities such as 'social facts' could be posited and endowed with a sufficient degree of stability and permanence to allow for operations involving various statistical computations and the extraction of data by means of surveys and statistical samples.[12]

Thus social scientists, creating and demarcating boundaries of disciplines emerged in institutional and professional zones distinct from those of government bureaucracy proper and from social and political movements. Links to bureaucracy and movements were often crucial but it would be misleading to reduce the agential capacities of social scientists to the assumed systemic properties of state apparatuses or political movements proper. A social theory of sheer differentiation is unable to make this history intelligible. However, nor does, what Shinn and Joerges (2002) label, the 'new orthodoxy' in social studies of science, i e an account positing discursive and scientific practices as just part of a 'seamless web' of human interactions, driven by a more or less constant set of agential attributes referring to economic and political rewards, yield an adequate understanding of the history of social science in its contexts.

In many ways, the history of the social sciences in the twentieth century must be seen in terms of developments before and after the broad professional breakthrough of the social science disciplines in the 1950s and 1960s. This breakthrough produced a kind of, if not consensus, then at least focal points concerning debates about texts that came to be regarded as 'modern classics'. Simultaneously it stimulated developments and training in the field of research methods. However it did not, in most disciplines, entail the emergence of one single strict methodological canon. In this respect, also disciplinary developments from the 1970s up to the present have been marked by the co-existence of a variety of research traditions. Three key features stand out in the development of the social sciences in recent decades.

Firstly, their professional consolidation has proceeded. Their institutional position in the modern mass higher education systems seems relatively secure. Never before have there been so many social scientists in the world, and never before have they been so well organised in professional terms, well trained in technical terms, and internationally linked through journals, networks, and research conferences. Many social science disciplines – and political science, sociology as well as educational research, business administration studies and public policy research are good cases in point – have to a large extent emerged and evolved as confederations of different practices. They have been held together by a common concern with broad substantive themes rather than by a core of theoretical assumptions.

Secondly, it is remarkable to what an extent truly innovative research contributions have resulted from work in scholarly settings outside of the structure of regular disciplinary university departments. There are three types of such environments that seem to have been particularly important to the intellectual advancement of the social sciences. Firstly, especially in the major American research universities, the system of so-called organised research units has been of great importance in providing settings outside of the departmental structure designed to be hospitable to research of a transdisciplinary nature. Secondly, research institutes, also normally of a transdisciplinary nature, have been created both outside and inside of the structure of regular universities. Thirdly, the role of so-called institutes for advanced study has been steadily increasing during the second half of the twentieth century. These different developments are reflected in the recommendation, some time ago, of the so-called Gulbenkian commission (Wallerstein et al. 1996), that every university create a transdisciplinary institute for advanced study, a recommendation that, lacking all other than persuasive force, now slowly seems to correspond to an emerging institutional reality.

Thirdly, the demands upon the social sciences, from politicians, administrators and the public at large, have not subsided but increased. Sometimes such demands have resulted in the emergence of new proto-disciplines. Generally, however, social scientists face a growing number of 'competitors', policy experts, management consultants, trend forecasters, modelling specialists but also representatives of new transdisciplinary fields such as that of the cognitive sciences. Even the most consolidated fields of social science, such as economics, are subject to critical inquiry into their boundaries whether from the point of view of theories of physics and computation (MacKenzie 2001), or history and anthropology, or theoretical and evolutionary biology.

Institutionally, the social sciences operate in the diverse and contested environment that Nowotny, Scott and Gibbons (2001) prefer to label – partly descriptively but, at least by implication, sometimes also normatively – 'mode 2'. This is an institutional environment characterised by disciplinary transgressions, responsiveness to external demands and assessments and more fluid boundaries than that of an imagined world of 'traditional' disciplinary science. Even if it is the case, as argued at some length in this essay, that there never has been a pristine sphere of disciplinary social science, 'mode 1', the problem remains that the tendencies identified by Nowotny, Scott and Gibbons are at work in the contemporary academic landscape.

Ultimately, this poses the problem to what extent epistemic and institutional autonomy is at all possible in the contemporary world. An earlier functionalist orthodoxy tended to see this as assured as a result of a secular process of differentiation. The 'new orthodoxy', to use the terminology of Shinn and Joerges, rejects the possibility of such autonomy already in the agential presuppositions of the orthodoxy itself. However neither of these two orthodoxies has provided a historically compelling argument. The intellectual autonomy of discursive practices has historically, as indicated at some length above, always been contingent and contested. The same is true of the existence of institutional spaces that allow for such practices. The institutional subordination and heteronomy of universities and other institutional spaces has always been a matter of agential imagination and contestation. If so, social studies of science should take the agential nature of human beings and the contingent nature of societal institutions seriously. In order to do this successfully, social studies have to engage directly and intensely with action-based social theory.[13]

Social studies of science ultimately rest upon assumptions of social theory and every such theory is related to an understanding of the rise and transformation of modernity as a global condition for all our actions in the contemporary world. Social theory is no mere interesting addition to the field of social studies of the natural and social sciences.[14] Social theory is constitutive of the key assumptions of social studies of science. Our ability to reflect critically on such assumptions decides the direction of social studies of science and ultimately our understanding of the world of modernity in which we all live.

There is no reason to assume that different traditions in different parts of the world will just disappear and be replaced by a single encompassing, worldwide civilisation. Modernity as such is not a civilisation. It is a set of conditions with global extension that now affect our actions, interpretations and habits, across nations and irrespective of which cultural roots we may have or lay claim to.

Humanity is in many ways constituted by the weaving together of a variety of histories of cultural and commercial interactions, entanglements and connections on a trans-regional and indeed a global scale. Science and technology is integral to this global connectedness. In order to arrive at an understanding of the contours of this global connectedness it is necessary to have as a starting point a mode of analysis that links up social theory, global history and social studies of science. I have outlined some of the conditions for this to occur.

Swedish Collegium for Advanced Study, Uppsala, Sweden

NOTES

[1] Wittrock (2000). See also Eisenstadt, Schluchter and Wittrock (2000).

[2] See, e.g., Heilbron (1995), Heilbron and Wittrock (1998), but also Brian (1994), Fox, Porter and Wokler (1995), Lepenies (1988), and Wokler (1987).

[3] The argument about that the formation of modernity entailed that a universalistic commimtent inherent in the Enlightenment was abandoned has in recent years been pursued perhaps most vigoursly by Robert Wokler, e.g. in Wokler (1998).

[4] A particularly succinct essay on Ranke's formative role is Gilbert (1990).

[5] Wittrock and Wagner (1992) and (1996). See also Rothblatt and Wittrock (1993), Geiger (1986) and Jarausch (1983).

[6] The argument in this section is largley based on a long-term research programme with results published in Heilbron and Wittrock (1998); Wagner et al. (1990); Wagner, Wittrock and Wollmann (1991); Wagner (1999a and b); Wittrock (1999); Wittrock and Wagner (1992) and (1996). An analogous line of argumentation is pursued in Wallerstein et al (1996). For a discussion of the role of universities in this process see Rothblatt and Wittrock (1993). For another overview see also Dierkes and Biervert (1992).

[7] This also entailed that an alernative, more historical, interpretation as propounded by e.g. by Reinhard Bendix came to paly a less prominent role as did Bendix' own magnum opus, *Kings or People* (1978).

[8] For an overview of these developmentes see Wagner et al. (1990) and Wagner (1999).

[9] An exemplary introduction to debates on globalisation is Held and McGrew (2000).

[10] An interesting critical discussion of network models in science and technology studies is that of Knorr-Cetina and Bruegger (2002).

[11] For an interesting recent overview see Buss (2000).

[12] Desrosières (1991) is an excellent discussion of the dilemmas involved in this construction of social facts.

[13] In recent years contributions by Hans Joas (1996, 2000a, 2000b) have been central to this debate. An ambition in his works is to elaborate a conception of action that is more encompassing than that of rational action theory. In so doing Joas draws on the traditions of both pragmatic and hermeneutic philosophy. Needless to say, some of the most prominent efforts to introduce agency and intentionlity into the analysis of texts and contexts is that of the so-called Cambridge school, drawing on works in the tradition of speech-act theory. In another context (Wittrock 1999), I have tried to indicate some of the dilemmas involved in transposing speech-act theory to textual analysis. They have to do both with problems of imputing intentionality and with linking the analysis of individual speech acts to macro-societal developments. In recent years there have been some specific efforts to establish a close link between the analysis of action and the analysis of networks. Outside of the field of social studies of science, maybe the most ambitious effort is that which tries to establish such links not just metaphorically – and I have already and appreciatingly referred to the critical notes by Knorr-Cetina and Bruegger (2002) concerning metaphorical uses of

the concept of network – is that which attempts to explicitly elucidate the nature of linking mechanisms. An excellent overview is provided by Hedström and Swedberg (1998). These efforts have however still largely been limited to analysis of action in the tradition of rational action theory. In principle, however, there seem to be no compelling reasons why the analysis could not be extended to include also a wider and more encompassing conception of action. Finally, it should be pointed out that in philosophical research, the long-standing tradition of philosophy of action has in recent years been increasingly complemented by research in the field of logic of action. Some of this research involves direct collaboration between logicians, cognitive scientists and computer scientists. It still seems to be an open question to what extent the results of this collaboration – in contrast to the field of philosophy of action where direct links to social theory have long existed – may or may not be conveniently linked up to the advances in action theory that have occurred in social theory.

[14] A different but analogous argument has recently been advanced by Judy Wajcman (2202).

REFERENCES

Baker, Keith M. (1994), 'Enlightenment and the institution of society: Notes for a conceptual history', in W. Melching and W. Velema (eds.), *Main Trends in Cultural History: Ten Essays,* Amsterdam: Rodopi, pp. 95–120.

Bendix, Reinhard (1978), *Kings or People: Power and the Mandate to Rule,* Berkeley, CA: University of California Press.

Brian, Eric (1994), *La Mesure de l'Etat: Administrateurs et Géometres au XVIII Siècle*, Paris: Albin Michel.

Buss, Andreas (2000). 'The evolution of Western individualism: To the memory of Louis Dumont', *Religion* 30: 1–25.

Collini, Stefan, Donald Winch and John Burrow (1983), *That Noble Science of Politics: A Study in Nineteenth Century Intellectual History,* Cambridge: Cambridge University Press.

Desrosières, Alain (1991), 'How to make things which hold together: Social science, statistics and the state', in P. Wagner, B. Wittrock, and R. Whitley (eds.), *Discourses on Society: The Shaping of the Social Science Disciplines*, Dordrecht, The Netherlands: Kluwer, pp. 195–218.

Dierkes, Meinolf and Berndt Biervert (eds.) (1992), *European Social Science in Transition: Assessment and Outlook*, Frankfurt a. M.: Campus; Boulder: Westview.

Dyson, Kenneth (1980), *The State Tradition in Western Europe*, Oxford: Robertson.

Eisenstadt, Shmuel N., Wolfgang Schluchter and Björn Wittrock (eds.) (2000), *Public Spheres and Collective Identities*. New Brunswick, NJ: Transaction Press.

Fox, Christopher, Roy Porter and Robert Wokler (eds.) (1995), *Inventing Human Science: Eighteenth Century Domains*, Berkeley, CA: The University of California Press.

Furner, Mary O. (1975), *Advocacy and Objectivity: A Crisis in the Professionalization of American Social Science 1905–1965*, Lexington, KY: The University Press of Kentucky.

Geiger, Roger L. (1986), *To Advance Knowledge: The Growth of American Research Universities 1900–1940*, New York: Oxford University Press.

Gilbert, Felix (1990), *History: Politics or Culture? Reflections on Ranke and Burckhardt*, Princeton, NJ: Princeton University Press.

Graham, Loren, Wolf Lepenies and Peter Weingart (eds.) (1983), *Functions and Uses of Disciplinary Histories*, Dordrecht, The Netherlands: Reidel.

Haskell, Thomas S. (1977), *The Emergence of Professional Social Science*, Urbana, IL: University of Illinois Press.

Hedström, Peter and Richard Swedberg (eds.) (1998), *Social Mechanisms: An Analytical Approach to Social Theory*, Cambridge: Cambridge University Press.

Heilbron, Johan (1995), *The Rise of Social Theory*, Cambridge: Polity Press.

Heilbron, Johan (1998/2001), 'French moralists and the anthropology of the modern era: On the genesis of the notions of "interest" and "commercial society"', in J. Heilbron, L. Magnusson and B. Wittrock (eds.), *The Rise of the Social Sciences and the Formation of Modernity*, Dordrecht, The Netherlands: Kluwer, pp. 77–106.

Heilbron, Johan, Lars Magnusson and Björn Wittrock (eds.) (1998; Paperback edition 2001), *The Rise of the Social Sciences and the Formation of Modernity: Conceptual Change in Context, 1750–1850*, Dordrecht, The Netherlands: Kluwer.

Held, David and Anthony McGrew (2000), *The Global Transformations Reader: An Introduction to the Globalization Debate*, Cambridge: Polity Press.
Jarausch, Konrad (1983), *Transformations of Higher Learning*, 1860–1930, Stuttgart: Klett.
Joas, Hans (1996), *The Creativity of Action,* Cambridge: Polity Press.
Joas, Hans (2000a), *The Genesis of Values*, Cambridge: Polity Press.
Joas, Hans (2000b), *Kriege und Werte. Studien zur Gewaltgeschichte des 20. Jahrhunderts*, Weilerswist: Velbrück.
Knorr Cetina, Karin and Urs Bruegger (2002), 'Inhabiting technology: The global lifeform of financial markets', *Current Sociology* 50(3): 389–405.
Kelley, Donald R. (1990), *The Human Measure: Social Thought in the Western Legal Tradition*, Cambridge: Cambridge University Press.
Koselleck, Reinhart (1987), *Critique and Crisis: Enlightenment and the Pathogenesis of Modern Society*, Oxford: Berg.
Lepenies, Wolf (1988), *Between Literature and Science: The Rise of Sociolog*, Cambridge: Cambridge University Press.
MacKenzie, Donald (2001), 'Physics and finance: S-terms and modern finance as a topic for science studies', *Science, Technology, & Human Values* 26(2): 115–44.
Manent, Pierre (1994), *An Intellectual History of Liberalism,* Princeton, NJ: Princeton University Press.
Manent, Pierre (1998), *The City of Man,* Princeton, NJ: Princeton University Press.
Manicas, Peter T. (1987), *A History and Philosophy of the Social Sciences*, Oxford: Blackwell.
Nowotny, Helga (2000), 'Transgressive competence: The narrative of expertise', *European Journal of Social Theory* 3(1): 5–21.
Nowotny, Helga, Peter Scott and Michael Gibbons (2001), *Rethinking Science: Knowledge and the Public in an Age of Uncertainty*, Cambridge: Polity Press.
Raeff, Marc (1983), *The Well-Ordered Police State: Social and Institutional Changes through Law in the Germanies and Russia 1600–1800*, New Haven, CT: Yale University Press.
Ringer, Fritz K. (1969), *The Decline of the German Mandarins*, Cambridge, MA: Harvard University Press.
Ross, Dorothy (1991), *The Origins of American Social Science,* Cambridge: Cambridge University Press.
Rothblatt, Sheldon and Björn Wittrock (eds.) (1993), *The European and American University Since 1800: Historical and Sociological Essays*, Cambridge: Cambridge University Press.
Rueschemeyer, Dietrich and Theda Skocpol (eds.) (1996), *States, Knowledge and the Origins of Social Policies*, Princeton, NJ: Princeton University Press,
Rueschemeyer, Dietrich and Ronan Van Rossem (1996), 'The Verein für Sozialpolitik and the Fabian Society', in D. Rueschemeyer and T. Skocpol (eds.*), States, Social Knowledge, and the Origins of Modern Social Policies*, Princeton and New York: Princeton University Press, Russell Sage Foundation, pp. 117–162 .
Shinn, Terry and Bernward Joerges (2002), 'The transverse science and technology culture: Dynamics and roles of research-technology', *Social Science Information* 41(2): 207–51.
Schluchter, Wolfgang (1996), *Paradoxes of Modernity: Culture and Conduct in the Theory of Max Weber*, Stanford, CA: Stanford University Press.
Soffer, Reba N. (1978), *Ethics and Society in England: The Revolution in the Social Sciences 1870–1914*, Berkeley, CA: The University of California Press.
Smith, Rogers M. (1997), 'Still blowing in the wind: The American quest for a democratic, scientific political science', *Daedalus* 126(1): 253–87.
Tocqueville, Alexis de (1964), *Oeuvres Complètes, Tome XII, Souvenirs*, Paris: Gallimard.
Tribe, Keith (1988), *Governing the Economy: The Reformation of German Economic Discourse, 1750–1840*, Cambridge: Cambridge University Press.
Tribe, Keith (1998/2001), 'Natural law and the origins of Nationalökonomie: L.H. von Jakob', in J. Heilbron, L. Magnussen and B. Wittrock (eds.), *The Rise of the Social Sciences and the Formation of Modernity*, Dordrecht, The Netherlands: Kluwer, pp. 189–205.
Wajcman, Judy (2002), 'Addressing technological change: The challenge to social theory', *Current Sociology* 50(3): 347–63.
Wagner, Peter (1990,) *Sozialwissenschaften und Staat. Frankreich, Italien, Deutschland 1870–1980*, Frankfurt a.M.: Campus.
Wagner, Peter (1994), *A Sociology of Modernity: Liberty and Discipline*, London: Routledge.

Wagner, Peter (1999), 'The twentieth century – The century of the social sciences?', in A. Kazancigil and D. Matkinson (eds.), *World Social Science Report 1999*, Paris and London: UNESCO and Elsevier, pp. 19–41.

Wagner, Peter, Carol H. Weiss, Björn Wittrock and Hellmut Wollmann (eds.) (1991), *Social Science and the Modern State: National Experiences and Theoretical Crossroads,* Cambridge: Cambridge University Press.

Wagner, Peter, Björn Wittrock and Richard Whitley (eds.) (1991), *Discourses on Society: The Shaping of the Social Science Disciplines*, Dordrecht, The Netherlands: Kluwer.

Wallerstein, Immanuel et al. (1996), *Open the Social Sciences: Report of the Gulbenkian Commission on the Restructuring of the Social Sciences*, Stanford, CA: Stanford University Press.

Wittrock, Björn (1999), 'Social theory and intellectual history: Rethinking the formation of modernity', in F. Engelstad and R. Kalleberg (eds), *Social Time and Social Change: Historical Aspects in the Social Sciences*, Oslo: Norwegian University Press. pp. 187–232.

Wittrock, Björn (2000). 'Modernity: One, none, or many: European origins and modernity as a global condition', *Daedalus* 129(1): 31–60.

Wittrock, Björn (2001), 'History, war and the transcendence of modernity', *European Journal of Social Theory* 4(1): 53–72.

Wittrock, Björn and Peter Wagner (1992), 'Policy constitution through discourse: Discourse transformations and the modern state in central Europe', in D.E. Ashford (ed.), *History and Context in Comparative Public Policy*, Pittsburgh, PS: The University of Pittsburgh Press, pp. 227-46.

Wittrock, Björn and Peter Wagner (1996), 'Social science and the building of the early welfare state: Toward a comparison of statist and non-statist Western societies', in D. Rueschemeyer and T. Skocpol (eds), *States, Social Knowledge and the Origins of Modern Social Policies,* Princeton, NJ: Princeton University Press. pp. 90–113.

Wokler, Robert (1987), 'Saint-Simon and the passage from political to social science', in A. Pagden (ed.), *The Language of Political Theory in Early Modern Europe*, Cambridge: Cambridge University Press, pp. 325–38.

Wokler, Robert (1998), 'The Enligthenment and the French Revolutionary birth pangs of modernity', in J. Heilbron, L. Magnusson and B. Wittrock (eds.), *The Rise of Social Science and the Formation of Modernity*, Dordrecht, The Netherlands: Kluwer, pp. 35–76.

CHAPTER 6

TERRY SHINN

THE 'TRIPLE HELIX' AND 'NEW PRODUCTION OF KNOWLEDGE' AS SOCIO-COGNITIVE FIELDS*

Assertive demands, originating in government and to a lesser extent in industry, for readjustments in science/enterprise/state relations were set in motion by the energy crises of the 1970s and again, by the economic slumps of the 1980s and early 1990s. Science and technology were bannered as a solution to overdependence on foreign energy supplies and panacea to economic sluggishness and sharply rising unemployment.[1] Sociologists, economists, political scientists and science policy makers have responded to the ensuing emergence of a new set of political, industrial and social expectations, discourses and projects in a variety of ways. One reaction has consisted of studies of the interplay between research, business and the government, and suggestions of conceptual frameworks to explain the observed changes. Other scholars have proposed a blueprint for transformation, and have become partisan activists in the promotion of change. Both these reactions have their place within the confines of particular cognitive and social fields.

This essay will explore this reaction with reference to the fields associated with the 'Triple Helix' and the 'New Production of Knowledge'. *The New Production of Knowledge*, published in 1994 and authored by Michael Gibbons, Camille Limoges, Helga Nowotny, Simon Schwartzman, Peter Scott and Martin Trow (Gibbons et al. 1994), argues that the way in which scientific knowledge, technical practices, industry, education and society at large are organized and function today lies in sharp contrast with the relationships in earlier times. The authors speak of two distinct modes of knowledge production. Mode 1 is characterized by a cleavage between academia and society. Academia revolves around an autonomous university, self-defined and self-sustained scientific disciplines and specialties, and the determination by scientific peers of what does and does not constitute science and truth. Here there is allegedly no interaction between academia and industry. By contrast, Mode 2 knowledge production perceives the weakening or collapse of the modern university, the disappearance of scientific disciplines

* This text was first published in French under the title 'Nouvelle Production du Savoir et Triple Hélice. Tendances du Prêt-à-Penser les Sciences', *Actes de la Recherche en Sciences Sociales*, Mars 2002, 1441–2: 21–30, and subsequently appeard in English as 'The Triple Helix and New Production of Knowledge. Prepackaged Thinking on Science and Technology', *Social Studies of Science*, August 2002, 32–4: 599–614.

B. Joerges and H. Nowotny (eds.), Social Studies of Science and Technology: Looking Back Ahead, 103–116.

and the atrophy of peer control over the direction and content of research programs. Mode 2 science is characterized by interdisciplinarity, by the fluent movement of short term task force teams of experts to problem domains, and by the primacy of social and economic problems in establishing what spheres of knowledge should be developed. Society thus denies the legitimacy of science's prerogatives, its institutional autonomy and cultural identity.

By contrast, the Triple Helix stresses historical continuities. Earlier relations between the university, industry and government persist. But to these models of learning and action, is now added another, which has been called the 'Triple Helix'. Unlike the New Production of Knowledge, the Triple Helix does not possess a charter document in the form of a single extremely visible book which can readily be cited. Rather than repudiating institutional differentiations, this perspective identifies the birth of a supplementary layer of 'knowledge development', a layer in which specific groups inside academia, enterprise and the government meet in order to address new problems arising in a deeply changing economic, institutional and intellectual world. The Triple Helix is intended to be a sociological expression of what has become an increasingly knowledge-based social order.

These two perspectives prompt five questions. Questions one and two are reflexive: First, how much attention has each perspective received? Second, who actually 'use' these two perspectives? What national, institutional and disciplinary roles do they play? Stated differently, and using the language of Pierre Bourdieu, what is the structure of their socio-cognitive field? (Bourdieu 1975). By 'socio-cognitive field' is meant the interdependencies between the historical, institutional, and political elements structurally integral to knowledge, and the rational and rhetorical components that are equally determinant. A 'socio-cognitive field' is a domain where cognitive and social elements are simultaneously self-referencing and combinatory as they interact to form a whole. Third, what are the characteristics of the claims, data, methodology, and theory of the two perspectives? Are their profiles similar, or do they operate in different intellectual and institutional socio-cognitive fields? Fourth, what changes of analysis, diagnosis or prognosis (if any) have occurred in these two perspectives over the last several years? Lastly, what are the domains of internal inconsistency, inadequacy or malfunction of each approach ?

A PINCH OF REFLEXIVITY

Sociologists of science and technology routinely do bibliometric studies of scientists working in the physical and life sciences in order to learn about the productivity of a research school, about those who occupy it, and those who constitute its audiences. To examine the 'impact' of the Triple Helix and the New Production of Knowledge and the institutional loci of the people who incorporate these perspectives (or at least who refer to them), I have drawn on the *Social Science Citation Index* for the period 1995 to 2000, as well as information available on the Internet. An exact comparison of the two encounters serious methodological snags, so the picture that I present is an approximation, although I believe it is sufficiently rigorous to suggest important differences between the two communities.

The New Production of Knowledge: The Dynamics of Science and Research in Contemporary Societies constitutes a charter document and a group manifesto (Gibbons et al. 1994). A second treatise, *Re-thinking Science: Knowledge Production in an Age of Uncertainty,* by Helga Nowotny, Michael Gibbons and Peter Scott, has recently appeared (Nowotny, Gibbons and Scott 2001). For my study, I limited quantitative tabulations to references made to the first book, by Gibbons et al. For the period 1995 to June 1999, a total of 98 references were made to the book: 1995 (8), 1996 (16), 1997 (15), 1998 (30), 1999 (29). The number of references for the first six months of 1999 already equals those for 1998. The citing journals include sociology, the sociology of science and technology, science policy, psychology, social psychology, and education. References in education reviews are particularly abundant. Indeed, both the quantity and range of citations to the *New Production of Knowledge* are impressive.

By contrast to the *New Production of Knowledge*, the Triple Helix does not possess a charter document.[2] This suggests an important structural difference between the two socio-cognitive fields. The Triple Helix has emerged gradually, and its textual architecture takes the form of numerous interlocking pieces, such as introductions and conclusions of collective works, chapters, articles or unpublished conference papers. To make my bibliometric task more manageable, I tabulated the references made to six defining or descriptive empirical or theoretical pieces published by Loet Leydesdorff or/and Henry Etzkowitz (the founders of the triple helix perspective) between 1995 and 1998; and to the pieces published in two collective volumes edited by Leydesdorff and Etzkowitz in 1997 and 1998 (Etzkowitz and Leydesdorff 1996, 1997, 1998, 1999; Leydesdorff and Etzkowitz 1996, 1998a,b,c,d).

As measured by the *Social Science Citation Index*, the impact of the Triple Helix perspective is negligible – almost non-existent. For the period 1996 to mid-1999, I found only one reference to the extensive corpus to which I extended my bibliometric search. I also examined the place occupied by the two perspectives on Internet, where the story is rather similar. Using the *altavista.com* search engine, I discovered about thirty web pages referring to the combined word sequences 'New Production of Knowledge' and 'Mode 2'. The theme of education is again paramount; for example, in relation to designing the 'new' university, the 'cyber-university' etc. However, 'New Production of Knowledge' web sites even extend to psychoanalysis of organizations! (The International Society for Psychoanalytic Study of Organizations Symposium 1999). On the Internet, this perspective is a good example of the band wagon effect, as individuals and groups from a variety of disciplines and occupations jump aboard. Apparently, the terminology ('New Production of Knowledge') resonates in a number of areas, where it can be seen to engender a community, whose shared feature consists of invoking a set of words. Whether the shared vocabulary is rooted in structured concepts or is merely a felicitous phrase remains to be seen. Durkheim demonstrated the power of metaphors and their centrality to sociology, but is the 'New Production of Knowledge' a metaphor, or just a catch phrase?

The Triple Helix is not prominent in either the *Social Science Citation Index* or the Internet. On the basis of the search engine *copernic*, I have identified eight sites.[3] These include information on the three international Triple Helix conferences (Amsterdam 1996; New York 1998; and Rio 2000), as well as a general purpose information site and discussions forum. These sites appear to be updated and reasonably active. Of course, as everyone knows, the Internet is a place for self-promotion, so it must be treated with caution as an impact indicator. The hand of Loet Leydesdorff is discernible in most of these sites, and this contrasts with the 'New Production of Knowledge' sites where the authors of the seminal book appear to play only a marginal role.

How is it that the impact indicators of the Triple Helix remain so low, both in the *Social Science Citation Index* and on the Internet? What does this mean? How does the low impact indicator for the Triple Helix square with the reality that its analytical focus has spawned three large international conferences, triggered scores of articles, and raised considerable amounts of public and private money? Immediately, the modest impact indicator in the *Social Science Citation Index* and on the Internet suggests that there are reasons to rethink the relationship between the growth of ideas and reputations, on the one hand, and what citation counts tell us, on the other. Indeed, what, if anything, do citation counts tell us about the precision, rigour, and documentation of claims? In the social sciences, abundant citation may be a better gauge of evocative language, speculation, and far reaching generalization than of prudent measurement or careful hypothesis testing.

The geographic composition of the communities drawing upon the Triple Helix and the New Production of Knowledge also differs significantly. The geographic distribution of the Triple Helix audience is very broad, and includes large numbers from Latin America, Asia and Africa. On the other hand, the New Production of Knowledge audience is concentrated in Western Europe and the United States and Canada. Based on the institutional affiliations of citing authors and individuals attending relevant meetings, over 90% of the New Production of Knowledge audience is based in the North, as against about 65% of the Triple Helix (the latter count is based on participation in conferences). The Triple Helix thus enjoys a sizable following among the developing countries. As we shall see, there are clear-cut reasons for this.

METHODS, CLAIMS AND CONCEPTS

The intellectual structures of the Triple Helix and the New Production of Knowledge differ significantly. The nucleus of the New Production of Knowledge is located in a single volume, in which are expressed claims about the demise of universities, scientific disciplines and academic laboratories; and a rise in inter-disciplinarity, economically and socially-relevant research themes, and the appearance of perpetually fluid business-linked research task-forces, in the framework of a new kind of socially useful epistemology. The book may in this sense be read as a compact and articulate charter that reflects the interests of scholars and others concerned with global business, learning and social relations.

The reader is struck by three features. *The New Production of Knowledge* raises few questions about the evolution of science and technology and about changes in their relations with enterprise and society. Instead, it offers a number of prefabricated indications about where science has putatively come from and where it is allegedly going. No questions, but lots of answers. On a parallel plane, almost no concrete evidence is given for the assertions advanced; and no provision is made for future empirical historical or sociological work. While the absence of data in the book is distressing, people interested in this approach and desiring to explore its possibilities might hope for precise information in subsequent works. However, this hope has so far not been realized: the six authors have not embarked on confirming empirical projects; neither have colleagues come up with corroborating data.

The perspective's empirical underdevelopment is particularly regrettable, for one can easily imagine that its tenets could prove highly instructive for the study of a series of important but little explored domains – such as the fluid and multi-determinant relations of knowledge orientation, production, application and evaluation in/for communities of the physically handicapped or ill.[4] But to my knowledge, research along these lines has not been inspired by the New Production of Knowledge. Moreover, the systematic empirical work that has been carried out with reference to the New Production of Knowledge (perhaps in fields less conducive to its predictions), has, by and large suggested that the claims run counter to available evidence, or that the claims are at best not clearly validated by available facts (Pestre 1997; Weingart 1997; Godin 1998; Shinn 1999; Godin and Gingras 2000). In short, there appears at present to be a pocity of careful control studies in selected spheres to identify the strengths and limiting conditions of the sweeping generalizations expressed in the book. The book lacks the methodological motor that is necessary to drive any research programme forward.

These programmatic and methodological difficulties may be a consequence of the fact that the approach lacks a theoretical referent. It is not specifically connected to any conceptual framework – for example to that of Durkheim, Weber, Parsons, Bourdieu, Habermas, or Luhmann. The *New Production of Knowledge* does not work out or define its key sociological concepts. One is tempted to ask, does it possess any? The answer is complicated. It does incorporate concepts, but whether the concepts are rigorously sociological is doubtful.

That said, the approach is 'anti-differentiationist', as it seeks to minimize or to deny demarcations between academic, technical, industrial, political and social institutions. It thus dismisses boundaries and divisions of labour. The perspective rejects the notions of specific forms of knowledge and specific social constituents in favour of undifferentiated knowledge and undifferentiated social ensembles, where even the distinctions between nature and culture disappear.

However, this radical anti-differentiation line is never buttressed with sociological theory, concepts or models. Instead, it stands as a free-floating, unintegrated component. This situation is permissible if the perspective is interpreted not as a sociological research programme linked to the study of the production, diffusion and use of knowledge, but instead as a performative discourse. This postmodernist penchant y is coupled with what appears to be a muted preference for globalization. Instead of theory or data, the New Production of Knowledge – both book and concept – seems tinged with political commitment. The authors appear to be true believers in a new cognitive and social order. They work actively in its favour and to persuade others to think likewise. One has to wonder if the perspective is not more a social platform than a serious, systematic framework for scholarly inquiry.

Of course, ambitious ventures involve a measure of self-promotion and propaganda. However, a research programme must exhibit a cognitive plan in addition to a set of professional strategies. The question becomes: what is the intellectual 'project' of the New Production of Knowledge? But the same token, one must ask the same question of the Triple Helix: what is its intellectual project and agenda?

The horizons of the Triple Helix are four-fold. First, it has developed an empirical base, in the form of multiple case studies of changing relations between university, industry and the state. Many of the articles appearing in the February 2000 issue of *Research Policy* document the changing relations between the three strands composing the Triple Helix with respect to specific economic sectors and cognitive fields – biotechnology, aeronautics, computers and instrumentation. Implicitly, these studies raise a central question: what is the purview of the Triple Helix? Does it apply only to a narrow range of economic, cognitive, technical and governmental configurations, and if so which? The centrality of empirical data goes some way toward neutralizing the normative propensity associated with sociological models. Inclinations towards generalization are in part balanced by analyses of concrete events. As will be shown below, a healthy attention to matters of fact has recently led to some key changes in the model's focus, prescriptions and conceptual thrust. This structured inclusion of detailed information in the modeling process, and in evaluating the model's utility, would appear to contrast with the 'New Production of Knowledge', which remains thin on data and prone to sweeping generalizations.

Second, the Triple Helix explicitly addresses concrete and pressing problems of government, academic and industrial policy. Here, the model's authors are sometimes involved in stimulating entrepreneurs, university administrators and public figures to rethink policy and conduct in response to changing cognitive, technical, economic and international trends. The mode of action is not that of a lobby or movement, but rather that of a mindful 'think-tank'. Additionally, the spate of publications and meetings associated with the Triple Helix helps policy makers keep abreast of changing environments – and when possible, to anticipate change. This is shown by the involvement of the National Science Foundation, the Centre National de la Recherche Scientifique, NATO, the European Commission, and political and academic authorities in Brazil and other developing countries, in reflection on the Triple Helix.

Often implicitly, and sometimes explicitly, the authors of the triple helix ask whether the putatively new arrangements emerging between academia, industry and the state are relevant to the actual process of economic, intellectual and political emergence underway in Latin America, Africa and Asia. Are the institutional realities of developing countries sufficiently similar to those in northern nations (on whose experiences the model is based) to make the Triple Helix analysis relevant? If there is a mismatch, is it possible, or desirable, to use the Triple Helix as a guideline to transformations in university/industry/state relations in order to reinforce processes of emergence? Given the size of the delegations that participated in the Rio conference in April 2000, the hope that this may be true runs high. This aspect of the Triple Helix contrasts with the perspective given by the New Production of Knowledge, which speaks more to institutionally and technologically advanced environments.[5]

Third, the analytic thrust of the Triple Helix is opposite to that of the New Production of Knowledge, in which distinctions between (or differentiations between) science and technology, industry and academia, society and knowledge are denied. By contrast, the Triple Helix pursues a neo-differentiation strategy (Shinn 1999, 2002; Shinn and Joerges 2002). Etzkowitz and Leydesdorff argue that while throughout much of the nineteenth and the first half of the twentieth century the related yet distinct strands occupied by the university, industry and state functioned effectively, internal events in each, and changing relations between them, is giving rise to still another differentiated unit – one which fuses the three strands in a historically unique way, the Triple Helix. This emergent entity constitutes a new synthesis between government, academia and enterprise. But unlike the New Production of Knowledge, this synthesis does not efface previous discontinuities, but rather constitutes an additional, new discontinuity – the 'triple helix' (as opposed to three related single helix).

This neo-differentiation perspective generates many questions: precisely what concrete entities does it comprise? What mechanisms have led to its emergence? What new functions does it perform? How can we know that the Triple Helix is a "new" differentiation rather than a readjustment that has modified environments without imperiling the established institutions? Are entities like incubators, start-ups, and other new forms of government/enterprise partnerships the deciding examples on which the validity of the Triple Helix depends? This would appear to be an empirical question, although a difficult one!

Fourth, the Triple Helix is accompanied by a theoretical framework that takes the form of self-organization and co-evolutionary theory. The standard references here are Humberto Maturana and Niklas Luhmann (Maturana 1980; Luhmann 1996; see also Krohn, Küppers and Nowotny 1990). The key claims of this theory are the following: (1) Under specified conditions, institutional and cognitive structures become ill-adapted to current situations and unstable. (2) Several structures evolve, and this co-evolution generates a historically new institutional and/or cognitive structure. (3) *Time* constitutes a foundational dimension in this dynamic process. (4) Co-evolutions temporarily resolve problems of mismatches in the complexities of earlier systems. (5) In time, the new layers of complexity are themselves accompanied by fresh mismatches (institutional and/or cognitive) and this breeds further cycles of co-evolutions.[6]

Many interlocutors express unease with the theoretical pronouncements surrounding the Triple Helix, and are even bewildered by them. This may derive in part from difficulties with mathematical formulations associated with the theory. It may also stem from difficulties in penetrating the theory's insider terminology (expressions such as "lockin" and "overlays"). If a theoretical message is being transmitted, it is not intelligible to numerous in the audience. An incomplete understanding of the theory tied to the Triple Helix potentially hinders a full appreciation of the model and its inherent possibilities. The theoretical message that accompanies the Triple Helix must be made intelligible. If not, the co-evolutionary concept will likely become viewed as irrelevant or wrong, and detached from the empirical, diagnostic and prognostic components of the Triple Helix. The result would be a less ambitious, versus a strongly predictive model integrated into a general social theory.

The fit between theory and the empirical data constitutes another problem: co-evolution theory describes structures and transformations on a meta-level and in macroscopic terms. The search for appropriate units of analysis takes place at a high level of aggregation, generalization and abstraction. However, such analytic operations and postulates may not sit comfortably with the rich, interesting and insightful empirical studies promoted by the Triple Helix. Intermediary mechanisms must be specified to link well-established institutional, economic and cognitive changes to the co-evolutionary theory in an unambiguous and corroborative way.

It is fair to ask whether co-evolution constitutes the only or the best theoretical frame for the Triple Helix. Do other systems, such as Durkheim's functionalism, or Bourdieu's concepts of "field" and "habitus", provide alternatives? If not, then why? Does the probable objection that alternative systems do not include the *time* dimension, or do not do so in a systematic fashion, constitute a sufficient argument to lock the Triple Helix to co-evolution?

SOCIOLOGY VERSUS INTROSPECTION

Until recently, the message of the *New Production of Knowledge* has remained constant. Although some of its authors (Michael Gibbons and Helga Nowotny, in particular) have often spoke in defense of their ideas, nothing new has been forthcoming in print. Fortunately, Nowotny, Gibbons, and Scott's *Re-thinking Science* (published in 2001) allows us to examine the issue of change in the New Production of Knowledge perspective (Nowotny, Gibbons and Scott 2001). What is their message today, and where does it stand with respect to 1994?

Like many, the authors talk about a 'new contract' between society and science. They also write about 'society learning to speak to nature' – by which they mean the eclipse of the science referent as the basis for scientific legitimacy, and its replacement by a purely social referent. Their book introduces a number of new terms – or at any rate, existing terms in unfamiliar usage. Many are worthy of further reflection.

1. For example, the authors argue in favour of 'contextualization', by which is meant the need to 'de-differentiate' between science and society. While Modernity has been characterized by distinctions between nature/culture and science/ society, this demarcation, they argue, no longer holds. The postmodern order is to be 'heterogeneous', and heterogeneity does not entail differentiations. Heterogeneous audiences address knowledge producers in the 'agora' (ibid., 47–9).
2. Postmodern society is characterized by 'reverse communication' – that is, by communication from society toward knowledge producers, and not vice-versa. Society decides on what knowledge is to be. Knowledge producers accept and follow. Knowledge is socially relevant learning, and is generated in fluid relations between the state, markets, and industry. Knowledge production is represented as a transition phase operation (ibid., 110).
3. 'Experts' operate in a framework of 'socially distributed knowledge'. Learning is no longer set off from society, and the activities of experts ensure that learning addresses social needs. Successful enterprise in a global economy constitutes one such paramount need (ibid., 226).
4. The authors of *Re-thinking Science* insist that the New Production of Knowledge involves a new epistemology, which they label 'socially robust knowledge.' They argue that advances in physical theory have come to a stand-still. Possibilities of fresh conceptualization are exhausted. 'Socially robust knowledge' refers to scientists' putative inclination to formulate promising research problems in terms of the 'novelties' that they may engender. 'Socially robust knowledge' consists of the systematization of application oriented, mission-driven research and learning (ibid.,

167). The rigours of relevance become the epistemological yardstick of the New Production of Knowledge, and its measure will be taken by experts functioning in the system of socially distributed learning. This, the authors emphasize, is what is entailed by Mode 2.

The message conveyed in *Re-thinking Science* is similar to that advanced almost a decade ago in *The New Production of Knowledge*. Both point to the end of disciplinary science, universities, laboratory-rooted research, and differentiations between scientific knowledge *per se* and society. Both accord primacy to the social relevance of learning, and in particular, to the demands of enterprise. The 1994 book hints at a new epistemology: the latest book develops the theme and identifies it as a mainstay of Mode 2. Like the 1994 volume, the latest book is silent when it comes to giving systematic evidence, or for suggesting directions for research. The book appears to be mainly a product of introspection and a basis for future reflection.

Finally, one can not help wondering if *Re-thinking Science* does not open the way to, or perhaps even legitimate, a neo-corporatist vision of the world. By collapsing the specificity and institutional setting of science, of technology, of industry and of politics, do not Nowotny, Gibbons and Scott inadvertently lay the foundations of a social amalgam whose direction and details can be readily imposed by an authoritarian political force?

This static stance contrasts with the mobile, fluctuating analyses advanced by the architects of the Triple Helix. At a meeting in June 1998 to compare the Triple Helix and the New Production of Knowledge, I suggested that the two approaches exhibit a degree of inverse symmetry. I argued that the New Production of Knowledge constitutes a radical 'anti-differentiation' message, and that the Triple Helix constitutes a radical 'neo-differentiation' message. During the meeting, I asked proponents of the Triple Helix two questions. First, precisely which institutions and initiatives embody the allegedly emergent synthetic 'triple' strand, as opposed to the three separate historical strands which compose it? Second, whatever its exact form, precisely how 'radically differentiated' is the putative 'neo-differentiation'? No clear-cut answers were forthcoming. However, I consider that now (some four years later), Etzkowitz and Leydesdorff have provided replies to my queries. Indeed, perhaps too many replies for the sake of clarification. Be this as it may, Triple Helix- thinking has evolved considerably since its initial formulation.

In their overview of the Triple Helix appearing in *Research Policy*, Etzkowitz and Leydesdorff give voice to two ideas (Etzkowitz and Leydesdorff 2000).

1. The Triple Helix is centered inside the traditional university. Etzkowitz and Leydesdorff insist that discipline-based departments are converging in new ways, and while maintaining traditional lines of research, are also turning to industrial research and intermediary forms of research. Universities are thus generating a variety of midwife institutions that link them to economic and social concerns. The university constitutes a privileged site in which discourses come together, merge and give rise to new forms of discourse and action. In this publication, the highly illusive Triple Helix has at long last been identified.

2. In the same publication, comes the term 'endless transition'. This is an important addition. An emphasis on repeated co-evolutions attenuates the focus on a single landmark. We no longer have to search for a single macro-entity which embodies a dramatic three-strand confluence. The model now becomes compatible with much smaller changes and co-evolutions occurring inside one of the three strands. People interested in the Triple Helix approach are now free to search for small variations and variants (endless transitions on a micro-level). This re-ordering within the model in favor of finite and micro-level transitions dovetails with the recent attention given to the changes taking place inside the university strand mentioned above.

So what has the Triple Helix become? According to this overview, the university is sustained. Along with its traditional functions of teaching, certification and fundamental research, recent cognitive and economic changes have merely added new functions. The historical role of universities is preserved, and extended to accommodate changing circumstances. Here, the university has been cannily slipped through the backdoor as the decisive element in contemporary cognitive and economic dealings. Concomitantly, the novelty of the metaphor "triple helix" discreetly recedes. The disturbing dislocations of the Triple Helix are being supplanted by evolutionary, transitional and incremental adjustments within academia!

However, Etzkowitz and Leydesdorff seem to disagree with this reading. In his keynote address to the third international Triple Helix conference, held at Rio in April 2000, Etzkowitz declared that the Triple Helix is embodied in 'incubators'. But this claim conflicts with the position taken by himself and Leydesdorff in *Research Policy.* Even greater confusion ensued when Etzkowitz insisted that the Triple Helix is to be found in the 'second academic revolution', a stance at least consistent with the *Research Policy* affirmations. But one wonders what kind of thinking underpins such claims. What evidence can be mustered in support of either of the two propositions?

An examination of *sociofile* and of several Internet search engines for 'incubator' and for 'second academic revolution' reveals that little has been written on these topics – indeed so little that it is unclear to exactly what (sociologically and cognitively) the two terms refer. Before crediting either incubators or the second academic revolution with being the cornerstone of the Triple Helix, it is first necessary to determine their cognitive, economic, and social properties.

However, if one adopts the 'Endless Transition', what is to be made of the formerly revolutionary Triple Helix? Because, if, as admitted by the authors, the key co-evolutions can arise within just one of the three constitutive strands, then the status of the alleged historically new and unique macro-entity (the 'tripled helix') must provoke reservations. In view of these refinements, can one say that the model still exists, or is it not rather giving way to a new formulation, more in harmony with empirical evidence collected in the course of careful case studies – a formulation which has little to say about 'radical neo-differentiation'? Whatever its future, the triple helix has prompted many to consider how science/government/industry relations have changed, are changing, and are likely to change. This orientation will evolve as it draws on case studies and debate.

CHALLENGES

The size and disciplinary/professional diversity of the New Production of Knowledge's audience, stems in part from the fact that its argument touches on many spheres, extending from education to research, business, politics and the organization of contemporary democracy (Horelli 1997). Many think of it as a skeleton key, a magic potion, for understanding an array of problems. The New Production of Knowledge is not a research school, since it does not articulate a research programme. *The New Production of Knowledge* and *Re-thinking Science* do not define questions, set-forth a methodology, provide reasoned answers, or set limiting conditions. On the contrary, they can be likened to political manifestos, whose expository form is rhetoric.

The socio-cognitive field of the Triple Helix is very different. When measured in citations, its audience is negligible. But if gauged by reference to international meetings and developing nations, the Triple Helix mobilizes a large number of followers. The Triple Helix may or may not constitute an analytic model, but it does constitute a serious research school with an empirical and conceptual agenda.

Both of the approaches can be faulted for failing to take into account two important aspects of knowledge and artefact production and diffusion. The first deficiency lies in their failure to recognize that the university, business and government all function in a national setting. Even scientific disciplines and specialties operate differently in different national institutions, and this is also true for business. This does not diminish the importance of multi-national, trans-national or global phenomena. Globalization is on the rise; but at least at present, the de-nationalization of science is not eclipsing the national component of the organization and work of scientific teaching and research (Crawford, Shinn and Sörlin 1992). Evidence abounds that Richard Nelson's concept of 'national systems of innovation' still accounts for much of science/industry/government dealings (Nelson 1993; D'Iribarne et al. 1998; Shinn 2000).

I suggest that we move away from an 'either or' scenario, in which mainly global or mainly national practices and arrangements are emphasized – each to the exclusion of the other. It is clear that a host of changes are sweeping science, industry and government. However, these are being played out in the framework of the nation state. This is the case even in Europe, where, despite serious efforts at a common science and economic policy, there is no decline in national science initiatives. An example of this is the bitter dispute over the design, construction and site of the third generation of cyclotrons – shall it be in England or France? It has been decided that each nation will have its own cyclotrons.

One way of reformulating this issue is to ask: which economic sectors and cognitive fields are most strongly tied to national systems, and which operate outside national constraints? (Hage and Rogers Hollingsworth 2000). A sociological approach to changing science/industry/state relations should combine concerns expressed in the New Production of Knowledge and the Triple Helix, with schemes that factor in national traditions, laws, employment and career patterns, and national institutions and cultures. To remain silent on these entrenched and enduring particulars is surely misguided.

A second problem facing the two approaches relates to the way in which they deal with one of sociology's key concepts; namely, 'differentiation'. The New Production of Knowledge analysis suggests that 'differentiations' and divisions of labour are a thing of the past. It is a failing that this startling claim is not girded by discussion of how 'differentiations' have operated in the past, how and why they have allegedly eroded, and what their putative demise implies for sociological theory. The Triple Helix is equally unhelpful, although from a different point of view. The perspective retains classical concepts of differentiation and integration. But in practice, this simply entails the projection of longstanding cycles of integration, neo-differentiation and neo-integrations through an infinity of co-evolutionary iterations. Regrettably, this contributes little.

Despite their many important differences, the two approaches constitute a shared search for what might best be called "transversalitty" – transversality that crosses cognitive, technical, economic and societal boundaries. Significantly, neither the new production of knowledge nor triple helix has examined the historical record for the existence and actions of communities rooted in transverse actions; instead working under the assumption that transversality is resolutely a functional product of our time and culture. Nevertheless, some recent sociology of science and technology suggests that for almost two centuries, first in Europe and then in the U.S., Japan and the USSR, small but often influential groups have embodied the very form of transverse operations more or less alluded to by the triple helix and new production of knowledge (Shinn 1993, 1997, 1998, 1999, 2000, 2002; Joerges and Shinn 2001; Shinn and Joerges 2002). This frequently informal and unofficial group of practitioners, dubbed 'research-technologists' has historically generated generic devices, artefacts and methodologies. Such devices (for example the ultracentrifuge, rumbatron, Fourier transform spectroscopy, stereo-comparator, laser etc.) take the form of open-ended instrumentation that spread across the boundaries of science and engineering, academia and industry, metrology and state service, as specialists in particular niches adapt and integrate them. Research-technologists and research-technology operates at the interface between established organizations and institutions. It stands "in-between" orthodox professions and bodies, and is thereby interstitial. It both sustains instituted differentiations and divisions of labor and violates them. Research-technology, an ordinary albeit often un-noticed feature of near history and contemporary life, is in some key respects emblematic of what the triple helix and new production of knowledge seek in their different ways to grasp. The challenge for both perspectives is to identify and probe similar transverse entities, to show how they function, and to indicate their potential in a social, economic, cognitive and technical order experiencing rapid change.

The social studies of science and technology are a privileged vantage point from which to analyze the changes that are re-configuring and fusing cognitive, artefactual and social factors. Since changes on such a grand scale prove elusive, there is a tendency to simplify, and to deal in headline-grabbing metaphors. Utmost care must be taken to ground concepts and conclusions in empirical studies, to establish the limiting conditions of their results, and to exercise prudence when articulating models so that they critically reflect the constraints of data. Such guidelines should be kept in mind in the New Production of Knowledge and Triple Helix perspectives, to ensure that they contribute something enduring to scholarship and practice. The sad alternative is that they simply nourish an unproductive frenzy, and be little more than fleeting fads.

Maison des Sciences de l'Homme, CNRS, Paris, France

NOTES

[1] For an analysis of the critical paths along which science and technology led society since 1945, see Jean-Jacques Salomon (1999).

[2] Neither does the Special Issue of *Research Policy*, XXIX (2), (February 2000), function as a doctrinal statement.

[3] On 20 February 2000, website addresses included:
http://www.itoi.ufrj.br/thelix.htm,
http://www.itoi.ufrj.br/general.htm,
http://www.itoi.ufrj.br/cfocus.htm,
http://www.chem.uva.nl/sts/loet/th2/respol.htm,
http://www.chem.uva.nl/sts/loet/th2/ihe98.htm,
http://www.chem.uva.nl/sts/loet/th2/papers/th2ley.htm,
http://www.chem.uva.nl/sts/loet/th2/bookabs.htm,
http://www.sura.org/~ghb/talks/triple_h/tsld001.htm,
http://vest.gu.se/vest_mail/0650.html,
http://vest.gu.se/vest_mail/0474.html,
http://vest.gu.se/vest_mail/sci-tech_1996/subject.html,
http://vest.theorysc.gu.se/vest_mail/1621.html,
http://vest.theorysc.gu.se/vest_mail/1339.html,
http://platon.ee.duth.gr/data/maillist-archives/th/threads.html,
http://platon.ee.duth.gr/data/maillist-archives/th/,
http://www-diotima.math.upatras.gr/mirror/mailbase.uk/lists/eurocon-knowflow/1998-07/0027.html,
http://www-diotima.math.upatras.gr/mirror/mailbase.uk/lists/eurocon-knowflow/1998-09/0002.html.

[4] A group headed by Joske Bunders at the Free University of Amsterdam is undertaking promising research in health care and biotechnology using hypotheses drawn from the New Production of Knowledge.

[5] An important corrective to this generalization is in the research of Roland Waast of the Institut de Recherche sur le Développement, which sees in the 'New Production of Knowledge' a fruitful approach to studying transformations in the organization and work of science and technology in post-1960s Africa.

[6] This brief summary does not do justice to the details, intricacies and complications of the theory. My present level of understanding does not allow me to move beyond these general precepts, and I suggest that this may be symptomatic of a broader problem and even malaise.

REFERENCES

Bourdieu, Pierre (1975), 'The specificity of the scientific field and the social conditions of the progress of reason', *Social Science Information* (Paris), XIV (6): 19–47.

Crawford, Elisabeth, Terry Shinn and Sverker Sörlin (1992), 'The nationalization and denationalization of the sciences: An introductory essay', in Elisabeth Crawford, Terry Shinn and Sverker Sörlin (eds.), *Denationalization Science. The Contexts of International Scientific Practice*, Dordrecht: Kluwer Academic Publishers, pp. 1–42.

D'Iribarne, Alain et al. (1998), *The French Science and Technical System Between Societal Constructions and Sectorial Specificities*, Research proposal for the 14th seminar of the European Group for Organizational Studies: Stretching the Boundaries of Organization Studies into the Next Millenium (9–11 July 1998), Maastricht University, Faculty of Economics and Business Administration.

Etzkowitz, Henry and Loet Leydesdorff (1996), 'The future location of research : A triple helix of university-industry-government relations II', *EASST Review* XV (4): 20–5.

Etzkowitz, Henry and Loet Leydesdorff (eds.) (1997), Universities and the Global Knowledge Economy: A Triple Helix of University-Industry-Government Relations, London: Cassell.

Etzkowitz, Henry and Loet Leydesdorff (1998), 'The endless transition: A 'triple helix' of university-industry-government relations', *Minerva* XXXVI: 203–18.

Etzkowitz, Henry and Loet Leydesdorff (1999), 'The future location of research and technology transfer', *Journal of Technology Transfer* XXIV (2/3): 111–23.

Etzkowitz, Henry and Loet Leydesdorff (2000), 'The dynamics of innovation: From national systems and 'Mode 2' to a triple helix of university-industry-government relations', *Research Policy* XXIX (2, February): 109–23.
Gibbons, Michael, Camille Limoges , Helga Nowotny, Simon Schwartzman, Peter Scott and Martin Trow (1994), *The New Production of Knowledge: The Dynamics of Science and Research in Contemporary Societies*, London: Sage.
Godin, Benoît (1998), 'Writing performative history: The New Atlantis ?', *Social Studies of Science* XXVIII (3): 465–83.
Godin, Benoît and Yves Gingras (2000), 'The place of universities in the system of knowledge production', *Research Policy* XXIX (2): 273–8.
Hage, Jerald and J. Rogers Hollingsworth (2000), 'Idea innovation networks: A strategy for integrating organizational and institutional analysis', *Organization Studies* XXI: 971–1004.
Horelli, Liisa (1997), 'A methodological approach to children's participation in urban-planning', *Scandinavian Housing & Planning Research* XIV-3: 105–15.
International Society for the Psychoanalytic Study of Organizations 1999 Symposium (1999), Issues in *Management Research*: www.sba.oakland.edu/ispso/html/1999Sympo...on James 1999.htm
Joerges, Bernward and Terry Shinn (2001), *Instrumentation Between Science, State and Industry*, Kluwer Academic Publishers, Dordrecht.
Krohn, Wolfgang, Günter Küppers and Helga Nowotny (eds.) (1990), *Selforganization. Portrait of a Scientific Revolution,* Dordrecht: Reidel.
Leydesdorff, Loet and Henry Etzkowitz (1996), 'Emergence of a triple helix of university-industry-government relations', *Science and Public Policy* XXIII: 279–86.
Leydesdorff, Loet and Henry Etzkowitz (eds.) (1998a), *A Triple Helix of University-Industry-Government Relations: The Future Location of Research?*, New York: Science Policy Institute, State University of New York.
Leydesdorff, Loet and Henry Etzkowitz (1998b), 'Technology innovation in a triple helix of university-industry-government relations, Asia Pacific Tech', *Monitor* XV (1): 32–8.
Leydesdorff, Loet and Henry Etzkowitz (1998c), 'The triple helix as a model for innovation studies', *Science and Public Policy* XXV (3): 195–203.
Leydesdorff, Loet and Henry Etzkowitz (1998d), 'Triple helix of innovation: Introduction', *Science and Public Policy* XX (6): 358–64.
Luhmann, Niklas (1996), *Social Systems*, Stanford, CA: Stanford University Press.
Maturana, Humberto (1980), Autopoiesis and Cognition: The Realization of the Living, Dordrecht: Reidel.
Nelson, Richard R. (1993), *National Innovation Systems: A Comparative Analysis*, New York: Oxford University Press.
Nowotny, Helga, Michael Gibbons and Peter Scott (2001), *Re-Thinking Science: Knowledge Production in an Age of Uncertainty,* Oxford: Polity Press.
Pestre, Dominique (1997), 'La production des savoirs entre académies et marché', *Revue d'Economie Industrielle* LXXIX: 163–74.
Salomon, Jean-Jacques (1999), Survivre à la science. Une certaine idée du futur, Paris: Albin Michel.
Shinn, Terry (1993), 'The Bellevue grand électroaimant, 1900–1940: Birth of a research-technology community', *Historical Studies in the Physical Sciences* 24 (1): 157–87.
Shinn, Terry (1997), 'Crossing boundaries: The emergence of research technology communities', in H. Etzkowitz and L. Leydesdorff (eds.), *Universities and the Global Knowledge Economy: A Triple Helix of University-Industry-Government Relations*, London: Cassell Academic Press, pp. 85–96.
Shinn, Terry (1998), 'L'effet pervers de commémorations en sciences', in P. Abir-Am (ed.), *La mise en mémoire de la science*, Paris: Les éditions des Archives Contemporaines, pp. 225–47.
Shinn, Terry (1999), 'Change or mutation? Reflections on the foundations of contemporary science', *Social Science Information* XXXVIII (1): 149–76.
Shinn, Terry (2000), 'Axes thématiques et marchés de diffusion', *Sociologie et Société* XXXII (1): 43–69.
Shinn, Terry (2002), 'Intellectual cohesion and organizational divisions in science', *Revue Française de Sociologie* 43: 99–122.
Shinn, Terry and Bernward Joerges (2002), 'The transverse science and technology culture: Dynamics and roles of research technology', *Social Science Information* 41 (2): 201–51.
Weingart, Peter (1997), 'From 'Finalization' to 'Mode 2': Old wine in new bottles?', *Social Science Information* XXXVI (4): 591–613.

CHAPTER 7

SABINE MAASEN

THE CONUNDRUM OF CONSCIOUSNESS: CHANGING LANDSCAPES OF KNOWLEDGE AT THE TURN OF THE MILLENNIUM

BECOMING CONSCIOUS OF CONSCIOUSNESS

"The past decade has seen a rising tide of interest in consciousness, accompanied by a surge of publications, new journals and scientific meetings" (Zeman 2001, 1264) – today, this is perhaps the best way to open the floor to any review of the field of consciousness. The sheer amount of activity in consciousness research is stunning, indeed, particularly in view of the fact that all studies and approaches are united by one conviction only: "At our present state of the investigation of consciousness we don't know how it works and we need to try all kinds of different ideas" (John Searle). In brief: Consciousness has become 'an issue' in academic discourse. This is not to say, of course, that consciousness is altogether new on the agenda. On the contrary, consciousness has been an object of contemplation for centuries (e.g., in philosophy and theology); it has been dealt with by political economists (e.g., Karl Marx) and by early psychologists (e.g., William James); it can be found in esoteric speculation as well as in social scientific studies on e.g., feminist, racial, or environmental consciousness. Another concern is directed toward the (deep-rooted) notion of self-consciousness. In other words: Consciousness has been a matter of diverse, yet systematic, contemplation and concern for centuries.

What is new, however, is that it has become the focus of a concerted *scientific* effort. Most notably, the cognitive neurosciences are currently investigating the 'brainy bases' of consciousness. Inspired by the 'cognitive turn' in psychology, the 'decade of the brain', and the institutionalization of a 'cognitive science', neurobiologists, cognitive psychologists and cognitively-inclined philosophers seem to hold the promise of finally coming to terms with an issue that is considered to be one of humanity's last mysteries. In so-doing, they fly in the face of our long-cherished intuitions about what consciousness is; this 'inner feel', this special quality accompanying perception, reflection, and action. Instead, and most provocatively, cognitive neuroscientists straightforwardly deny any mysterious experiential aspect, that is, they deny the existence of 'qualia', of 'what it is like' to be or feel (Nagel). Rather, they insist on consciousness being nothing but the

B. Joerges and H. Nowotny (eds.), Social Studies of Science and Technology: Looking Back Ahead, 117–153

name for an interplay of various regions in the brain upon which cognitive functions operate. Other than that, talking about experiential or phenomenal aspects of consciousness is relegated to folk psychological wisdom: at its most extreme, its function is claimed to be neither a physiological nor a cognitive, but a purely social one; consciousness is an institution invented to attribute causation and responsibility to a self – it is 'I-talk' (see Kusch 1997).[1]

Such claims are bold, yet not very surprising given the enthusiasm in cognitive neuroscience. Titles such as "Neuroscience: breaking down scientific barriers to the studies of brain and mind" (Kandel and Squire 2000) are programmatic. Neuroscience is about the scientific study of brains *and* minds. Consciousness, it seems, is the right issue at the right time. Entering the new millennium, we can finally get rid of antiquated customs and approach consciousness anew. Now that heuristics, techniques, and instruments are available, the 'issues proper', that is, the neural correlates of consciousness (NCC), can and should be addressed. Against sceptics who view cognitive neuroscience as no more than a new phrenology, neuroscience has news: Progress has been made in analyzing cognition at multiple levels, from systems, to cells, to molecules; thus, there is more than correlation of neural firing and experience. Neuroscience, equipped with fMRI and the ability to genetically engineer conditional knockout animals (mice, rats, and monkeys) is on the verge of bridging correlative and causal understandings. These techniques are regarded as "powerful additions to the armamentarium of cognitive science in the 21st century" (Miyashita and Farah 2001, 147). No wonder then that neuroscientists are looking to the future quite optimistically.

The sceptics, it seems, are not yet convinced. Are we indeed, and do we just act upon, an assembly of neurons and cognitive functions? In fact, as most philosophers keep reminding the neuroscientists, we still discuss the 'mind-body problem', an age-old, self-insistent paradox: While we believe ourselves to be physical beings, we also believe that we are conscious beings whose awareness of physical beings cannot be physical. It thus seems that we are both physical and non-physical. The problem is, how do the physical and non-physical relate; more specifically, how does the physical lead to the non-physical, or how does the non-physical lead to the physical? By definition, the physical and the non-physical have no means of 'communication', hence the problem. Sociologists (without arousing much interest) regard consciousness as a social construction, and anthropologists emphasize the cultural variability of consciousness. Transpersonalists, in turn, insist on pure consciousness, something beyond human grasp – a view that artists, for the most part, agree with when it comes to explaining aesthetic experience. Last but not least, our (Western) folk psychology fundamentally rests on the notion of being conscious individuals (reflective, responsible, hence: sociable beings).

This *tour d'horizon* may suffice to give a first impression of what is happening in the study of consciousness: As scientific paradigms multiply and philosophical wisdom is challenged, social scientific approaches are struggling against oblivion, and folk-knowledge may soon be shaken as well. A new order of knowledge about consciousness is about to emerge, bringing certain disciplines to the fore, relegating others to the back. The new order is not so much a matter of *one* discipline taking the lead, but rather, is about various *co-operative* efforts. Cognitive neuroscientists agree: "The 'problem of consciousness' has been identified as an outstanding intellectual challenge across disciplines" (Zeman 2001, 1264). Indeed, recent attempts to understand consciousness are all characterized by 'building bridges'. Interestingly enough, in the course of this, *various* multi-, inter- and transdisciplinary perspectives have emerged, with all the paraphernalia of institutions (meetings, journals, associations).[2]

These moves in the field of consciousness studies, and thus the thesis of this paper, have emerged from the underlying tension between what are conceived of as the physical and the phenomenal aspects of consciousness, discernible in virtually every single account. Even those who deem the experiential perspective irrelevant talk about 'it'. Consequently, despite what most cognitive neuroscientists would have us believe, this tension does not disappear by virtue of a scientification of scientific approaches to consciousness. Neither are qualia, other than in Daniel Dennett's opinion, so easily eliminated, sophisticated philosophical reasoning notwithstanding. Moreover, there are (and presumably will always be) those who reject the notion of consciousness being 'just another mental capacity' residing in the brain. They regard this epistemic shift as impoverished at worst, or one-sided at best, and hence, explore alternative ways to capture consciousness as both a physical and a non-physical phenomenon. By way of focusing on e.g., spiritual, social scientific, and/or aesthetic experiences, still more interdisciplinary perspectives are introduced, convening (different assortments of) expertise across the scientific cultures. Ultimately, this turns both the concept of consciousness and the field of consciousness studies into hybrids – a state which, in turn, provokes further debates and alliances.

In brief: Consciousness has become a field of epistemic combat that is likely to concern the participants (both present and future) for quite some time. Consciousness should also concern science studies. Notably, it should concern a type of science studies that is interested in the 'shifting and sifting' of knowledge.[3] Based on a sociology of knowledge inspired by Foucault (e.g., 1974) and Luhmann (e.g., 1980), one would start from, yet go beyond, the (re-)emergence of a grand topic and look for its (re-)productions in local discourses, scientific and extrascientific. Why? Firstly, such grand topics are made (and remade) in a variety of more or less local discourses. Secondly, the making of knowledge is discourse-specific. Thirdly, discourse-specificity is not to be equated with arbitrariness: individual discourses process knowledge according to their vocabulary, methods, world views, etc., and in turn, these discourses, by connecting to a certain piece of knowledge, have to process its pre-existing meanings, uses, etc. Fourthly, local specificity and global significance do not contradict each other: rather, the price for achieving global significance is to be paid in locally specific adaptations. All of this is true for consciousness as well.

Indeed, from a sociology of knowledge point of view, 'consciousness' is but a piece of knowledge circulating in society. Consciousness does not 'belong' to any special discourse, neither to philosophy, nor to meditative explorations of inner experiences, nor to the forefront of brain research. Rather, it occurs in various contexts, thereby assuming various meanings or shades of meanings. These different meanings are partly dwelling side-by-side, happily ignoring each other, but also partly interacting, and partly competing. A sociology of knowledge approach, interested in the making of knowledge in society, seeks to reveal the multi-discursive constructions of terms or concepts. Therefore, sociology of knowledge is not, and cannot be, about definitions. Rather, it is about tracing *the emergence of consciousness as a heterogeneous concept.* I will ask questions such as: In which discourses does consciousness occur and when? What does the respective discursive context do to the term and, conversely, what does the term do to its respective contexts of application? Do the various shades of meaning ignore, interact, or compete with one another? Asking these questions implies conceiving consciousness as site and medium of meaning production. Understanding the dynamics of knowledge is not a goal in itself, however. Ultimately, it is about understanding the systematics of issues and how they become heterogeneous and even contradictory. This may be of particular importance once certain debates transcend the academic sphere of interdisciplinary discourses and cause ethical or political conflict. In such cases, understanding the (necessarily) selective processing of knowledge may prove a sound basis for, e.g., social scientific counseling.

But how should those selective processes and largely incremental twists and turns in the re-makings of knowledge be rooted out? To this end, the following study rests on a method called metaphor analysis. The idea is that, by way of interacting with novel contexts, consciousness assumes novel (shades of) meanings. In a way, that is, consciousness acts like a metaphor: A metaphor is a term or phrase that is somewhat unfamiliar with respect to its poetic context. Both unfamiliar phrase and poetic context need to interact, thereby producing an innovative reading, in order to 'make sense'. Likewise, consciousness, formerly to be found in non-scientific contexts only, finds itself used in various scientific fields: Interacting with them, it undergoes shifts in meaning and use, which themselves become the object of affirmation and critique in other discourses.

After a brief introduction to the issue of metaphor and the method called metaphor analysis, I will present the outline of the broader study. This outline starts by presenting a discursive map and calendar concerning the discourse on consciousness. Based on ISI databanks (Science Citation Index and Social Sciences Citation Index), I will first inquire into the quantitative dimensions of consciousness research: How many contributions do we have, from which disciplines, what are their prime concerns, do they change over time? This will help to go beyond initial impressions (see above), and strictly includes only those discursive contributions that are explicitly concerned with consciousness (as indicated by mentioning the term in either the title and/or the abstract). These data will lead to more detailed observations, which will then be confronted with views from within cognitive neuroscience. Thereafter, the analysis will be complemented by a brief discourse analysis of a recently established journal, namely, the 'Journal of Consciousness Studies'. Its policy is to re-open the scientific debate to phenomenal approaches of various kinds. An overview of the themes published throughout its 8 years of existence will also be confronted with views from within the emerging scientific-cum-humanist community.

The rationale for this sequence of studies, switching between (quantitative) description and analyses of views from within, is to disclose more precisely the intricacies of the wheres and hows of major, possibly paradigmatic, changes in consciousness (studies).

METAPHORS

Metaphors of Consciousness / Consciousness as Metaphor

As we turn to metaphor analysis, the reader may well wonder whether this study is about investigating 'metaphors of consciousness'? Although it is not, it might be instructive to take a quick look at some of these metaphors, in order to learn more about how metaphors work:

It has frequently been noted that theories of consciousness abound with metaphors. In the domain of psychology, pertinent writings of its founding father, William James, are evidence of this. His *Principles of Psychology* (1890), alongside *Does 'Consciousness' Exist?* (1904), demonstrate how figurative language directs James's study of consciousness. For James, writes Kress (2000), metaphor does not just describe, but also constructs consciousness. It governs his conflicting theories of mind through a series of rhetorical configurations and provokes the continual reconstruction of his ideas of human subjectivity. His attitudes toward metaphor are mixed, however: On the one hand, James presupposes a clearly delineated concept around which one might wrap a verbal expression; on the other hand, he also seems painfully aware that every new metaphor launches an entirely new theory. James therefore creates a series of metaphors out of which consciousness materializes: liquid metaphors, nuclear metaphors, ethereal metaphors, corporeal metaphors. The ultimate principle of Jamesian consciousness seems to be its creative capacity, which he translates into his notion of 'pure experience', while simultaneously remaining committed to figures of speech (especially: 'stream of consciousness').

Both before and after James, authors have invented a host of further metaphors of consciousness in order to gain access to an ever-elusive concept. Among many other things, consciousness has been likened to a spotlight, a recursive loop, a stage; likewise, the specific quality of consciousness has been described as a state, an act, input or output, as a unity, a diversity, or *a unitas multiplex* (see Bruner and Fleisher Feldman 1990, 230). An important basic metaphor of consciousness is nakedness ('bare consciousness', 'mere sensing'), put forward by Aristotle (*sensus communis*) and John Locke ('primary qualities'). These metaphors all advance a reproductive or passive view of consciousness: Certain ideas are somehow pushed or pulled into the mind. A more productive or active view of consciousness has been advocated by scholars like Kant or Leibniz: From their angle, "the contents of consciousness are what mind creates rather than what mind encounters" (Bruner and Fleisher Feldman 1990, 232). Both views, however, distinguish between an inside and an outside of consciousness, detailing the various functions of consciousness, such as the function of 'shielding' itself from the demands of both the external world and the unconscious (e.g., Freud). Information-processing models of contemporary cognitive science are recent sources for metaphors of consciousness. It is conceived as the 'readout' at the end of 'filtering' processes. These filters change over time, depending on what one knows about perceptual processes and depending on the machine analogues (see Bruner and Fleisher Feldman 1990, 234).

The difference in reproductive versus productive views of consciousness has been more clearly delineated by metaphors of cognition. On the one hand, there are those which picture the mind as a 'mirror of nature', on the other, there are metaphors of illumination (e.g., searchlights, spotlights), creative synthesis (e.g., mental chemistry, construction), assignment of meaning (e.g., semantic networks, encoding/decoding), or topography (e.g., cognitive maps, schema). Another set of metaphors centers upon symbolism: Symbolism advances language (icons, indexes, symbols) rather than physics as the most appropriate source for metaphors of cognition. These metaphors, according to Bruner and Fleisher Feldman, highlight what consciousness *does* instead of what it *is*, and ask about the uses to which it may be put in constructing meaning and in assigning interpretations. Now, whether or not one agrees with this assessment, one may nevertheless feel that the basic message is well argued: The study of consciousness (and cognition) abounds with metaphors and thus, is indicative of the psychology involved, the questions it asks, the conceptions it arrives at.[4]

As regards the neurosciences, the theater metaphor has been suggested by classical and current authors, Pavlov and Crick being pertinent examples. Today, the theater metaphor branches off into a network of sub-metaphors: It has been suggested that consciousness might be associated with cortical convergence zones, and theaters exist to allow numerous convergent influences to shape a coherent performance on stage (see Damasio 1989). However, the theater metaphor – as any metaphor – has met with criticism as well: Dennett and Kinsbourne, for instance, deem it Cartesian and thus misleading (see Dennett and Kinsbourne 1992). Others hold that computers could simulate virtually all hypotheses deriving from the theater metaphor. Still others claim that some aspects of consciousness are not amenable to explanation, neither with nor without metaphors, that is. Only few scholars seems to be of the opinion that metaphors can be productive, more precisely, that certain metaphors can indeed be productive for certain scientific problems. Bernard J. Baars pleads energetically for the value of the theater metaphor when it comes to scientific reasoning about consciousness. For him, "the fundamental function of the theatre architecture is to make possible novel, adaptive interactions between the sensory inflow, motor outflow and a range of knowledge sources in the brain" (Baars 1988, 59, 1). Generally speaking, "the criteria for productive metaphors are the same as for other scientific ideas: they should help organize existing evidence, yield testable hypotheses and suggest conceptual clarifications" (Baars 1988, 59, 2).[5]

Rather than going into any of these suggestions in any depth, however, I would like to focus on the issue of metaphors in a more general sense. Reviewing the points made by the psychologists and neurobiologists just mentioned, we can broadly summarize them as follows: Metaphors

- seem *inevitable*; certainly, they are ubiquitous in scientific as well as in other discourses;
- are *intricate*, that is, irritating and always contested elements of (scientific) discourse[6];
- have all kinds of impact on (scientific) reasoning, *innovative* ones included.

Indeed, by confronting consciousness with various metaphors, some of its aspects, functions, and modes of operation have been highlighted, whereas others have been shaded, or have turned out to be neutral (see Hesse 1974). Every new metaphor provokes new ideas as well as criticism; sometimes both innovators and critics co-establish a new, elaborated language game, which then guides further research for a while. This is the general message as regards metaphors of consciousness: Whatever consciousness is supposed to mean at a given point in time is the result of an encounter with the discourses (and their preferred metaphors) interacting with it. True to its Greek origin (*meta-phorein*=transfer), a metaphor is about transferring and shifting meanings.

The aforementioned mechanism is the point of departure for a more radical idea: One could apply the notion of metaphor to any term or concept that apparently undergoes significant changes. This idea contends that these changes come about due to new discourses interacting with a particular term, thereby instigating shifts and changes in the term's uses and meanings. Why not suppose, along with Mary Hesse and others ('all language is metaphorical,' see Hesse 1974), that this is how discourses evolve – by way of terms, concepts, vocabularies, and world views interacting with one another? Adopting this stance, I therefore suggest re-conceptualizing the notion of metaphors as methodical devices. *Hence, instead of analyzing metaphors of consciousness I suggest to treat consciousness itself as a metaphor*. The study will follow the term consciousness to those discourses in which it happens to occur, and will look for the respective interactions. This type of method has been termed metaphor analysis[7]. It aims to provide a rich account of major instances of knowledge dynamics by way of looking at metaphors as units of knowledge and discourses as their environments of knowledge, both selecting against each other.

On Metaphor Analysis

In order to grasp fully the way in which metaphors can function as sites and media of knowledge transfer (see Bono 1990), one needs to take stock of basic metaphor theoretical insights and to transfer them into a sociology of knowledge framework. Semantic, pragmatic, and constructivist theories emphasize different aspects of metaphors.

The *semantic* view of metaphors has been advanced by Max Black (Black 1962) and Mary Hesse (Hesse 1972). According to them, metaphors serve to re-describe a phenomenon of a primary system in terms of a secondary system and in so-doing, interact with one another. According to Black, however, this interaction will ultimately lead to a mutual transfer of meanings or aspects of them. *Thus, Semantic theories are about shifting meanings in both metaphor and context.*

Pragmatic theories stress the importance of processing a metaphor within the importing discourse. Donald Davidson advances the idea that a "metaphor belongs to the domain of use. It is something brought off by the imaginative employment of words and sentences and depends entirely on the ordinary meanings of those words and hence on the ordinary meanings of the sentences they comprise" (see Davidson 1981, 202). *Thus, Pragmatic theories are about new meaning(s) as a product of the interaction between the literal meanings of both metaphor and context.*

Thus, when reformulated in *constructivist* terms, an unfamiliar element (that is, a metaphor), according to Rorty, not only forces us to process both metaphor and its immediate context, but also provokes an entire 'new vocabulary' (Rorty 1989) in which both the metaphor and its context make sense. On a less radical note, one might learn from 'ethnotheories' in the area of scientific innovation: these regard metaphorical transfer of knowledge as a regular occurrence, since the concepts or models that become analogized have already figured as 'solutions' in other fields. From this perspective, and opposed to Rorty's view, a metaphor is something already known or familiar with respect to a context that is novel or unfamiliar. In both versions, however, *constructivist theories are about inventing a new language game capable of integrating metaphor and context in a meaningful way.*

Enlightened by the accounts just given, I suggest metaphorizing the concept of metaphors: A discourse-analytical approach conceives of metaphors as single discursive elements within a broader discursive context. Whereas discourses are language games that have become thoroughly familiar, metaphors are (freshly imported) terms or concepts that are (still) unfamiliar to their context (see Davidson 1981; Rorty 1989). Thus, speaking of metaphors means always to be speaking of a *relation* between a concept and a discourse. Secondly, it always implies an aspect of *transience*.[8]

Both the relational and the transient aspects of metaphors turn them into ideal candidates for observing the invention, circulation, and modification of knowledge: From a discourse-analytical level, the discursive processing of a metaphor can be described as one of those mechanisms by which the production of knowledge is controlled, selected, organized, and channelled (see Foucault 1974). That is, if an already established discourse (i.e., cognitive psychology) selects *for* a certain metaphor (i.e., 'consciousness'), it necessarily varies the meaning of consciousness by interpreting it discourse-specifically, and perhaps it retains this new meaning. Note that this capacity of generating new meaning is not a *feature* of a metaphor but the *result* of an interaction of a metaphor and its discursive environment.

As to the analytical procedure, these are steps to take:

- Step one reads: 'Pick one!' – I pick 'conscious/ness'.
- The second step reads: 'Draw a discursive map!' and: 'Reconstruct a discursive calendar!' With the aid of bibliometric methods I follow the term or concept chosen through all scientific and philosophical discourses in which it appears. The results of this step provide a fair approximation as to the whens and wheres of the metaphor in question. In other words, the dynamics of knowledge are thus represented in quantitative terms: One reveals both the increasing or decreasing occurrences of a certain concept over time and discourses.
- The third step reads: 'Discourse analysis!' Only after the occurrences of a certain metaphor are available, I select a set of publications that seems most productive and enlightening as regards the dynamics of consciousness. Discourse-by-discourse, I look for specific interaction with the term and try to confront it with literature commenting on the history and/or systematicity of any discursive developments (e.g., reviews).
- The fourth step reads: 'Local specificity – global significance?' Do the specific shades of meaning converge on a heterogeneous topic (be it an issue, a paradigm, a cultural matrix, ...)? For this is what metaphor analysis is about: It looks for *both* locally specific processing of metaphors and for ways in which they – gradually – produce (heterogeneous sets of) meanings across various (types of) discourses for a given period of time.

So, what are the dynamics of consciousness, now that consciousness itself has become such a highly volatile concept, interacting with cognitive and neurosciences, with philosophy, the social sciences, ...?

A METAPHOR ANALYSIS OF CONSCIOUSNESS

"Consciousness is a word worn smooth by a million tongues. Depending upon the figure of speech chosen it is a state of being, a substance, a process, a place, an epiphenomenon, an emergent aspect of matter, or the only true reality" (George Miller 1962, 25). On a very general level, one can isolate a social aspect of the term, that is, joint knowledge shared by a community of people (e.g., Marxist, feminist consciousness) and a mental sense. The latter can be subdivided into the notion of a state concomitant of all thought, feeling, and volition (transitive: conscious of), or as the normal condition of a healthy waking life (intransitive).[9] Thus, stating that consciousness is not one phenomenon today borders on a truism (Chalmers 1996, 7). Still, it might be worthwhile looking at the most comprehensive data bank (the ISI: Science Citation Index/SCI and Social Science Citation Index/SSCI), which lists journal articles over a broad range of disciplines and covers about 30 years, in order to get a more precise picture:

- How often do the terms 'conscious' and 'consciousness' appear in titles of articles?
- What is the range of disciplines covered by these articles? Does the attention shift from one set of disciplines to another?
- What are the themes associated with consciousness?

To emphasize: At this stage of analysis, the occurrences indicate when, how often, and in which disciplines authors deal with the subject. These observations will be confronted with a view from within the scientific branch of consciousness studies later on.

Conscious/ness in SCI / SSCI

Absolute and Relative Occurrence

Throughout a period of 26 years (1975 – 2000), the terms conscious/ness occur with considerable frequency: Looking at the publications in absolute numbers, the Science Citation Index reveals an increase by a factor of two (1974: 210 – 2000: 435) and the Social Science Citation Index an increase by almost a factor of three, although on a lower level (1974: 113 – 2000: 293).

The relative frequency of publications paints a different picture: In the natural sciences it remains by about 0.5‰, after a peak of about 0.9‰ in the mid-1980s. In the social sciences, the data show an increase (1974: 1.25‰ – 2000: 1.96‰).

Taken together, we find noticeable growth in academic publications explicitly dealing with conscious/ness. On the basis of these data one can only speculate, however, about why the natural sciences apparently began to refrain from using the term during the 1990s, despite the fact that even higher activity on this topic would seem to be more plausible during the 'decade of the brain'. One reason might be the internal differentiation within the field. When psychologists, philosophers, social scientists, theologians, ethnologists, and scholars in law and politics all write about conscious/ness, scientists

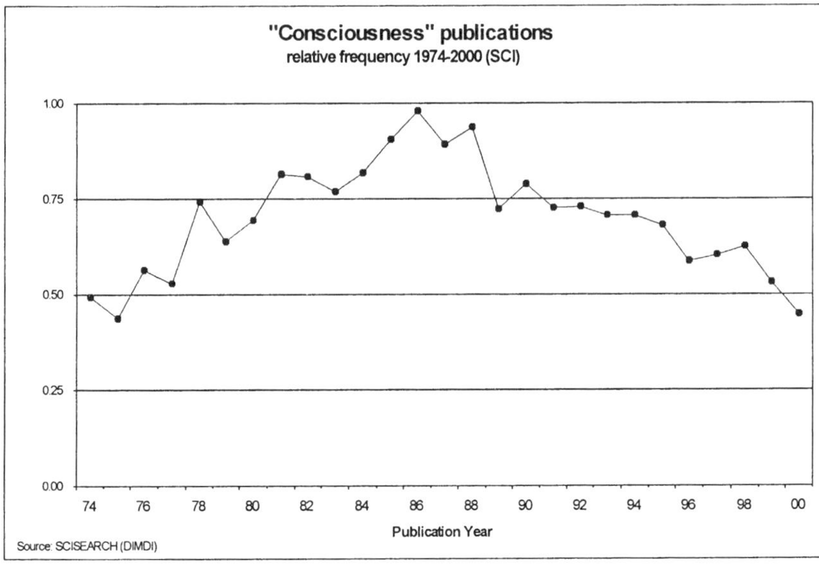

Figure 1. Relative frequency of 'consciousness' in titles of articles. Data are based on the Science Citation Index (SCI) 1974–2000.

may deem it advisable to differentiate, not only in approach and outlook, theory and method, but also in terms of labels.[10] We can count on the fact that scholars working in the field do use more specific terms in the titles of their publications in order not to become confused with other (that is, non-scientific) approaches.[11]

Fields

Focusing on selected years (1975, 1980, 1985, 1990, 1995, 2000), I will now investigate the fields in which consciousness studies are conducted, research activities, and those journals that publish the articles we are concerned with here. This line of investigation pursues the breadth and internal structure of scientific interest: Which *disciplines* or *fields* participate in research activities on conscious/ness? I will look at the SCI first, then the SSCI.

In the SCI, the neurosciences become the most important field dealing with conscious/ness, followed by physiology, pharmacology & pharmacy. Biology, including cell biology, covers middle ground. Medicine, general and internal, seems to be less important over the years, as opposed to those branches of medicine that are oriented toward experimental research.

The SSCI also reveals a broad array of *fields* dealing with conscious/ness, ranging from psychology, psychiatry and philosophy, through the social, political, and historical sciences to law, anthropology, education, area studies, language and linguistics. Yet here as well, we find hierarchies of interest in the issue among the disciplines. The leading field throughout the years is psychology, ranging from 15,66% in 1975 to 56,7% in 2000: While this includes sub-fields such as clinical, developmental, and social psychol-

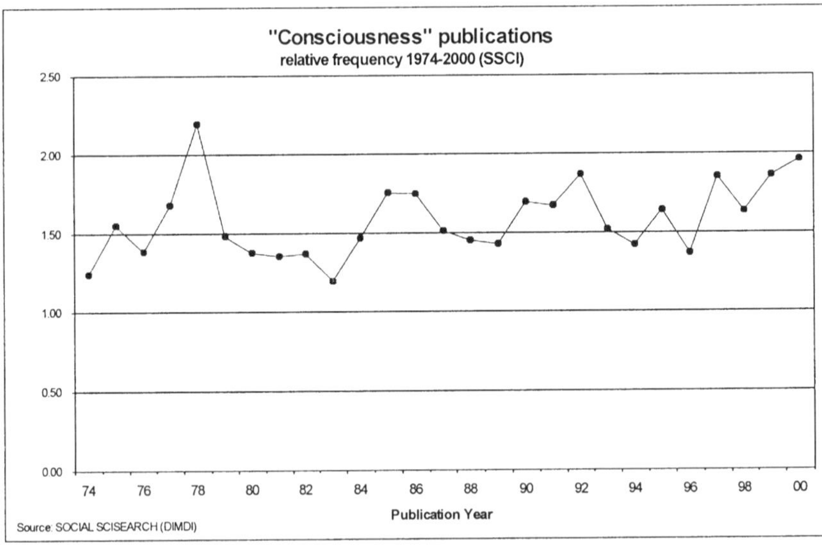

Figure 2. Relative frequency of 'consciousness' in titles of articles. Data are based on the Social Science Citation Index (SSCI) 1974–2000.

ogy, there is a noticeable increase in experimental psychology dealing with conscious/ness (1985: 2.29%; 1995: 8,23%; 2000: 33.84%). At the same time (1985, 1990, 1995), the field of neuroscience contributes 8–15% of the publications on conscious/ness. In summary: Throughout these years, the academic occupation with conscious/ness becomes more experimental, more cognitive, more 'neuro', and, most of all, more pharmacological.

But what about philosophy and the social sciences, the two prominent contributors to consciousness throughout the 1970s and 1980s? Philosophy, although under-represented in the ISI data[12], shows a considerable decrease in publications on conscious/ness: Initially providing about 9% of the articles (1975), they range on a 1,21%–2,13%-level in 1995/2000. A similar trend can be found in political science and history: Starting on the same level, they decrease rapidly in the 1990s (to about 1%). Other fields dealing with socio-cultural aspects of consciousness, such as law, anthropology, and religion, mostly contribute about 1% or 2% of all articles per year. Sociology, however, maintains a steady level of about 10%, with the exception of 1995 (5.57%) and a noticeable increase in 2000 (17.37%). Despite the latter finding, it is most obvious that philosophical and social scientific treatment of conscious/ness are a minority issue, especially in view of the fact that the percentages have to be seen in relation to the absolute numbers of publications covered by the SCI and the SSCI, respectively. (The SSCI contains only half as many articles published on consciousness as the SCI!). Although research on consciousness in the social sciences exists, this is evidently not 'where the iron is at its hottest'.

Journals

When it comes to the leading journals that publish scientific articles on conscious/ness, their primary concern with physiology and pharmacology is evident. Journals such as the *British Journal of Pharmacology*, the *American Journal of Pharmacology*, as well as *Circulation*, for instance, publish the major share of articles on conscious/ness. Yet another community, the neurosciences, figures less prominently: *Behavioral and Brain Sciences*, *Brain Research*, *Neuropsychologia* and *Neurophysiology* publish 1,27 (1980), 2,2% (1985), 4,99% (1990), and 4.31% (2000) of all articles mentioning conscious/ness in their title. One exception is the year 1995, when more than 10% of all publications appear in neuroscientific journals. Based on these data, one can conclude that within the natural sciences, research on conscious/ness is primarily a physiological phenomenon, addressed in a variety of different communities, yet most prominently in the context of pharmacological application.

As regards social scientific articles on conscious/ness, they are scattered over a broad spectrum of different journals that publish one or two articles at most (in 1975 and 1980). Since 1985, the *Behavioral and Brain Sciences* has been listed in the data and dominated the field (1985: 6.42%, 1990: 9.27%; 1995: 18.8%) from the start, only to be overtaken by *Consciousness and Cognition* in 2000 (28.67%).[13] Next come journals that belong more specifically to the area of psychology. Since 1990, a few journals have been founded which bridge philosophical and psychological concerns with conscious/ness: the *Journal of Mind and Behavior* and *Philosophical Psychology*. Yet not only are they insignificant with respect to those journals that attempt to establish closer links between the neurosciences and cognitive psychology (see above), but furthermore, journals trying to establish even broader interdisciplinary perspectives have come into being, most explicitly, the *Journal of Consciousness Studies*. Today, in fact, these journals seem to be the proper place for non-scientific contributions to the study of conscious/ness to appear. A considerable decrease is to be found in, for instance, *Die Deutsche Zeitschrift für Philosophie*, the *Journal of Asian Studies*, or the *Journal of Economic History*: in 1975/1980, one could find 3-7 articles per year in philosophical, social scientific and other journals, whereas journals publishing such pertinent articles are hardly to be found any more (most notably since 1990).

Leaving pharmacologically-oriented research on consciousness aside, the data on the growth, breadth and disciplinary distribution of research activities show that conscious/ness has increasingly become a matter of experimental neurocognitive science. In the next step, I will look at the titles of articles published, in order to obtain an overall impression of which particular issues they address, and to see whether they vary among and between disciplines, as well as over time.

Titles/Themes

Regarding the specific issues addressed, I will select a few disciplines and examine the lists of full-length titles for the years 1978, 1988, and 1998:

- In *anthropology, sociology*, and *law,* the term consciousness is regularly connected to certain groups or strata, such as class, ethnic group, or gender. It is a political, historical, or moral /religious concept: national consciousness, epochal consciousness, or Jewish consciousness being pertinent examples. Finally, the term refers to specific practices and technologies: For example, the titles address sorcery as a practice of consciousness, or the technocratic consciousness.

- In *philosophy* we find the accounts of metaphors on consciousness used by various philosophers throughout the centuries (Brentano, Descartes, Hobbes, Kant, Husserl, Heidegger, Hegel, ...). Moreover, a number of articles, predominantly published in 1978, deal with political topics, such as social structure and consciousness, or the formation of a Socialist consciousness; that is, specifically Marxist accounts of consciousness. Alongside some esoteric listings, one also finds a slightly increasing number of neuro/evolutionary topics, such as 'consciousness & the brain', 'scientific approaches to consciousness'.
- In *psychology* the titles can be attributed to various fields of studies, too. One group is concerned with the field of social science/psychology and deals with questions of identity formation, self-constitution, self-perception, and self-regulation. Another group addresses various states of consciousness, such as multiple/divided/altered states (sometimes discussing therapeutic interventions), as well as consciousness while sleeping, in coma, or under hypnosis. Still others address basic phenomena and provide conceptual elaboration, for instance, they discuss the theory-of-mind approach to consciousness. A minority group focuses on general questions such as 'consciousness in psychology' or 'the history of consciousness'.
- In the neurosciences, the situation seems less diverse. A notable group of titles refers to studies on patients in coma, that is, in non-conscious states. Another discusses altered states of consciousness (pathologies, drugs, etc.). Various titles focus on the issue of attention, awareness, and alertness. Finally, and on a more general level, some titles address 'the function of consciousness', or methods '. . . to study consciousness scientifically.' If titles referring to 'conscious' only are included, one finds a huge amount of publications dealing with methodical issues such as a particular subject (rat, dog, sheep, patient, child, male/female) and a particular method (EEG, fMRI, etc.).

Frequency of Words

Because full-length titles can be very suggestive when one attributes both meanings and frequencies to them, it is highly advisable to take a step back and look at single words and phrases only, as they occur in titles. The frequencies of single words and phrases (pairs of words) may tell a different story.

As far as the frequency of single words is concerned, there are, of course, no surprises at the top: While conscious can be found in 3,933 cases, consciousness occurs 1,781 times in both the SCI and the SSCI. Also, immediately below in the ranking, one finds a clear predominance of terms that relate to neuroscientific research, mostly related to pharmacology. The terms refer to the subjects of study, most often animals, clearly headed by rat(s) (1,425), followed by dog(s) (835), rabbits (260), sheep (91), etc. Secondly, they refer to the state of the subject, e.g., whether under sedation (279), anaesthesia/ anaesthetized (108), or unconscious (68). Thirdly, they refer to the locus of research, such as the brain (156), or the heart (128), and fourthly, to all kinds of processes, functions, reactions (inhibition: 127; antagonist: 93), or failures (64). Finally, they postulate effects (693), model(s) (100), theory (ies) (83).

Only considerably lower in the hierarchy of terms do we find 'self-consciousness' (122), the most frequent term not pertaining to a scientific discourse. Next comes the term 'social' with 59 listings only, and the 'self' with 55. Even more infrequent are terms such as identity (41), national (31), class (35), public (34), political/politics (52), black (24), gender (22), culture (21), or Australia (2). It is worthy of note that the terms just mentioned virtually exhaust the list of words pertaining to non-scientific (i.e., social scientific and/or philosophical) contexts, which is not the case with terms pertaining to a clearly scientific (i.e., predominantly neuroscientific/physiological) context. The latter outnumber non-scientific words by far.

This line of investigation emphasizes that the types of consciousness connected to social scientific and philosophical concerns have become so infrequent that it seems as if they are on the verge of becoming lost, whereas physiological shades of meaning (such as awareness, for example) are coming to dominate the academic discourse on consciousness.

- national c. (27), modern national c. (14), class c. (22), mass c. (10), political c. (10)
- environmentally conscious (34)
- historical consciousness (25)
- private self consciousness (22), self consciousness (17)
- social (16), public (8), popular (6) consciousness
- Palestinian identity (14), ethnic (9), African American (9), black (9), ethnic self (6) consciousness
- body consciousness (12)
- feminist consciousness (12)
- phenomenal consciousness (11), consciousness explained (10), moral consciousness (10)
- cost consciousness (6)

Figure 3. Co-Word Analysis (selective; considering social scientific co-words only).

Reviews

Yet another indicator concerns a certain type of publication: the review. The review is designed to inform colleagues of the author's own and/or neighboring discipline about the state of the art as regards a certain problem or area of study. By implication, there has to be a considerable body of work to write about, which is in all probability controversial and/or in need of further investigation, and which is looked at from the reviewer's point of view. Idiosyncratic assessments aside, reviews are indicative of a topic that has a certain stage of vividness and is regarded as 'an issue'. Moreover, reviews contribute to the delineation of fields, they make claims as to which kinds of research activities should prevail and finally, they address – and thereby constitute – scientific communities.

Between 1993 and 2001, the SCI and SSCI list 55 entries for reviews on consciousness. Only eleven of them do not belong to the neurocognitive domain. As to cognitive aspects, the reviews deal with attention, episodic memory, as well as temporal binding/binocular rivalry, etc. In the neuroscientific realm, scholars review the thalamic contributions to attention and consciousness, they inquire into the links between consciousness, pain, and cortical activity, or whether acethycholine is a neurotransmitter correlate of consciousness, etc.

Of the reviews on non-scientific contributions to consciousness, there are three contributions on legal issues, three in the realm of race/politics and two in parapsychology, as well as one contribution each to health costs, environmental ethics, and psychology.

The last mentioned review on 'consciousness and the psychological science' (Pickering 1999) is the only one that explicitly deals with the growing body of experimental neurocognitive research on consciousness and its impact on the science of psychology. All other reviews refer to consciousness as a façon de parler (cost consciousness) or to its political connotations. In sum, a look at the reviews published since 1993 not only suggests the predominance of neurocognitive approaches once more, but also considerable internal differentiation regarding issues addressed and/or methods used. Last but not least, the SCI/SSCI show no entries for philosophical studies on consciousness.

Most Often Cited Authors

While the data examined thus far did not reveal any special role for philosophers in the field of consciousness studies, the apparent invisibility of philosophy changes dramatically when authors are ranked according to how frequently they are cited in publications addressing consciousness. Again, the dates refer to the SCI and SCCI (1973 until today).

The remainder of the extended list (covering the 'top hundred') includes representatives of philosophy, as well as neuroscientists and cognitive psychologists, yet no social scientist.

While other indicators did show some activity in the social scientific domain, it borders on insignificance once one looks at data indicating the impact of contributions by single authors from the respective fields. Quite the opposite is true for philosophy: Although the group of scholars contributing to the field is much smaller, they have a considerable impact nevertheless, as can be shown by the frequency of quotations pertaining to their field. Yet another pattern is revealed for the neurosciences: Here, there seems to be a pair of star authors (Crick and Koch) and a host of scholars who are cited moderately often. Cognitive psychology seems to have no such stars, yet is represented by a considerable number of scholars who are cited quite often.

Although one should be careful when working on the level of counting and categorizing titles, frequencies of words, reviews and highly cited authors, one can nevertheless obtain a reasonable impression of the diversity of issues discussed within and among the various fields. Based on these quantitative data, one might extract some major trends that need further investigation: All indicators considered signal the prevalence of pharmacological studies. Only way below these studies in terms of ranking do we see the beginning of a trend toward a more scientific, especially more cognitive and more neuroscientific approach to consciousness. While other notions are present as well: pol-

Author	Year	Title	Cita-tions
Dennett, Daniel	1991	Consciousness explained, Boston: Little, Brown and Co., 1991.	153
Dennet, Daniel Kinsbourne, M.	1992	Time and the Observer: The Where and When of Consciousness in the Brain, Behavioral and Brain Sciences, 15 (1992), 183–247	39
Paxinos, G.	1986	The rat brain in stereotaxic coordinates / George Paxinos, Charles Watson. Sydney ; Orlando : Academic Press, 1986.	140
Crick, F. & Koch, C.	1990	Towards a neurobiological theory of consciousness. Seminars in Neuroscience, 2:263-275	64
	1995	Are we aware of neural activity in primary visual cortex? Nature 375:121-123.	42
Searle, John	1992	The rediscovery of the mind, Cambridge: MIT Press, 1992.	76
	1983	Intentionality. An essay in the philosophy of mind Cambridge, New York : Cambridge University Press, 1983	26
Fenigstein, A.	1975	Public and private self-consciousness: Assessment and theory. Journal of Consulting and Clinical Psychology, 1975.	76
James, William	1890	The principles of psychology, New York, H. Holt and company, 1890.	76
Baars, Bernard J.	1988	A cognitive theory of consciousness, Cambridge: Cambridge University Press, 1988	64
Nagel, Thomas	1974	What Is It Like to Be a Bat? Philosophical Review, 1974, 83, 4, 435-457.	51
...			
Block, Ned	1995	On a confusion about a function of consciousness: Behavioral and Brain Sciences. 1995, 18, 2, 227-247	40
Chalmers, David J.	1996	The conscious mind. In search of a fundamental theory. New York : Oxford University Press, 1996.	29

Figure 4. List of highly cited authors (selective).

itical, moral, philosophical, educational, they are concerned with different issues, mostly pertaining to collective representations of various kinds. Consciousness research as a whole is characterized by a notable division between scientific accounts and 'others'.

This division, however, not only separates the sciences from the humanities and the social sciences, it also cuts through individual discourses, as is apparent in philosophy and psychology. There is a slightly growing body of studies within philosophy that actively contribute to the neurocognitive debate, thereby significantly rephrasing what was previously referred to as the 'mind-body problem', without special reference to neural processes: "The key philosophical question posed by consciousness concerns its relationship to the neural processes which correlate with it. How do the events which register in our experience relate to those occurring in our brains?" (Churchland after Zeman 2001, 1282). Likewise, and much more strongly, psychology has become divided. A steadily growing faction within psychology is involved in producing functional theories of consciousness and trying to explain their relevance for producing fitting behavior: According to them, consciousness is mobilized whenever challenges or unpredictable events require special attention. Hence, consciousness studies seem to create a novel community, to which a certain faction of neurobiologists, psychologists and philosophers belong. Scholars studying collective approaches to consciousness neither seem to form their own community, nor to belong to the emerging one.

The view from outside, it seems, yields a fairly unequivocal result: The paradigm shift in consciousness studies has almost, if not already, happened. Consciousness has become an object of scientific study. This is also reflected in reports from within, that is, from those who now form the inner circle of consciousness studies, constituted by the neurosciences, cognitive psychology, and philosophy. It is to these views from within that I now turn, in order to substantiate or differentiate the picture obtained from the view from outside. How do representatives of the evolving neurocognitive field of consciousness conceive consciousness? The answers will differ. To put it boldly: While neuroscientists have simply conquered novel territory without further ado, cognitive psychologists have had to re-acknowledge consciousness as an object of scientific inquiry, and philosophers have had to face the risk of being used on a case-by-case basis, rather than systematically informing the experimental results.

On Scientific Approaches to Consciousness: Views From Within

Making Claims on 'Consciousness': Neuroscience

A recent review in the realm of the neurosciences is plainly entitled 'consciousness' (Zeman 2001). This title is a statement: The review should tell us what does and does not reside within the confines of consciousness research today, according to a neuroscientist.

Alongside introducing various concepts of consciousness (as the waking state, as experience, as mind) and self-consciousness (e.g., as self-recognition, as awareness of awareness, as self-knowledge), Zeman highlights two lines of investigation emerging within the field that he regards as 'sciences': the *science of wakefulness* and the *science of awareness*. The science of wakefulness predominantly rests on the study of electrical correlates of states of consciousness, and the discovery that critical structures in the brainstem, thalamus and basal forebrain regulate conscious states. Consciousness understood as wakefulness inquires into the associated neural, behavioral and psychological functions, and their – jointly or independently functioning – control systems. The science of awareness predominantly rests on the study of the visual system in animals and man and searches for increasingly fine-grained correlations between cerebral activity and experience, namely in the areas of (visual) perception, memory and action. As to visual perception, two types of studies are conducted: One type of study looks into visual experiences without any change in external stimuli (e.g., hallucination, ambiguous figures); complementarily, another looks at the difference between explicit neural processes, which give rise to conscious awareness (e.g., conscious vision), and implicit neural processes (e.g., blindsight), which allow visuomotor performance in the absence of awareness. Further studies focus on action: Changes in cerebral activity, as consciously directed action becomes automatized or otherwise divorced from conscious control, which may illuminate the neurology of consciousness. Again, pathologies and disorders which disrupt our sense of ownership of our action are promising sources of insight into the neurological bases of volitional or conscious action (e.g., alien limb syndrome).

While all the studies above operate on locally specific theories, there are nevertheless some that attempt to provide a more general account: In the next section of his review, Zeman differentiates neurobiological theories, information processing theories, and social theories. *Neurobiological theories* roughly revolve around the conviction that structures in the upper brainstem core play a critical role in arousal, and that thalamic and cortical activity supplies much of the content of consciousness. Moreover, they agree that the neuronal correlates of consciousness (for short: NCC) are, in all probability, a cell assembly[14]. Disagreement arises on questions such as how large an assembly must be, or which regions it has to comprise, in order to give rise to consciousness, etc. Basically, neurobiological theories reverberate between the twin poles of neural activity and experience. *Information processing theories* are concerned with the self in terms of the functions consciousness might have for an organism. In general, while many things that we do need little or no attention, consciousness is required whenever the situation involves challenges, is unpredictable or critical. These theories identify consciousness with a stage or aspect of information processing, which directs our waking behavior. Critics (e.g., Libet 1996; Velmans 2000) counter that consciousness rather serves its biological purpose on a longer, more reflective time scale: It provides us with insight into the minds of others, thereby illuminating the workings of our own (Humphreys 1978; Frith and Frith 1999). *Social theories*, therefore, suggest going beyond the brains and behavior of isolated subjects. Other than social scientific theories proper, however, 'social' does not refer to the level of collective representations (e.g., in the Durkheimian sense). Articulated within an individualist methodology, 'social' theories of consciousness are all about an organism's chance to survive with the help of higher-level information processing.

Significantly, Zeman only refers to philosophy after these grander-scale theories: This section is broadly ordered according to the relationship between conscious and neural events that each single theory postulates, yet is in no way connected to the aforementioned theories or findings.

- *Identity theories* hold that neural events are identical to the corresponding conscious events. They offer a reductionist and materialist, or physicalist solution to the mind-body problem – a solution which dates back as far as Lucretius and which was emphasized later on by Thomas Hobbes (today see e.g., Churchland 1983). Critics argue that reductionism leaves out the mind (e.g., Searle 1992) because, for instance, the blind student of the visual system could never gain a full understanding, as she would never know 'what it is like to see' (Nagel 1979; Jackson 1982).
- *Functionalist theories* claim that the essence of consciousness lies in the functions that it serves, i.e., in a certain subset of information regarding the inputs or outputs

our nervous systems achieve (most prominently, see Dennett 1991). Functionalism reinterprets our experience in terms of a series of judgements made in response to sensory information. Although this approach takes cognitive processes into account, it does not take qualitative properties of consciousness into consideration.

- *Dualist theories* concede a close relationship between neural and conscious events, yet consider them fundamentally distinct classes of phenomena. Dating back to Descartes' separation of material and spiritual substances (Descartes 1961/1968), today's dualisms refine this argument. Chalmers has proposed a 'naturalistic property dualism' that regards mental features of reality as properties of natural physical entities that bear a lawful relation to their physical ones. Explaining consciousness, on this view, requires a set of psychophysical laws that describe *the relationship between* conscious and neural events.[15]

Concluding, Zeman states:

> Future work in philosophy and science may change the ways in which we think about consciousness so radically that its reduction to physical process begins to look plausible. For the time being we have no alternative but to use all three vocabularies of biology, behavior and experience to understand the mind (Zeman 2001, 1284).

Zeman's turn of phrase is revealing: The conundrum of philosophy cannot be resolved without confronting it with different kinds of 'vocabularies'. These assume the role of a heuristics eventually leading to a scientific, brain-based, explanation of the mind. Particularly, wakefulness and awareness are the aspects of consciousness to study. In fact, they *are* what consciousness is about. This is a decisive shift in reasoning about consciousness, that merely grants philosophy a commenting role and largely excludes the social sciences (see above).[16]

Rediscovering/Reclaiming Consciousness: Psychology and Philosophy

The disciplines considered thus far all have different histories, connected in some way to consciousness: While neurobiologists are just conquering novel territory, the social sciences, philosophy and psychology already have a history of their dealings with consciousness. Most interestingly, psychology, as a scientific discipline, has an entire history concerning the question whether or not to study consciousness at all.[17] In a nutshell, this history can be told by sketching William James' attitudes and approaches throughout his scholarly life: He switched from an introspective stance when he discovered consciousness as an entity, yet ultimately, denounced consciousness altogether. This same pattern has re-occurred within the discipline of psychology throughout the last 100 years:

In the beginning of psychology as a discipline, introspectionism attempted to give a full description of the mental landscape as it appeared to the subject. Scholars like Külpe or Titchener, having a scientific project in mind, adhered to nothing less than the atomic table of the human mind (Külpe 1901; Titchener 1915). But as they accredited the word of the introspecting subject with ultimate authority - those words that produced different results in different labs – introspectionism itself became discredited (Güzeldere 1995a, 38). The emerging positivist attitude thus did away with consciousness as a scientific issue: "Behaviorism claims that consciousness is neither a definite nor a usable concept"

(Watson 1970, 2). Having been an influential paradigm for psychology for over half a century, behaviorism removed the mentalistic vocabulary from the psychological agenda – consciousness can only recently be found once again in textbooks and dictionaries (e.g. Kiefer in Muesseler and Prinz, forthcoming). Yet consciousness was always present as a hidden variable: "Behaviorism owes its *ism* to consciousness. And what would it be without its *ism*? Well it would be physiology" (Boring 1963, 275). Julian Jaynes adds: "Behaviorism was only a refusal to talk about consciousness" (Jaynes 1976, 15).

With the advent of cognitive psychology in the seventies, consciousness became a kind of component or aspect of information processing models, thereby regaining scientific status. Mandler and Shallice, for instance, insisted on the significance of consciousness research: "Theoretical developments in cognitive psychology and the increasing use of introspective methods require a rationale, and that this should involve consideration of consciousness ... the common preposition driving the cognitivist research on consciousness was that the basic phenomenological concept – consciousness – can be mapped onto the information-processing concept" (Shallice 1972, 383). This, however, did not occur without objections, such as Neisser's:

> The treatment of consciousness as a processing stage ... does not do justice neither to the usage of the word 'consciousness' in ordinary discourse nor to the subtleties of experience. A better conception of consciousness, which has been suggested many times in the history of psychology, would recognize it as an aspect of activity rather than as an independently definable mechanism (Neisser 1976, 104–5).

Yet another aspect of rediscovering the issue of consciousness was connected to psychology's rediscovery of the *un*conscious: Other than the Freudian unconscious, which is not inaccessible given appropriate therapeutic treatment, cognitive psychology posits a *cognitive* unconscious. It exists due to the way our perceptual/ cognitive system is constituted and is, in principle, inaccessible. This notion has instigated a whole research industry.

> Theorists of diverse persuasions have been led to propose cognitive dichotomies, which have given a rather bewildering array of labels: unconscious vs. conscious, procedural vs. declarative, automatic vs. controlled, reflexive vs. reflective, and many others. These distinctions do not always divide cognition along the same lines ...[but] there are tantalizing similarities among the proposed dichotomies. In particular, the first member of each pair is generally viewed as involving unconscious mental processes, a topic that has seen a recent resurgence of interest among experimental psychologists (Holyoak and Spellman 1993, 265).

Thus, as far as cognitive psychology is concerned, consciousness has regained the status of a scientific object and is currently being investigated by various strands of study: It finds itself re-framed by an impressive number of dual conceptions (see above).[18]

As cognitive psychology is mainly driven by a causal intuition, however, little emphasis is generally placed on the phenomenal account (Güzeldere 1995, 49). This tendency became even more acute once psychology came to form part of the cognitive sciences, which have proved inspiring, not only for the empirical disciplines involved (e.g., neurology, cognitive psychology, and the information sciences), but also for analytical philosophers. Their reasoning on classical problems, such as intentionality, subjectivity, and consciousness, has been enriched with a huge heuristic potential stemming from empirically-based theories of information processing in (complex) systems, human and non-human – at least in the view of those (few) philosophers who feel 'cognitively inclined'.

Cognitive science has thus been a major driving force, rendering consciousness studies not only scientific but interdisciplinary as well. Some, however, find this type of interdisciplinarity far too narrow (see below). A good example is Yanina Shapiro's review of 13 (of about 45) books on consciousness by well-known scholars in the field of the biological and neurosciences, as well as in philosophy throughout the first half of the 1990s. It is entitled: *The Consciousness Hype: What Do We Want Explained?* (Shapiro 1997). Shapiro's review sketches a story according to which consciousness, as soon as it became 'discovered' by science and analytical philosophy, received special treatment by each author who cared to deal with the issue. Above all, its history is one of theories mutually ignoring each other. If anything, scholars like Edelman (1989) and Calvin (1996) established that consciousness was no longer an outcast of science but one of its greatest mysteries: Consciousness, via neuronal group selection, emerges from a Darwinian machine called the brain; it can best be studied via computational models of visual perception, the best known of the neuronal pathways; and it can be described as a set of necessary and sufficient conditions (see Shapiro 1997, 840). Yet here is where things start to drift apart. Questions such as

> Should a theory of consciousness be consistent with physics? What is the philosophy (e.g., reductionism, functionalism, dualism,) the theory should espouse? Should it accept the unconscious as indisputable fact that explains consciousness? What are the necessary and sufficient conditions of consciousness? What can be done about that problem of problems, self-consciousness...? (Shapiro 1997, 841)

seem to allow no consensual answers. Why?

Shapiro's basic critique is that scientists, as yet, have only bits and pieces of scientific facts pertaining to the brain. They glue them together with just the kind of philosophy that seems right to them, whereas the philosophers select neurobiological and cognitive facts as they need them. However, on her view, there is more to be criticized than just this 'quilt-a-theory' approach. In the end, the discussion as a whole suffers from a major flaw in that it disembodies consciousness. "... so far, the discussion has not been about conscious bodies and their thoughts, but about how the unconscious thoughts and feelings become conscious in the brain that is not attached to any body" (Shapiro 1997, 855).

From a discourse analytical point of view, one cannot help but observe that Shapiro, too, is involved in suggesting the right questions in consciousness research; she, too, makes judgements about what is/is not scientific, and finally, she is not at all critical as regards the current breadth of interdisciplinarity. Social, political, and moral aspects, so much seems to be clear by now, do not play a role. From a Kuhnian perspective, this situation is intelligible in terms of field formation. While struggling about the right questions to ask, and to determine whether or not being a science is common procedure in pre-paradigmatic states, consciousness studies seems to be on the brink of 'normal science' as far as the accredited participants are concerned.

This is one of the major challenges the 'Journal of Consciousness Studies' (JCS) faces. On the basis of the renewed interest in consciousness, it wishes to re-open and broaden the forum by welcoming humanists, transpersonalists, and social scientists back into the fold.

JCS: AN EXERCISE IN EXTENDING & INTEGRATING THE DEBATE ON CONSCIOUSNESS

The Legitimating Narrative for a Multidisciplinary Effort

> The 1990s have been christened the 'Decade of the Brain' due largely to the remarkable developments in brain research and brain-scanning technology which are set to revolutionize[19] our understanding of cognitive processes. Our purpose in launching the *Journal for Consciousness Studies* is to argue that it should also be the 'decade of the mind', and to point out that the two things are not necessarily synonymous (Editorial Introduction 1994, 4).

While the editors acknowledge that neuroscientists have returned consciousness to the academic agenda after nearly a century in exile, they deem it high time to approach consciousness in all its aspects.[20] The scientification of consciousness, on their view, has its downside: The threat is that we are but 'a pack of neurons' (Crick). But will the mystery surrounding consciousness come to an end once we know all about its neuronal correlates and have solved the 'binding problem', that is, the synchronization of different neuronal inputs to form the single unified perception we commonly experience? Or, alternatively, once we know the natural history of the mind, that is, the evolution of consciousness as something conferring a clear evolutionary advantage? The JCS wants its heterogeneous community to acknowledge both neuroscientific details as well as the experiential concerns regarding consciousness. It calls for *ontological rapprochement* between physicalists and phenomenalists, as well as for *epistemological reconsiderations*: an observer-related epistemology might overcome the fruitless opposition between reductionist and dualist accounts. Only then, according to the JCS, may we hope to solve what has been termed the 'hard problem' (bridging the gap between neuronal activity and phenomenal experience).[21]

Last but not least, the editors pursue a social, political, and ethical agenda: What follows for politics, for ethics, for social change, if we are but a 'pack of neurons'? Ethical relativism, uncertainty, and alienation, are the concerns they mention explicitly. Most acutely, they hint at the consequences of treating consciousness as nothing but some neuronal activity in the anterior cingulate sulcus: "Our brave new world might wish to develop a form of 'consciousness engineering' to parallel the new technologies in genetics" (Editorial Introduction 1994, 9).[22]

In conclusion, the editors insist on their decision *not* to define what consciousness is. In a way, their legitimating narrative goes like this: While peers watch over quality and the editors[23] over diversity, the discourse on consciousness watches over itself, eventually allowing interdiciplinarity/ies to emerge. So what is consciousness about, according to eight years of multidisciplinary discourse in the JCS? By way of an overview, the following section will focus on the issues addressed and approaches suggested, particularly by way of emphasizing the different types of multidisciplinarity. Secondly, I will analyze two instances of meta-discourse held by the editors of this journal: How do they assess values and shortcomings of their efforts and the reception of their readers, respectively? In addition to this, the analysis will include reports on the biannual conferences organized by the Association for the Scientific Studies of Consciousness (ASSC), published in the JCS. Taken together, these articles will present us with another 'view from within'. This time, we look at those in consciousness research who actively promote a multidisciplinary approach *beyond* the cognitive neuroscientific alliance.

Themes and Approaches in Eight Years of the JCS

Until today, 40 volumes of the JCS have been published, the big issues being the 'hard problem', 'Zombies', neuroscience of consciousness, quantum theories of consciousness, the self. Another bias is toward the basic cognitive functions upon which consciousness operates, such as perception, memory, and awareness, including a discussion of general cognitive theories of consciousness (e.g., 'Global Workspace Theory' by Baars 1997). Moreover, throughout the years, various articles and issues draw connections between cognitive processes and mysticism, particularly Buddhism. Special issues address empathy, morality, and art. Numerous single articles concern themselves with the epistemological foundations of consciousness research, with social, emotional, and corporeal aspects of consciousness, and a variety of specific problems (e.g., consciousness in plants, poetry, politics and psychoanalysis). With one exception (Güzeldere 1995a,b), there is no paper that gives a general overview of consciousness studies, and even the latter is supposed to be an introduction to one of the 'hot' topics (the 'hard problem'), rather than a state–of–the–art account. Highlighting a few issues may show how the issues addressed diverge with respect to the scope of interdisciplinary perspectives they imply or invite:

A considerable portion of articles has been devoted to the *hard problem* (see fn. 21). Although this problem can be said to be mainly driven by the philosophical discourse, the editors still reinforced a broader debate by initiating 'symposia', to which philosophers of various creeds as well as a few neurobiologists, cognitive psychologists, and computer scientists contributed their views. This procedure of focusing the debate, however, yielded even more contributions: From 1994 until 1997, the JCS published six symposia on the hard problem; in 2000, there is one other symposium on the mind-body problem (resulting in a total of 35 contributions). Alongside these symposia, there are a host of individual articles dealing with single problems, such as epistemological stances toward the hard problem (idealist, functionalist, postfunctionalist, etc.).

Somewhat related to the hard problem is one symposium and three individual articles on *Zombies*, again predominantly treated by philosophers: Of course, this problem is not about Haitian folklore, the living dead, but about the philosopher's zombies. It is a thought experiment concerning a being for whose cognition a functional description could be exhaustive. According to the experiment, "given any functional description it will still make sense to suppose that there could be insentient beings that exemplify the description": Zombies (Moody 1994). Although cognitive psychologists, neurobiologists and a number of philosophers should, theoretically at least, be keen on this thought experiment, there is not a great deal of inter-disciplinary exchange. Prominent philosophers such as Thomas Nagel consider it unsolvable, in principle. Meanwhile, the debates on Zombies are fervently continuing, largely engaging philosophers. This has led to a significant characterization: These are 'qualia wars'.

Not surprisingly, the *neuroscience of consciousness* (18 contributions, including one on neurosurgery) is another 'big issue' the JCS. Unlike the aforementioned debates, however, the pertinent approaches, methods, and insights are covered by individual articles only – thus far, it seems, no real multidisciplinary debate, focusing on specific topics, has emerged. One of the few exceptions concerns the cortical dysfunction called 'blindsight'[24], which engages philosophers as well. Recently, however, neuroscientific contributions have appeared that try to dampen the general optimism with regard to neuroscientific accounts of consciousness, in spite of their intricate research instruments. According to Revonsuo (2001), for instance, consciousness, or the phenomenal level of organization, probably resides at higher levels of complex electrophysiological and bioelectrical phenomena in the brain. However, not only do single cell recordings, PET, fMRI, EEG, and MEG pick up different kinds of signals from different levels of organization in the brain[25], but also, we lack, as yet, established theories in neuroscience of how these signals relate to each other. Revonsuo's critique is bold: The correlations that have been detected so far do not allow a reconstruction of the nature or organization of conscious processes on the phenomenal level – while supporting the neuroscientific project, he complains about a lack of deeper theorizing in such an experimentally-based discipline.

One particular approach, however, aroused some interest among both cognitive scientists and philosophers alike for quite some time, presumably simply because it makes broader theoretical claims: The *quantum-theoretical approach* toward consciousness (9 contributions), championed by Roger Penrose and Stuart Hameroff. According to this approach, consciousness should be viewed as an emergent property of physical systems. While most approaches remain speculative, Hameroff, e.g., postulates that cytoskeletal microtubules within neurons could be a possible site for quantum effects. Consciousness is re-defined as an "emergent macroscopic quantum state driven or selected by neurobiological mechanisms (neural networks, attentional scanning circuits, coherent firing of distributed neurons) with origins in quantum coherence in cytoskeletal microtubules with the brain's neurons" (Hameroff 1994, 106). Many contributions are highly critical of this approach: In the JCS, Scott, for instance, holds that distinctive features of pertinent brain activity (global coherence, binding of cell assemblies, etc.) are more typical of classical dynamics than of quantum dynamics, which is a linear theory. Therefore, natural scientists should turn to hierarchies of non-linear classical systems (see Scott 1996, for example). Despite these controversies, quantum theoretical approaches have been (and, to some extent, still are) a striking example of a highly controversial, multidisciplinary debate.

There is another frontier which seems to attract even more interest, however: *transpersonalist* accounts of consciousness. For a long time, they considered consciousness to be a 'pure experiential event', something beyond communication. Yet nine articles and one symposium on *Cognitive Models and Spiritual Maps* try to bridge this type of gap as well. It is the gap between constructivists (religious experience is wholly constructed from the fabric of pre-existing materials) and so-called perennial psychologists (mystical experiences are defined by some common underlying experiential cores, notably the so-called Pure Consciousness Event). Consciousness, in this volume, is thus meant to be the mediating term between *qualia*, or felt experience, and the 'hard' reality we refer to as the external world. Since perceptual systems continually translate between the two systems, it is time to ask how this is accomplished (see Andresen and Forman 2000, 7f.).

Only recently, the JCS fostered a controversy over the issue of *art* (one symposium, one target paper plus commentaries, 2 single contributions). Not unlike spiritual, meditative, or psychedelic experiences, for example, aesthetics seem to transcend consciousness as well, and hence may be another litmus test for empirical investigations pursued from a third-person perspective. The editor of the symposium on art lays down the challenge: "Nearly all of whatever brain activity it is that corresponds to aesthetic experience is unconscious ... some scholars hold that art is to merge subject and object in an ecstatic epiphany that transcends individual consciousness" (Gogue 2000, 13). Still, given this preconception cherished in the art community, Gogue and some of his colleagues are of the opinion that methods such as fMRI and theories of embodied consciousness are most promising for yielding further insights into the process of aesthetic experience.[26]

With about the same degree of coverage, one can find contributions on *evolutionary* and *ecological* aspects of consciousness (5 articles and a two–part symposium on evolutionary origins of consciousness[27]); only one article concerns itself with consciousness in non-human animals. The symposium on morality, however, does not seem to make any particular effort to attach itself to the consciousness debate: Rather, it remains a self-contained issue discussed primarily among evolutionary ethnologists and psychologists who deem morality an evolved capacity. The term consciousness does not even appear once in the introduction to this symposium, thus forcing readers to draw the connections for themselves.

The cross-roads at which virtually all the above-mentioned aspects and perspectives meet is the issue called *self*. Inquiries into the nature of the self have a long tradition. We have learned to frame it both within the ancient tradition of Socratic self-knowledge in the context of ethical life, and within contemporary discussions of brain function in cognitive science. The variety of responses to the problem of self includes the assertion that there is no self; that the idea is a logical, psychological, or grammatical fiction; that the sense of self is properly understood and defined in terms of brain processes; that it is the center of personal and public narrative; or that it belongs in an ineffable category all of its own. Accordingly, approaches involve philosophy, psychology, neuroscience, theories of embodiment, and artificial intelligence, all of which try to explore, and, at times, relate various ideas. They address questions of whether or not the self is enduring, is applicable to computer models, can be reduced to neural mechanisms, is a linguistic entity and/or a social phenomenon (see Gallagher and Shear 1997, 399ff.). The JCS devotes 4 volumes (2 of which are double issues) to symposia on *Models of the Self* (edited by Gallagher and Shear), plus 8 individual articles: While encompassing most perspectives mentioned above, there is one in particular which is lacking: the social perspective.

The *social* foundation of consciousness is only explicitly addressed on one occasion. Three further articles (since 1998) deal with loosely related issues. The one article targeting the social construction of consciousness deserves closer scrutiny as its authors, Burns and Engdahl (1998a,b), are themselves stunned by the fact that the new research industry on consciousness is proceeding without sociology (and vice versa, one might add[28]). This is all the more surprising as scholars like Durkheim, Schütz, Mead, Bateson as well as Berger/Luckmann, Goffman and many others have made significant contributions. Granted that consciousness is a matter of physical or material processes as well as of sentience and phenomenal processes, socially based cognitions, representations, reflective processes and discursive reflections nevertheless need to be accounted for as well. The main thesis is that individual reflectivity, as a defining form of consciousness, *derives from* collective processes. Individuals learn a language, language-mediated collective representations, and a cultural cognitive frame. Being active cognizers, structuring data and organizing perceptions within established cultural frames, human beings utilize their ability to discriminate, categorize, and react to environmental stimuli and signals. Moreover, they reflect upon, judge, and alter their modes of discrimination, categorization, and reactions to external stimuli. Awareness of self-processes requires a representation of self–states and processes, and thus depends on memory, including collective memory, that stabilizes representations, cognition and memories.

Consciousness in this sense of reflectivity is a matter not only of awareness that we experience feelings and perceptions, etc., but of the fact that the latter have meaning and can be understood in a larger context. Language and verbal thought enable elaborate articulation, focusing, and awareness of awareness itself; they enable continual and elaborate self-referencing and reflectivity, which is, as is argued here, the characteristic feature of consciousness proper.[29]

In closing, Burns and Engdahl concede that this additional level of analysis may complicate consciousness studies – as yet, however, the multidisciplinary community does not seem to be much disturbed. Consciousness, despite proclamations to the contrary, is not the subject of *one all-encompassing interdisciplinarity*. Rather, consciousness, according to the results of this analysis, is the subject matter of a *specific* interdisciplinarity which itself consists of various, even *more specific* interdisciplinarities. The most active interdisciplinary level is delineated by all those discourses that engage themselves with the neuro, cognitive and experimental challenge (be it affirmatively or critically); there is a less active inter-, or perhaps as yet only multi-disciplinary, level, upon which one finds other frontiers and other common grounds as well. In one camp, the experiential account talks back (uniting phenomenologists, mysticists, art theorists and deconstructivists). In another camp, one finds self–accounts uniting all perspectives that reside within the confines of an individualistic methodology (including social psychology, theories of mind). Far less multidisciplinarily organized are debates concerning the hard problem or zombies – philosophical debates that dominated the discussion until 1997. A host of specific issues combine various perspectives in themselves, like the science historical study of researches into plant consciousness, for instance, yet they remain unconnected to one of the more general discussions – a fate that may be predetermined for sociological accounts as well. The latter's differentiation of awareness and reflectivity (hence, 'higher-order consciousness') does not concern natural scientists interested in the 'basic (hence brainy) issues' of consciousness.

Finally, a look at both the editorial evaluations of the JCS (1996, 1998), as well as at three reports on conferences organized by the scientific association on consciousness studies (conference II–IV, held at Tucson), confronts the observations just made with reference to reflections from within. These are indicative of a renewed effort to both broaden and integrate the knowledge space *(epistemological function)* and at the same time, continually remind the emerging hybrid community to cooperate, epistemological and metaphysical gaps notwithstanding *(social function)*.

On Scientific-Cum-Humanist Approaches to Consciousness: Views From Within

Editorials

The first evaluative editorial, issued in 1996 after three years of publishing the JCS, is mixed. It is quite positive with regard to the stimulating controversy of quantum theoretical approaches toward consciousness. Also, in some cases, cross-talks way beyond the published manifestations have been instigated (e.g., between philosophy and computer theory). More critically, the editors firstly remark on the abundance of philosophy articles, and secondly, on a shortage of solid experimental work in psychology and neurobiology. This, according to their assessment, is partly due to the fact that the hard problem has been widely misconceived as being a predominantly philosophical problem. To set this balance right, the editors announce the publication of more on NCC (a special issue on blindsight). Thirdly, they note a striking lack of contributions by the arts and the humanities, especially on ethics and social issues (see 1996). Fostering a broader dialogue, it seems, did have a promising, if slightly humble, start, in need of active guidance on the part of the editors.

A second evaluative editorial appeared one year later (1997). In the meantime, as is evident from the outset, there has been much discussion on the part of the editors and other guiding groups as to what a scientific study of consciousness should look like. Although everybody emphasizes that there is no general theory as yet, an increasing number of authors warn against a chaotic, all-encompassing type of interdisciplinarity. Thomas Metzinger, for instance, argued for the debate to be limited to a narrow band of specialists. Particularly, he challenged the interdisciplinary enterprise by stating that he was looking forward to the establishment of a dominant (information processing) paradigm, and the abandonment of all first-person approaches to the 'New Age' Journal (sic!). This leaves, once again, only cognitive psychology, neuroscience and (analytical) philosophy as legitimate candidates for scientific dialogues regarding consciousness. The editorial response neither sides with anti-scientism nor with scientism. The message is: Non-scientific accounts have to be included, yet they have to follow a standard approach as well, that is, a hermeneutic one.

Moreover, the editors want to prevent a balkanization of consciousness studies, because even the scientific branch of consciousness studies is less unified than it may seem. "In practice this means that the 1997 Montreal meeting ... was attended mainly by neuroscientists, ASSC Claremont ... by psychologists, New School (NY) by philosophers etc.... given the above-mentioned trends towards premature closure ... our only defense must be a pluralistic one" (The Editors 1997, 388).

The whole discussion, in summary, is about struggling with the *right type* of interdisciplinarity. In particular, the debate is about *accepting phenomenal accounts*. Do they enrich (as say the editors) or rather disrupt (as say the critics) further insights into the truth of consciousness?[30] Needless to say, the social sciences are not part of this discussion.

Conference Reports (ASSC)

Not surprisingly, the reports on the Tucson-conferences show a similar pattern. Starting out with an enthusiastic embrace of a broad type of interdisciplinarity (1996 on Tucson II)[31], the next report (1998 on Tucson III) hailed most sessions as very multidisciplinary, yet at the same time, criticized a lack of dialogue among the speakers. In some cases, particular representatives were considered to be missing, such as philosophy in a session on NCC. While certain themes are now ranked as 'standard issues' (e.g., explanatory gap, implicit processes) and other themes have been strongly emphasized (especially cross-cultural and aesthetic approaches), other issues were deemed under-represented, such as the emotional aspects of consciousness. However, there is a reason (and hence an apology) for this lack: Due to the cognitive and computational bias in cognitive psychology, pertinent studies are rare (except for, e.g., Damasio and LeDoux).

One observation the report particularly dwells upon is resistance to multidisciplinary communication: Sutherland attacks the division of the conference into two opposing camps (materialists-reductionists versus transpersonal-experientialists). Interestingly enough, in his view, most materialists take experiential aspects of consciousness seriously, whereas experientialists despise materialist accounts in all forms and fashions: For the first time, the experiential faction is criticized as being unwilling to cooperate. Nevertheless, according to Sutherland, the meeting and the journal, as well as the association, aim toward an era in which the transpersonal and humanistic dimension will naturally be included as part of the mainstream science of consciousness (rejections by natural scientists notwithstanding), without the need for some special label.[32] From the outside, it seems as if both ends of the continuum are still hesitant to rethink their ontological, epistemological, and pre-theoretical creeds. In another article, Sutherland emphasizes his point:

> The separation between our subjective experience and our scientific descriptions of the world goes back to the writings of Bacon and Descartes. Without the 'Cartesian Cut' the enormous progress of science and technology would not have been possible. But this has also led to a science which, in the end, has to dismiss our most intimate and cherished feeling – consciousness, selfhood, and free will – as insupportable illusions. This is increasingly felt unacceptable, as it paints an impoverished picture of the human condition. Therefore, scientists and theorists are competing to develop a model which does full justice to all the elements in the 'naturalistic triangle' of psychology, neurology, and phenomenology (Sutherland 1998, 244) – has consciousness studies found its paradigm in the 'naturalistic triangle'?

In his report on Tucson IV, two years later, Reimer testifies to the breadth of issues addressed, which indeed cover all aspects of the naturalistic triangle. In particular, he emphasizes David Chalmer's remark that consciousness studies has gone through a transition from primarily comprehensive, incompatible, grand theories, to a spirit of incremental contributions which, according to him, has led to genuine progress in many areas (e.g., studies on NCC, achromats on color experience, blindsight, facial perception, selfhood, ...).

Most interestingly, he observes that transpersonalists have now struck back. During a session on meditation and consciousness, one scholar suggested that scientists studying neural correlates of meditative states should consider trained meditators as collaborators, not unacquainted subjects. He suggested that the concept of attention, which is most fundamental to both scientific attempts to understand consciousness and to various contemplative traditions, might serve as the best bridge between first- and third-person methodologies. He argued that both scientists and meditators might benefit from such collaboration, the former from sophisticated first-hand reports of mental states, and the latter from information about more natural categories and mechanisms that could result in better meditative techniques (see Reimer 2000). This plea for *mutual* enrichment was followed by a neuroscientific study. The scientist presented results from EEG studies of long-term meditators, a group of professional musicians, and a control group: They clearly indicated that the meditators showed significant differences in the dynamics of their attentional processing. Hence, the overarching naturalistic paradigm, proclaimed only two years earlier, seems to have yielded one instance of *inter*disciplinarity across the physicalist-phenomenalist frontier.

Meanwhile consciousness studies are continuing. Neither is there *one* concept of consciousness, nor *one* community, nor *one* field. At the same time, however, all studies converge on 'consciousness'. Perhaps concept, community, and field are best conceived of as hybrids.

CONSCIOUSNESS STUDIES: A HYBRID FIELD

This study attempted to reveal the dynamics of consciousness studies by way of, firstly, tracing the scientification of consciousness and by completing and commenting on this picture with disciplinary views from within, that is from within the neurosciences, cognitive psychology, and analytical philosophy. Secondly, the study examined the ongoing efforts of the JCS to foster bridges within a broader consciousness community (including phenomenal accounts) and scrutinized their assessments from within. All data were selected and united by the criterion that the articles deal with consciousness (as made explicit in the title, or by way of being published in the JCS). The sociology of knowledge underlying this approach is based on communications theory and discourse analysis. It is in communications that knowledge emerges, evolves, and at times, disappears, depending on whether or not, and if so, how, communications connect to preceding communications. Connecting to communication on consciousness today has become increasingly heterogeneous: The variations, however, are not arbitrary. Rather, if one applies the epistemology of the metaphor analysis, the variations occur via the interaction of the target term (conscious/ness) and the importing discourses. Both the unit of knowledge (conscious/ness) and the importing discourses or fields (e.g., neuroscience, philosophy, cognitive psychology) interact, that is: *they select against each other*.

In the present case, consciousness has been chosen as a unit of knowledge that has a long-standing tradition in itself. On the one hand, it essentially implies an experiential aspect, partly grounded in spiritual metaphysics. On the other hand, it implies social/sociological aspects. Both types of implication are thoroughly challenged by the cognitive and neurosciences. The responses to this challenge throughout the last decade have varied considerably:

- The cognitive and neurosciences have mainly approached consciousness in two ways; firstly, by denying its experiential aspects, and restricting themselves to phenomena, such as wakefulness or awareness, without making any further-reaching claims. Secondly, others have postulated and explored higher levels of complex electrophysiological and bioelectrical phenomena in the brain that would both ultimately and exhaustively explain the brainy bases of consciousness (including its experiential aspects, that is).

- Philosophy has responded in a split fashion as well. While a considerable number of philosophers have continued to work on standard issues (e.g., reconstructions of classical accounts), others have responded to the cognitive and neuroscientific challenge, either affirmatively or critically with regard to the different claims offered. Philosophers have thereby connected to a variety of accounts to which scientists, in turn, often refer only loosely – if they do so at all.
- Transpersonalists of various branches (religious, meditative, psychodelic) have insisted on the experiential aspect of consciousness. In some instances, they have even denied that experiences of 'Pure Consciousness' can be studied scientifically. Others, however, have pleaded in favour of the integration of the 'experiential experts' (e.g., long-term meditators) within the scientific study of consciousness, so as to grasp the neural correlates of such states.
- Social scientists have not denied the experiential aspect of consciousness, yet have emphasized that both contents and feelings associated with consciousness are socially co-constituted. While consciousness operates on brains, cognition, and individuals, it is a fundamental social phenomenon. This view, it seems, is a stand-alone, to which neither scientists nor transpersonalists, and only few philosophers, connect.

The interaction between consciousness and the different discourses mentioned above not only yields different results in and within different discourses. The interactions themselves also evolve, as neither consciousness nor the importing disciplines remain the same throughout the process. For instance, virtually every article today has to refer to neural aspects of consciousness, at least in passing – as indeed sociologists, philosophers and transpersonalists do, if in different ways. While most struggle with the experiential aspect, practically no one deals with sociological aspects of consciousness.

Yet at the same time, there is more than just diversity. Consciousness today may be justly described as a hybrid, a common ground, to which various discourses can attach themselves, although for different reasons, about which one can speculate. The interest of the *cognitive sciences* appears to be least obvious. While it is evident that modern research on consciousness is dominated by the cognitive sciences, notably neurobiology and cognitive psychology, the cognitive sciences, in turn, are by no means dominated by questions pertaining to consciousness. On the contrary, they are predominantly driven by their internal logic of research, given the instruments, techniques, and heuristics at their disposal. Interestingly enough, however, a fair amount of work today explicitly addresses what they deem to be the center of consciousness, namely its brainy bases, and is trying to finally solve the problem of the *causal* relation of neural firing and subjective experience. However, the question is not whether or not this goal will ever be achieved, but why such a highly diversified, busy and well-funded type of research attempts to explore consciousness at all. On the one hand, it seems plausible to assume that its attraction lies precisely in the opportunity to structure the diversity of the field. On the other hand, given the venerability of the topic, various strands of neuroscientific research can now be sold as solving one of humanity's last mysteries.

With a significant nuance, this mechanism pertains to *philosophy* as well: By attaching itself to experimental research (notably based upon brain-imaging, artificial neural networks, and real-world robotics) and applying the subtleties of philosophical expertise to this domain (affirmatively or critically), philosophy can regain broader attention.

Unlike the cognitive sciences, philosophy needs consciousness to keep its expertise involved in debates that transcend both the philosophical and the academic sphere. Participating in ethical concerns arising in modern society (e.g., bioethics, 'consciousness engeneering') or responding to cognitive science and its challenges to ancient problems of philosophy (mind-body, mind-brain) are promising ways for philosophy to achieve this goal.

For it is not unlikely that, in the near future, society will discuss more intensively the effects of neurobiological insights into conscious actions for, e.g., jurisdiction. Put differently, consciousness may become a key issue to which academic discourses, the general discussion, and societal practices (e.g., pharmacological, juridical) will ever more closely connect. That is to say, consciousness may become a transdisciplinary project. It is therefore all the more surprising that the *social sciences*, with few exceptions, do not yet see a market to which they could contribute an interesting product. Even if they did, they nowadays would have respond to the cognitive challenge as well, presumably by way of enlarging the knowledge space (from neuronal constructions of perceptions to social constructions of shared knowledge and vice versa).

This is different with respect to another push toward transdisciplinary perspectives: *Transpersonalists* (spiritual, parapsychological, religious) engage themselves in consciousness research by claiming privileged access to the 'real' phenomenon; pure consciousness. Thus, by connecting to a huge branch of transpersonal theories and practices, consciousness gains further attraction for a larger, mostly humanist, even New Age, branch of scholars. Most of them also self-consciously insist upon a non-scientific, phenomenal approach. Interestingly, for about the last three years, cognitive and neuroscientists have felt obliged to defend the scientific enterprise against what they consider to be phenomenalist intrusion.

Given the multitude and heterogeneity of knowledge today, it seems plausible to suggest that the well-introduced issue of consciousness assumes the function of an attractor for various disciplines, individually or jointly, and for various reasons. The current landscape of knowledge on consciousness, as shown, consists of constantly contested multidisciplinarities, and of various single disciplines (notably in the social sciences) which perform studies on consciousness, yet do not belong to any accepted community within the field. Pharmacology / neurobiology, cognitive psychology, philosophy / physicalist-phenomenalist / transpersonal accounts / the social sciences form clusters, each defining for themselves what they consider consciousness to be: a predominantly physical, neurocognitive, socio-political and/or spiritual matter. Yet, while nowadays most of those accounts are united by the fact that science may help to enlighten us on consciousness as well (if not exclusively), a considerable faction will not give up on phenomenal accounts altogether. Presumably, this tension will drive the hybrid engine called consciousness studies for quite some time. High time for science studies to learn how this affects the social significance of (a changing consciousness of) consciousness.

University of Basel, Switzerland

NOTES

[1] Although this threat to the ego has its own history by now, it does not lose its force throughout the disciplines involved – hard-nosed eliminativists like Dennett being the exception to rule (but do they really mean it while falling in love?).

[2] Definitions of multi-, inter-, and transdisciplinarity vary. This paper proceeds using the following: Both multi- and interdisciplinarity arise out of an intellectual problem, multidisciplinarity being a premature stage at which a common ground has not yet been established. Both types are not driven by solutions, in contrast to transdiciplinarity, which is driven by socio-economic problems, and is instigated more by general problems than by disciplinary questions: The context of application is the decisive frame of reference. The type of communication between the parties is characterized by consulting and negotiation. The organizational setting is flexible and transient. Accordingly, the level of institutionalization is low (see 'mode 2', Gibbons et al. 1994).

[3] Here, I fully accord with the strategic advice given by MaxiStrat™ (!), kindly forwarded to the Editiorial Board of Sociology of Science Yearbook by Stephen Hilgartner, see this volume.

[4] Bruner and Fleisher Feldman express themselves boldly: " ... at least where psychologists are concerned, it is indeed by their metaphors that we shall know them" (Bruner and Fleisher Feldman 1990, 237).

[5] A similar story could be told for the computer metaphor, according to which mental life can be mathematically formalized (Haugeland 1985). While computational accounts have been very inspiring, ranging from the speculative (e.g., Moravic 1988), through the innovative (e.g., Ansell-Pearson 1997), to the explicit (e.g., Johnson-Laird 1988), the computer metaphor has been criticized as well, e.g., by Pickering: "Thus, although it may have refined certain research topics, when generalized into a general theory of human cognition <including consciousness, SM>, the computational metaphor appears inadequate and restrictive" (Pickering 1999, 614).

[6] It is certainly fair to say that metaphors have always been regarded with suspicion and can be found in the midst of various dualistically structured debates: To mention but a few, they have been seen as ornamental, yet inessential; educational, yet lacking genuine insight; as economical carriers of complex meaning, yet easily misleading. Hence, metaphors have been troubling from the very beginning of the reflection on language. Max Black's seminal paper on "Metaphors" (Black 1962) eventually endowed the topic with new attraction. Until today, however, many scholars hold that metaphors, invisibly but nevertheless effectively carry meanings that can be disruptive to the truth of the topic in question. Thus, the analytical task is obvious: metaphors need to be revealed and erased. I will suggest otherwise: We should accept the inevitable and use these metaphors for our own ends, that is, for insights into the dynamics of knowledge (see below!).

[7] This approach has been developed together with Peter Weingart. For a broader introduction to theoretical underpinnings and methodical refinement, see Maasen and Weingart (2000).

[8] For instance, it could well be possible that, in course of interacting with the neurosciences, consciousness will acquire a different, physicalist meaning, or to put it differently, that neuroscientific efforts are considered to tell us something about, e.g., awareness, yet nothing about consciousness in its experiential aspect. The difference between the physical and the experiential could remain so important that consciousness sharpens its meaning *in contrast to* awareness, thus leading to a separation of terms, consciousness covering the experiential, awareness the scientific aspect.

[9] For a brief etymology, see e.g., Lewis (1960).

[10] This happened to chaos theory as well. Theoretically inclined authors increasingly reject the label for fear of being misrepresented as belonging to popular uses of the term. It should be noted, though, that earlier on they had been happy to become fashionable by way of piggy-backing on a popular term such as chaos, see Weingart and Maasen (1995).

[11] There is also a general proviso to be made as regards the indicator (consciousness occurring in titles): On the one hand, it does indicate those authors who explicitly deal with consciousness (even when they tackle very technical issues, which would not 'deserve' such a grand topic in the title). On the other hand, it does not cover those authors who study (some aspect of) consciousness, yet do not mention it in the title. The

choice of an approach accounting for only those articles that *explicitly* refer to conscious/ness is justified by discourse theory: Neither the participants of a discourse, nor the second-order observer, judge which contributions 'belong' to a discourse, but, in a non-essentialist way, the discourse itself (here: by way of using or not using conscious/ness). This perspective will be complemented later by views from within.

12 Amount of journals covering each field in the ISI 2000 (selective): anthropology: 51; neurosciences: 203; psychology general: 110, experimental: 67, social: 42; philosophy: 23; law: 105; sociology: 94.

13 These journals publish consciousness studies pursued in cognitive psychology. The latter is listed in both the SCI and the SSCI. Cognitive psychology is, in fact, the only significant overlap between both databanks.

14 Hebb (1949) defined the cell assembly as "a diffuse structure comprising cells in the cortex and diencephalon ... capable of acting briefly as a closed system, delivering facilitation to other such systems." On such a view, mental phenomena are subjective by virtue of involving relations to the self: we feel that it is ourselves that are doing the seeing. Other authors hasten to add: "This view is not homuncular, because we do not envisage that object representations are presented to organism representations, but rather that these two kinds of representations are linked in a neural network whose collective activity generates subjective experience" (Adolphs and Damasio 1995, 83).

15 Dual aspect theories (e.g., Nagel 1986) go even further: Some kinds of events might be inherently neural and mental, by necessity – a necessity that the human mind may not be equipped to understand. At present, however, we do not know what kind of properties could render physical events intrinsically mental and, as yet, speculations lead to unwarranted forms of panpsychism.

16 This shift, however, is not a 'conscious' one. Although Zeman claims in a footnote that consciousness and awareness are for him synonyms, in his text he seems to differentiate between the two. Generally as yet, there is no clear differentiation of terms, indeed, this is true for the social sciences as well: For Burns&Engdahl, for example, consciousness is 'awareness of awareness' (see below).

17 In the following, the presentation largely relies on Güven Güzeldere's account (Güzeldere 1995a,b).

18 This decidedly scientific-experimental stance, as opposed to phenomenal forms of investigation, is of special importance for psychology. Having virtually no 'natural history' of mental phenomena, in psychology the practices came first, whereas equipment and methods modeled on those in experimental physics only came later (e.g., Wundt, see Kusch 1995, 128). Not unlike other disciplines struggling for scientific status (e.g., economics), the natural sciences figured as models and sources of true and respectable knowledge – often by way of strictly excluding other ways of investigation, regarded as pre- or non-scientific. In a nutshell, this opposition is re-enacted in psychology's recent rediscovery of consciousness: Scientific approaches (in cognitive, partly social psychology) and non-scientific approaches, such as in psychoanalysis (let alone parapsychology), form two completely different clusters of research activities.

19 By the way, in the discourse on consciousness the revolutionary terminology is not all 'out' (see Hagner and Rheinberger, this volume).

20 Important ingredients of its novel success have been: Prominent scholars (Dennett, Crick and Koch, Penrose ...) addressing consciousness without terminological disguise (they call their respective topics 'consciousness' instead of 'anaesthesiology of cognition' ...), a host of conferences, scientific associations of or dealing with consciousness, the increase in publications, and the establishment of specific journals on consciousness.

21 This refers to a distinction of 'hard' and 'easy' problems in the study of consciousness introduced by David Chalmers in 1995: The easy problems refer to the empirical concerns of the cognitive neurosciences and their investigation of the cognitive and behavioral functions of consciousness. The hard problem is about the question of how any physical system could give rise to experience, to first-person perspective. Although we don't have all the answers, we know that easy problems (such as discriminatory abilities, reportability of mental states, the focus of attention, the control of behavior) can be approached scientifically. The hard problem, namely the problem of subjective experience, resists scientific explanation. Even more, Chalmers asks why all this information processing in the brain does not simply go on 'in the dark', free of any inner feel (see Chalmers 1995, 4–5). Evidently, the hard problem has its roots in the mind-body problem, dating back to Plato's distinction and reformulated by Descartes. As regards the so-called explanatory

gap, Güzeldere found an interesting quotation by the 19th century philosopher and psychologist Charles Mercier: "The change of consciousness never takes place without the change in the brain; the change in the brain never ... without the change in consciousness. But *why* the two are together, or what the link is that connects them, we do not know, and most authorities believe that we never shall and can never know" (Mercier 1888). Today, in spite of computer-based experimental design and instruments such as EEG, MEG, PET, fMRI, some authors still draw the same conclusion.

22 Thomas Metzinger, a cognitive philosopher, holds that the new relationship between brain and consciousness will soon change the ways in which we experience our everyday life. Cognitive science will lead to cognitive technologies (cyberspace) and consciousness technologies (psychoactive substances). Consequently, he suggests that continuous reflections on these issues should be institutionalized: an "anthropology assessment" is called for (Metzinger 1994, 16). Note that political and ethical issues are discussed as *consequences of*, not as *constitutive for* consciousness.

23 As well as a multidisciplinary advisory board, these include representatives of cognitive science, religion, cognitive studies, psychology, neuropsychiatry, biology, aneasthesiology, physiology, mathematics, sociology, and engineering.

24 This phenomenon can be found in patients with damage to their primary visual cortex. Despite this damage, patients retain the ability to detect, discriminate and localize visual stimuli presented in areas of their visual field in which they report that they are subjectively blind.

25 We look at neuroconsciousness (Revonsuo 2001, 7), for example, at both neuroanatomical and neurophysiological levels.

26 Ramachandran and Hirstein, e.g., suggest artists deploy certain rules or principles (8) to titillate the visual areas of the brain (Ramachandran and Hirstein 1999, 17) amenable to SCR testing (galvanic skin response) and single-cell recordings.

27 For many scholars, the evolutionary way is the one to go: Despite many scientific findings of neural and cognitive processes we, as yet, do not have more than a correlative, and no transparent theory of consciousness (Nagel). Such a theory, these scholars claim, would "need to explain how conscious experiences change behavior so as to provide the increased survival value that would have allowed evolution to select those brain mechanisms that underlie the conscious experiences" (Gray 1995, 9).

28 Hence, the anthropologist Cathrine Lutz writes: "Western social science has generally neglected to consider the cultural aspects of consciousness" (Lutz 1992, 64), and consequently pleads for an anthropology of knowledge. In a nutshell, her comparison between American and Ifaluk definitions of consciousness reveals that American definitions of persons have consciousness at their center (intrapsychic model), whereas Ifaluk focus on the obligations personhood imposes on others in the community (interpersonal model). That is to say, within an anthropological framework, any debate about consciousness is meaningful only insofar it "examines its cultural foundations and includes the breadth of human experience in its purview" (Lutz 1992, 86).

29 Burns and Engdahl do not fail to mention that this type of consciousness has its roots in the Western world (1998b. 182f.). Notably Christianity and modern therapeutic practices brought the self-reflective individual about (see also Maasen 1998).

30 The readers of the JCS have been mixed on this issue as well. The results of a poll revealed that about a third wished to have more scientific papers; 10% evaluated the papers published thus far as being too technical already, and 25% wished to read more on literature, art, and transpersonal accounts (Freeman 1997, 389).

31 "Perhaps the most remarkable and creative thing about the conference was the presence of people with starkly different intuitions and backgrounds debating the subject together: fMRI research talks to Eastern writings by Sufi divine Ibn Arabi. Monkey researchers talk to transpersonalists. The atmosphere as a whole was characterized as open yet rigid enquiry" (Sutherland 1996, 285).

32 And there is progress to be mentioned with regard to the institutionalization of this kind of multidisciplinary research: In 1997, the Fetzer Institute granted a \$1.4 million three-year-grant to establish a center for consciousness studies at the University of Arizona.

REFERENCES

Adolphs, Ralp and Antonio R Damasio (1995), 'Pain and the brain: Advances in pain research and therapy', *Consciousness and Neuroscience* 22: 83–97.

Andresen, Jensine and Robert K.C. Forman (2000), 'Methodological Pluralism in the Study of Religion', *Journal of Consciousness Study* 7 (11/12): 7–14.

Ansell-Pearson, Keith (1997), *Viroid Life: Perspective on Nietzsche and the Transhuman Condition,* London: Routledge.

Baars, Bernard J. (1988), *A Cognitive Theory of Consciousness,* New York: Josiah Macy Jr. Foundation.

Black, Max (1962), *Models and Metaphors: Studies in Language and Philosophy,* Ithaca, NY: Cornell University Press.

Bono, James J. (1990), 'Science, discourse, and literature: The role/rule of metaphor in science', in Stuart Peterfreund (ed.), *Literature and Science: Theory and Practice,* Boston: Northeastern University Press (pp. 59–89).

Boring, Edwin G. (1963/1932), 'The physiology of consciousness', in *History, Philosophy, & Science: Selected Papers* New York: Wiley & Sons, pp. 274–86.

Bruner, Jerome S. and Carol Fleisher Feldman (1990), 'Metaphor of consciousness and cognition in the history of psychology', in David E. Leary (ed.), *Metaphors in the History of Psychology,* Cambridge: Cambridge University Press, pp. 230–38.

Burns, Tom and Erik Engdahl (1998a), 'The social construction of consciousness. Part 1: Collective consciousness and its socio-cultural foundations', *Journal of Consciousness Studies* 5 (1): 67–85.

Burns, Tom and Erik Engdahl (1998b), 'The social construction of consciousness. Part 2: Individual selves, self-awareness and reflectivity', *Journal of Consciousness Studies* 5 (1): 166–84.

Calvin, William H. (1996), *How Brains Think: Evolving Intelligences. Then and Now,* New York: Basic Books.

Chalmer, David J. (1996), *Toward a Theory of Consciousness*, Oxford: Oxford University Press.

Churchland, Patricia S. (1983), 'Consciousness: The transmutation of a concept', *Pacific Philosophical Quarterly* 64: 80–95.

Damasio, Antonio R. (1989), 'Time-locked multiregional retroactivation: A systems-level proposal for the neural substrates of recall and recognition', *Cognition* 33: 25–62.

Davidson, David (1981), 'What metaphors mean', in M. Johnson (ed.), *Philosophical Perspectives on Metaphor,* Minneapolis: University of Minnesota, MN: Press, pp. 199–220.

Dennett, Daniel (1991), *Consciousness Explained*, London: Allan Lane.

Dennett, Daniel and Marcel Kinsbourne (1992), 'Time and the observer', *Behavioral and Brain Sciences* 15: 183–247.

Descartes, René (1968/1961), *Discourse on Method and the Meditations,* Harmondsworth: Penguin.

Edelman, Gerald M. (1992), *Bright Air, Brilliant Fire: On the Matter of the Mind,* New York: Basic Books.

Foucault, Michel (1974), *Die Ordnung des Diskurses,* Frankfurt a.M., Berlin and Wien: Ullstein.

Freeman, William T. (1997), 'What readers said', *Journal of Consciousness Studies* 4 (1): 7–9.

Frith, Christopher D. and Uta Frith (1999), 'Interacting minds – A biological basis', *Review of Science* 286: 1692–5.

Gallagher, Shaun and Jonathan Shear (1997), 'Editor's introduction' (to 'Models of the Self'), *Journal of Consciousness Studies* 4: 399–404.

Gibbons, Michael, Camille Limoges, Helga Nowotny, Simon Schwartzman, Peter Scott and Martin Trow (1994), *The New Production of Knowledge*, London: Sage.

Gogue, Joseph A. (2000), 'What is art. Editorial introduction', *Journal of Consciousness Studies* 7 (8/9): 7–15.

Gray, Jeffrey (1995), 'Guest editorial: Consciousness – What is the problem and how should it be addressed?', *Journal of Consciousness Studies* 2 (1): 5–9.

Güzeldere, Güven (1995a), 'Consciousness: What it is, how to study it, what to learn from its history', *Journal of Consciousness Studies* 2 (1): 30–51.

Güzeldere, Güven (1995b), 'Problems of consciousness: A perspective on contemporary issues', *Current Debates* 2 (2): 112–43.

Hagner, Michael and Hans-Jörg Rheinberger (forthcoming), 'Prolepsis. Considerations for histories of science after 2000', in Hans-Jörg Rheinberger and Bernward Joerges (eds.), *Social Studies of Science and Technology: Looking Back, Ahead.* Dordrecht: Kluwer.

Haugeland, John (1985), *Artificial Intelligence: The Very Idea,* London: Sage.

Hebb, Donald O. (1949), *The Organization of Behavior,* New York: Wiley.

Hesse, Mary B. (1972, 7th edition), 'The explanatory function of metaphor', in Y. Bar-Hillel (ed.), *Logic, Methodology, and Philosophy of Science,* Amsterdam: Elsevier, pp 249–59.
Hesse, Mary B. (1974), *The Structure of Scientific Interference*, London: Macmillan.
Hilgartner, Stephen (forthcoming), 'Insitutionalizing science & technology studies in the academy', in Hans-Jörg Rheinbergerand and Bernward Joerges (eds.), *Social Studies of Science and Technology: Looking Back – Ahead.* Dordrecht: Kluwer.
Holyoak, Keith and Barbara A. Spellman (1993), 'Thinking', *Annual Review of Psychology* 44 (51): 265–315.
Humphreys, Nicholas (1978), 'Nature's psychologist', *New Scientist* 78: 900–3.
Jackson, Frank (1982), 'Epiphenomenal qualia', *Philosophical Quarterly* XXXII: 126–36.
James, William (1890), *The Principles of Psychology,* New York: Henry Holt.
James, William (1904), 'Does consciousness exist?', reprinted in R.B. Perry (ed.), *Essays in Radical Empiricism,* New York: Dutton & Co., pp. 3–22.
Jaynes, Julian (1976), *The Origin of Consciousness and the Breakdown of the Bicameral Mind,* London: Penguin.
Johnson-Laird, Philip N. (1988), 'A computational analysis of consciousness', in A. Mandel and E. Bisiach (eds.), *Consciousness in Contemporary Science,* Oxford: Oxford University Press, pp. 357–68.
Kandel, Eric R and Larry R. Squire (2000), 'Neuroscience: Breaking down scientific barriers to the study of brain and mind', *Science* 2000 (290): 1113–20.
Kiefer, Markus (forthcoming), 'Bewusstsein', in J. Müsseler and W. Prinz (eds.), *Allgemeine Psychologie,* Heidelberg: Spektrum.
Kress, Jill M. (2000), 'Contesting metaphors and the discourse of consciousness', *Journal of the History of Ideas* 61 (2), 264–82.
Külpe, Oswald (1901), *Outlines of Psychology,* New York: The MacMillan Co.
Kusch, Martin (1995), *Psychologism: A Case Study in the Sociology of Philosophical Knowledge,* London: Routledge.
Kusch, Martin (1997), 'The sociophilosophy of folk psychology', *Studies in History and Philosophy of sciences* 28 (1): 1–25.
Lewis, Clive S. (1960), *Studies in Words*, Cambridge: Cambridge University Press.
Libet, Benjamin (1996), 'Neural processes in the production of conscious experience', in Max Velmans (ed.), *The Science of Consciousness,* London: Routledge, pp. 96–117.
Luhmann, Niklas (1980), 'Gesellschaftliche Struktur und semantische Tradition', in N. Luhmann (ed.), *Gesellschaftsstruktur und Semantik. Studien zur Wissenssoziologie der modernen Gesellschaft,* Frankfurt a.M.: Suhrkamp, pp. 9–71.
Lutz, Cathrine (1992), 'Culture and consciousness: A problem in the anthropology of knowledge, in F.S. Kessel, P.M. Cole and D.L. Johnson (eds.), *Self and Consciousness: Multiple Perspectives,* Hillsdale: Erlbaum, pp. 64–87.
Maasen, Sabine (1998), *Genealogie der Unmoral. Therapeutisierung sexueller Selbste,* Frankfurt a.M.: Suhrkamp.
Maasen, Sabine and Peter Weingart (2000), *Metaphors and the Dynamics of Knowledge,* London and New York: Routledge.
Mercier, Charles (1888), *The Nervous System and the Mind: A Treatise on the Dynamics of the Human Organism,* London: Macmillan.
Metzinger, Thomas (1994), 'Bewusstseinsforschung und Bewusstseinskultur', in H. Leuner and M. Schlichtung (eds.), *Jahrbuch des Europäischen Kollegiums für Bewusstseinsstudien,* Berlin: VWB, pp. 7–29.
Metzinger, Thomas (1998), 'Anthropologie und Kognitionswissenschaften', in Peter Gold and Andreas Engel (eds.), *Der Mensch in der Perspektive der Kognitionswissenschaften,* Frankfurt a.M.: Suhrkamp, pp. 326–72.
Miller, G A. (1963), *Language and Communication*, New York: McGraw-Hill.
Miyashita, Yasushi and Martha J. Farah (2001), 'Cognitive neuroscience at the turn of the millenium. Editorial overview', *Current Opinion in Neurobiology* 11: 147–9.
Moody, Todd C. (1994), 'Conversations with zombies', *Journal of Consciousness Studies* 1 (2): 196–200.
Nagel, Thomas (1986). The *View From Nowhere*. Oxford: Oxford University Press.
Nagel, Thomas (1988/1979), *Subjective and Objective, Mortal Questions,* Cambridge: Cambridge University Press
Neisser, Ulric (1976), *Cognition and Reality,* San Francisco: Freeman & Co.
Penrose, Roger and Stuart Hameroff (1994), 'Quantum coherence in microtubules', *Journal of Consciousness Studies* 1 (1): 91–118.

Pickering, John (1999), 'Consciousness and psychological science', *British Journal of Psychology* 90: 611–24.
Ramachandran, Vilanyanur S. and William Hirstein (1999), 'The science of art. A neurological theory of aesthetic experience', *Journal of Consciousness Studies* 6 (6/7): 15–51.
Reimer, Jacob (2000), 'Tucson 2000: A whirlwind tour', *Journal of Consciousness Studies* 7 (6): 70–80.
Revonsuo, Atti (2001), 'Can functional brain imaging discover consciousness in the brain', *Journal of Consciousness Studies* 8 (3): 3–23.
Rorty, Richard (1989), *Philosophy and the Mirror of Nature*, Princeton: Princeton University Press.
Scott, Alwyn (1996), 'On quantum theories of mind', *Journal of Consciousness Studies* 3 (5/6): 484–91.
Searle, John (1992), *The Rediscovery of the Mind*, Cambridge, MA: The MIT Press.
Shallice, Timothy (1972), 'Dual functions of consciousness', *The Psychological Review* 79 (5): 383–93.
Shapiro, Yanina (1997), 'The consciousness hype: What do we want explained?', *Theory & Psychology* 7 (6): 837–56.
Sutherland, Keith (1996), 'Toward a science of consciousness. 1996 Conference Report', *Journal of Consciousness Studies* 3 (3): 278–85.
Sutherland, Keith (1998), 'Review article: The mirror of consciousness', *Journal of Consciousness Studies* 5 (2): 235–44.
The Editors (1997), *Journal of Consciousness Studies* 4 (5, 6): 385–8.
Titchener, Edward B. (1915), *A Beginner's Psychology*, New York: Macmillan.
Velmans, Max (2000), *Understanding Consciousness*, London: Routledge.
Watson, John B. (1970/1924), *Behaviorism*, New York: Noton & Co.
Weingart, Peter and Sabine Maasen (1995), 'The career of chaos as a metaphor', *Configurations* 5: 563–620.
Young, Andrew W. and Ned Block (1996), 'Consciousness', in V. Bruce (ed.), *Unsolved Mysteries of the Mind*. Hove (UK): Erlbaum, pp. 149–79.
Zeman, Adam (2001), 'Consciousness', *Brain* 124: 1263–89.

CHAPTER 8

SHEILA JASANOFF

IN A CONSTITUTIONAL MOMENT: SCIENCE AND SOCIAL ORDER AT THE MILLENNIUM

NEW WORLDS TO ORDER

Strolling east along the splendid swath of Constitution Avenue, beyond the recessed lawns of the White House, diagonally across from the back entrance to the National Gallery of Art, the visitor to Washington, D.C. will be drawn to a building whose soaring Corinthian columns signal the presence of something exceptional within. Inside, under the hushed central rotunda of the National Archives of the United States, the now-curious visitor may take her place in the slow-moving line of tourists for a brief glimpse of the three documents that anchor the American state: the Constitution, the Bill of Rights, and the Declaration of Independence. Encased in thick, greenish, helium-filled, bronze and glass frames, elevated on a marble pedestal, the faded parchments are barely readable in the dim, protective light. It is hard to linger long enough to decipher the script; the guards take care to keep the line moving. But the entire setting – the heroic murals,[1] the sober display cases around the circular gallery, the inlaid floor and monumental architecture – encourages a feeling of reverence. This is no ordinary public space; it is the closest thing to a holy of holies in this brashly populist, secular republic.

Yet the ironic observer would note more than a touch of incongruity in the deferential encounter between the spectators and the objects of their veneration. The soft lighting and expensive, high-tech display (the documents descend into an impregnable vault by night) are theater at its postmodern best. The nation whose representatives stream by in all their heterogeneity of sex, race, religion, color, and attire is scarcely recognizable as the one whose blueprints the documents register. Fewer than half the visitors to the National Archives could have voted in the polity contemplated by the founding fathers who look gravely down from their painted murals (Keyssar 2000). Among the visitors are people whose admission to American citizenship would have occasioned a skeptical raising of the eyebrow, if not a shiver of fear, among the sedate gentlemen who wrote those grand, nation-building texts. Their genius, if we choose that term, seems in retrospect to have lodged in what they left unsaid – in a choice of words so flexible that it accommodated an onrush of diversity and change which the authors' imaginations could scarcely have apprehended.

B. Joerges and H. Nowotny (eds.), Social Studies of Science and Technology: Looking Back Ahead, 155–180.

While the written Constitution sits enshrined in glass and stone – and, fittingly, inert gases – it is the practices of government based on the founding texts that have done the most to ensure the republic's survival. For more than two hundred years, the thrust and parry of American political life have been directed toward filling the blanks left open by those historic documents. Judicial decisions played a central role. Early in the nineteenth century, Chief Justice John Marshall's bold decisions in *Marbury v. Madison* and *McCulloch v. Maryland* affirmed the power of judicial review and of Congress to make necessary and proper laws. In the process, Marshall helped establish the proposition that significant political innovation can occur without explicit constitutional reauthorization. Closer to our own day, other landmark rulings of the high court, such as *Brown v. Board of Education of Topeka* on school desegregation and *Roe v. Wade* on abortion rights, have struggled to redefine the rights of persons whose color, gender or, in the case of the unborn, physical invisibility deprived them of status in eighteenth century political life. Still more recently, some credit Chief Justice William Rehnquist with envisioning and carrying through a fundamental reorientation in favor of states' rights on issues such as sovereign immunity and separation of church and state. Between the legal milestones of judicial review and school desegregation, a bloody civil war was fought to cement the origin myth of a nation conceived with such brilliant foresight that its founding principles have survived intact, even though the texture of its public life has changed beyond recognition.

The reality of American politics belies the felt continuity of the founders' vision. Judicial creativity, however, has not been the only instrument of fundamental reform. Over two centuries, American state-society relations have been deeply affected by inventiveness in quite another quarter – science and technology. The principles underpinning the American state were drafted by representatives of a pre-industrial, almost pre-scientific, agrarian society. Since their day, revolutionary changes have occurred in the organization of commerce and industry (Ratner, Soltow and Sylla 1979), spurred by radical shifts in transportation, communication, medicine, finance, and manufacturing. Railroads once thickly webbed the country and then largely disappeared, replaced by highways and, eventually, the aerial routes of civil aviation. Telegraph and electric power lines (Hughes 1983) formed prototypes for today's virtual communication networks. Nuclear power came and went (maybe to return), while the Internet gave newspapers, telephones, and even television a run for their money. Holistic thinking about the environment became fashionable, linked to the growth of ecology, the discovery of chemical hazards, and the birth of commercial biotechnology. Meanwhile, in medicine, human beings were increasingly seen as composites of separately treatable body parts, mental states, genes, tissues, and organs. Technology revolutionized the conduct of war, as the United States, along with a handful of other nations, acquired the capacity to destroy humanity, and earned therewith a vastly stronger hold on the imagination of its citizens.

Since the scientific revolution, the legal and political institutions of the United States, as of other major powers, have changed in keeping with shifting perceptions of the natural world and the capacity to intervene in its processes. So, too, have citizens' self-awareness, expectations of each other, and the norms that reflect these altered perceptions. This coupled development of natural and social orders has, if anything, gained speed and salience as the world narrows. Science and technology, I will argue in this essay, are playing a constitutive role in determining how power will be exercised – and, equally important, constrained – in the emerging global order. In particular, the entanglement of science and technology in three processes of globalization are broad and deep enough, in my view, to merit the label 'constitutional': the redefinition of self, identity, and community; the appearance of the consumer as a political agent, asserting rights claims against commerce and industry; and the certification of 'global' knowledge for use in supranational governance. Through these three prisms, we can see how previously taken-for-granted roles of citizens, corporations, and social movements, as well as their relationships with governmental institutions, are being reconceptualized. Human engagements with science and technology, I suggest, are altering the very foundations of identity, citizenship, and sovereignty, thus silently laying the groundwork for constitutional governance in the 21st century.

Understanding these phenomena is critically important for legal and political theory, but standard analytic tools in these fields need to be supplemented by perspectives from the systematic study of science and technology. In place of the structural formalism and epistemological realism that have marked much orthodox legal and political scholarship, science studies offers a dynamic, constructivist, practice-centered approach that is better suited to analyzing the distributed processes of global constitution-making. Below, I begin by reviewing the contemporary discourses of constitutional change, noting their commonalities and differences. I then reflect on the understandings of law and lawlike processes in the science studies literature, and the treatment of science and technology in legal and constitutional doctrine. With these discussions in place, I turn to the three most prominent nodes at which science and legal order are joined in processes of co-production: identity-making; consumption and citizenship; and global knowledge production. In concluding, I address the implications of these developments for democracy in a post-national future.

HOW CONSTITUTIONS CHANGE

As the United States takes its first uncertain steps across the threshold of a new millennium, one senses that constitutional changes are abroad in the world. Some perceive it in the waning of federal authority and the growing popular alienation from institutions of national governance (Sandel 1996; Nye, Zelikow and King 1997).[2] The extraordinary U.S. presidential election of 2000 confirmed these fears for many: a polled-to-death public split its votes with such stunning accuracy that a statistician from Mars could plausibly have modeled the American electorate as millions of coins tossed in unison (Jasanoff 2001). Europeans deplore the 'democratic deficit' in European Union (EU) institutions (Eder and Kantner 2000) and growing voter apathy in national elections. Others worry that the rise of transnational legal regimes and bureaucracies will sap national autonomy and jeopardize national welfare (Darman 1978; consider also President George W. Bush's retreat from the Kyoto climate change accord so as to protect the American economy[3]). Theorists of sovereignty have focused on the displacement of the state by multinational corporations and global media that have rendered normal modes of participation and politics increasingly irrelevant. Still others see the issue in terms of the formation of new identities and grassroots alliances, creating social movements that do not respect the geopolitical claims of the nation-state. Indeed, as the attacks of September 11, 2001 in the United States so horrifically demonstrated, the devolution of power to non-state actors has shaken fundamental presumptions about what makes a state a state.

Does all this turmoil point only to the decay of existing orders or also to a more optimistic future for constitution-making? It is clear, particularly in the era of globalization, that it is not necessary to hold a formal convention in order to rewrite the fundamental presumptions that bind people to their political authorities. As U.S. experience indicates, radical shifts in social order do not have to originate with, or even be confirmed by, explicit constitutional amendment. Britain's 'unwritten constitution' offers an even clearer example. On the world stage as well, revolutions can and do happen without the benefit of constitutional ratification – as when the Cold War's bipolar order collapsed in 1989. But how do old constitutional dispensations change, and how do new ones emerge? Let us look at three sets of responses, at the levels of national, regional, and global politics. While each offers valuable insights into processes of constitutional change, I suggest that there is need for a fourth model, more suited to times of emergence and more respectful of human involvement with science and technology. This last, I will argue, is a bottom-up approach to constitution-making that has particular significance for global governance in the 21st century.

American history offers an instructive starting point for thinking about constitutional change. In more than two centuries, the U.S. Constitution has undergone only the barest modifications[4]; the text the founders wrote is still the text that schoolchildren study and the Supreme Court pays homage to. Constitutional rights are regarded as among the most cherished elements of national life – to be held, as far as possible, constant, and defended against the corrosive effects of time and social change. Not everyone, however, sees this continuity as real, let alone as reason for bemoaning the end of constitutional creativity. Bruce Ackerman (1983, 1991, 1998), in particular, has influentially argued that the apparent durability of the constitutional order is a major achievement of America's living democracy. At three 'constitutional moments' – the Founding, Reconstruction, and the New Deal – Ackerman believes that revolutionary reform was achieved through inspired but 'unconventional' adaptation of existing institutions and practices. Politicians actuated by transforming ideals successfully claimed to speak on behalf of 'the people' and repositioned the building blocks of government to fit their grandly unorthodox visions.

A different constitutional challenge confronts Europe. Here we have not so much the problem of an ossified, and ossifying, founding text as a search for doctrines to legitimate rule at a higher level than the nation-state. The European constitutional debate is, in this respect, particularly relevant to global constitutionalism. It springs from a similar well-spring of concern: the interests of peace, security, and economic stability seem increasingly to require stronger integration across countries, via new supranational institutions; yet, the cause of democracy seems to be better served through the communal structures long cultivated by nation-states, and hence to favor at best intergovernmental accords (Goldmann 2001, 1–3). This predicament has given rise to more than one school of thought. It has made some European constitutional scholars insist on a minimalist approach, based on carefully prescribed, formal powers delegated to the supranational organs by participating nation-states. This formula would locate the European constitutional framework within the treaties constituting the EU. Reluctance to expand constitutionalism beyond such intergovernmental arrangements stems from a perception, especially among German scholars, that 'Europe' remains for now a notional political space. It lacks a common language or traditions and common media of communication that would enable a fully *European* democracy to form. Pressing for a closer union strikes these observers as unrealistic and threatening to democratic principles (Grimm 1995).

Against the 'constitution deniers,' another school of thought favors a more positive approach to European constitutionalism. The aim of such an exercise, according to Neil Walker (1996), would be to seek out and address in a coherent fashion three sets of questions about the structure of the EU. These concern the logic of the Union (intergovernmental or supranational), the 'democratic deficit,' and the degree of allowable variation among member states. Walker (1996, 288) does not minimize the difficulty of this task: "How, then, is it possible to develop the agenda for the construction of a constitutional identity for a novel political order which draws upon structural principles which are equally capable of resisting reversion to the old sovereign state and progression to a new sovereign superstate?" The answer for him and other modest constitutionalists of the EU lies in developing a mid-range of concepts, such as subsidiarity, that will permit new ways of bridging law and politics. Empirical contexts for this discussion will be found in the intricacies of market integration, currency-making, border security, and comitology rules.

Despite its structural ambiguities and potential, the EU remains at bottom a union built on the sovereign will of nation-states. Globalization connotes something altogether different. It points to the emergence of an ordered political realm that, by definition, supersedes nations. If such a transcendental formation is indeed in the making, can it, too, be said to rest on constitutional foundations? Michael Hardt and Antonio Negri (2000) emphatically answer 'yes' in their ambitious thesis that what we are witnessing at the turn of the millennium is not globalization but the birth of 'Empire.' The order that is coming into being is not, on their account, a Bretton Woods-style piecing together of separate national sovereignties, with power consolidated in international institutions such as the United Nations system. Rather, it more nearly resembles the ancient Roman Empire, a world-formation with its force-fields lying outside the control of even the United States, the only genuine superpower.

The new global Empire, according to Hardt and Negri, has its own constitutional order, but its elements are far different from the institutional building blocks that U.S. or European politicians play with at Ackerman's 'constitutional moments.' Expressing the will of the people has little to do with the constitution of Empire. Hardt and Negri view the imperial constitution as something immanent and unstoppable, almost apocalyptic. It descends from the historian Polybius' theory of Roman government, which saw power as institutionally divided among three centers of action: force with the monarchy; justice and virtue with the aristocracy; and discipline and distribution with the people, or the demos. In modern constitutional systems, these authors argue, this ancient triad was replaced by the functionally tripartite structure of the executive, the judiciary, and the legislature. Now, however, we are caught in another transition – this time toward a 'hybrid constitution' (Hardt and Negri 2000, 316–19), in which the old governmental functions are so networked and distributed as to operate simultaneously everywhere, and therefore nowhere.

The constitutional theater for Hardt and Negri (2000, 319) is an 'imperial non-place.' The transformation they describe occurs almost independent of political will, in defiance of locality, and without discernible ties to social practice or agency. All this is as alien to the temper of Ackerman's analysis as it is to the debates about European constitutionalism. If Hardt and Negri, the theoreticians of Empire, take the high road of constitutional abstraction, the U.S. and European debates can be seen in some ways as taking the low road. Legal traditions, whether in Europe or the United States, view constitutional change in more situated terms, which are in principle more congenial to scholarship in science and technology studies: lodged in particular texts, mobilized by particular actors, and implemented by particular institutions of law and politics.

Yet in one respect all the constitutional discourses we have considered thus far have something in common that differentiates them from most work on the practices of science. Constitutional theorists mostly conceive of change as coming from above, whether through Hardt and Negri's imperial structural realignments, Europe's self-abnegating sovereign states, or Ackerman's inspired judicial and political innovators. But – paralleling Kuhn's (1962) famous argument about scientific revolutions – deep social change can also come about through countless smaller adjustments and accommodations that restructure the basic organization of power. Both kinds of change marked the end of the millennium: the former attracted more notice, but the latter may prove more democratic and durable. Many of these incremental movements were tied in crucial ways to developments in science and technology.

The role of science and technology in the making and unmaking of political order has been largely neglected by political theorists. To be sure, American conventional wisdom attributed the Soviet system's downfall to a form of technology policy: specifically, to President Reagan's determined pursuit of the Strategic Defense Initiative ('Star Wars'), and the resulting costly intensification of the arms race in the 1980s. But this ruinous story of a bankrupt Soviet state further impoverished in search of a bankrupt technology is just another change rung on the all-too-familiar theme of technological determinism: just as 'good' technology drives progress so 'bad' technology, in the Soviet case, brought disaster. Nor have attempts to explain the end of the Cold War in ideological terms proved more satisfying. The fall of the iron curtain was heralded as the 'end of history' (Fukuyama 1992), signifying the demise of the planning state (Scott 1998) and the universal acceptance of the market as the only viable template for economic and social organization. But the stark dichotomy between state and market drawn by many political commentators scarcely does justice to the incredibly complex infrastructures that grew up to undergird both in the modern era. Characteristically, too, deterministic, single-variable explanations of the Soviet system's collapse have offered few insights into the diversity of post-revolutionary orders that sprouted on the dust-heaps of an abandoned ideology.

Overall, political accounts of the Cold War's end have displayed little of the subtlety discernible in recent scholarship on technological change (see, for example, Smith and Marx 1994). Yet the tools developed for the study of large technological systems could profitably be adapted to the analysis of such massive social achievements as a state or an ideology, whether they succeed or fail (Latour 1990). Such constructs, no less than fighter planes or power plants, are pieced together from myriad material and social elements that work more or less well together; survival depends on the harmonious functioning of the parts with the whole. Either progressive innovation or radical breakdown could originate, on this account, at multiple points in the system, through agents who question power by political or other means (scientific claims, for instance). Of interest then are the circumstances that permit revolutionary ideas to form and to spread through systems built on other expectations – possibly achieving constitutional status.

What kinds of futures, or communities, are likely to be imagined, and by whom, as we move from a world organized along strict ideological divisions to one where the market now seems rampant? If commitment to nationhood was the glue that held together the dominant cultural identities of the nineteenth and twentieth centuries (Anderson 1991 [1983]), how will the socially and politically significant identities of the coming decades be organized and cemented? How will the new technologies of desire – biological, digital, miniaturized to the nano level – affect identity and agency in the societies of the twenty-first century? With threats of annihilation temporarily in abeyance, and with hunger and poverty receding in many parts of the world, on what basis will people articulate their needs and demands in relation to the formal structures of government? If unwritten and emergent rules of constitutional dimension are beginning to operate in the post-cold-war order, where can we observe these rules being crafted?

Clearly, it is not sufficient to look for answers only in the high politics of globalization, as enacted in international treaties or articulated in the expansion and management of global markets. To comprehend the new constitutional settlements that may be appearing at this historical moment, we must also focus on the ways in which civil societies are responding to novel, technologically mediated possibilities of prediction, manipulation, and interconnectedness. First, however, we must position this analysis with respect to two other relevant bodies of thought: the discussion of constitutionalism in science and technology studies, and the discussion of technoscientific change in legal scholarship.

SCIENCE STUDIES AND CONSTITUTIONAL LAW: MINDING THE GAP

Given the centrality of scientific knowledge and technological artifacts in contemporary life, it is reasonable to think that the basic ordering commitments of modern societies will be found not only in legal texts, but also, tacitly expressed, in the very organization of life around the products of human ingenuity and knowledge. Constitutional ideas should be embedded in people's collective imaginations and practices, built into their material culture, and worked out in innumerable daily routines that draw upon resources derived from science and technology. Order may emerge not merely, or even mainly, when positive law bestows it or a court affirms it, but also when people assume that they have the capacity and the right to change their behavior in fundamental ways, and act accordingly. Thus, there may be emergent, quasi-constitutional rights that no court has declared nor legislature has decreed, but that are created through altered popular conceptions of what sorts of people we have a right to be, or what we have a right to demand from our ruling institutions. In technologically advanced societies, such changed expectations are commonly associated with transformations in scientific knowledge and advances in technological capability.

To date, however, academic literatures have done relatively little to probe these structuring effects. The theme of scientific and technological progress is not new, of course, any more than its dialectical counterpart, the theme of humanity's problematic adjustment to its own inventions. Both have figured in significant streams of work –

analytic, imaginative, minatory, prescriptive – over at least two centuries. On the positive side, are innumerable biographies of inventors and scientists, celebrations of their discoveries, and tales of science's triumph over adversity and disease. On the negative side are many forebodings. Mary Shelley's 1816 'ghost story,' *Frankenstein*, brilliantly captured the terror of runaway scientific ingenuity; today, her concerns persist, under the headings of playing god, flirting with doomsday machines or surrendering to forces outside the creator's control. Philosophers and political theorists have warned of the dangers of hyperrationality and its corrosive effects on deliberation, civic engagement, and individual liberty (Habermas 1975 [1973]; Bauman 1991). The human sciences in particular, as Michel Foucault's oeuvre compellingly documents, can function as disciplinary tools by which governmental power is dispersed throughout society, although these sciences may also open up new possibilities for human creativity (Foucault 1971 [1966], 1979). Still another line of work reflects on the risks and uncertainties of new technologies, and how they permeate the social structures of modernity (Giddens 1991; Beck 1992 [1986]). From outside the western world have come denunciations of science and technology as instruments of dominance, even of violence (Shiva 1993, 1997; Visvanathan 1997). Together, these explorations have done much to destabilize the myth of scientific and technological progress, but in one respect they lack the power of the popular narrative: they reveal deep problems of governance in technological civilization, but they provide few insights into how good orders are achieved. Criticism undoes the complacency of unreflective optimism, but it does not rebuild confidence.

Work in science and technology studies has begun to change this picture by documenting – very generally – how the products of science and technology not only influence but are also shaped by human norms and institutions (Jasanoff et al. 1995). Technological objects, such as ozone holes, genes, smart bombs, computers, climate models, and Dolly, the category-defying sheep cloned from the cells of an adult 'mother,' are all seen by S&TS researchers as repositories of human commitments about what counts as 'good,' whether in reasoning, in making things, or simply in living with one another (Latour and Woolgar 1979; Bijker, Hughes and Pinch 1987; Latour 1988, 1993; MacKenzie 1990; Haraway 1991, 1997). The deployment of technological artifacts engages with and reshapes our perception of social order at many levels: for instance, by redrawing the boundaries between humans and non-humans or nature and culture (Callon 1986, 1987; Latour 1993); by altering fundamental notions of identity (Haraway 1997); and by challenging settled expectations of liberty and autonomy (Jasanoff 1995a).

Among S&TS scholars, Bruno Latour has been perhaps most explicit in calling attention to the constitutional dimension of these human accommodations with the products of science and technology. In his important 1993 monograph, *We Have Never Been Modern*, Latour described the considerable work that human societies do to 'purify' their world of technoscientific hybrids into separate spheres of nature and culture. He termed the resulting settlement 'constitutional' because it deals with one of the most fundamental divisions of social experience: that between 'us' humans and 'other' non-humans, be they animate or inanimate. For Latour the metaphysician, the world of objects is always full of social meaning and normative power. A mundane object such as a speed bump (or a 'sleeping policeman') performs, as its colloquial name implies, functions that are essentially human, albeit rendered without human consciousness. Yet, by locating this construct of earth and asphalt squarely in the domain of inert nature, we, the agents of modernity, set aside any need to reflect on the thing's moral status or the nature of its relationship with us. The world, Latour implies, would be an altogether different place, maybe more terrifying and less ordered, if technological objects were continually reinvested with human characteristics, as nature habitually was in pre-modern societies.

Provocative though these insights are, the regime of sharp demarcations that Latour attributes to modernity markedly contrasts with the fluidity and ambiguity of technoscientific constructs noted by other S&TS scholars (Cambrosio, Keating and Mackenzie 1990; Haraway 1991, 1997; Moll and Law 1994). Like any universalizing theory, Latour's notion of purification fails to account satisfactorily for the divergences one finds among quasi-constitutional understandings in different times, locations, and cultures. Nature is not perceived in the same way by all modern industrial or industrializing societies: different lines are drawn between humans and other species, and different assumptions are made about the degree of interconnectedness between environment and society. Not surprisingly, when it comes to accepting or rejecting particular technological achievements, disparate ideas of what is 'natural' or morally right drive social responses, leading to divergent appraisals of the same processes or artifacts (see, for example, Jasanoff 1995b; Gaskell et al. 1999). In short, the mechanics of demarcation beg for elucidation within specific social, political, and legal contexts. One may ask, in particular, what role prior institutional commitments (including those embedded in formal constitutional law) play in the simplification of hybrid networks into the reductionist framings of *social* and *natural*. Put differently, Latour's constitutional settlement plays itself out in a curiously ahistorical, unsituated, and impersonal space. His constitutionalism is a philosopher's abstraction. To obtain a more textured picture of the ordering power of technoscience, we need to undertake a more grounded inquiry.

Regrettably, legal studies do not instantly provide the hoped for solutions to questions opened by work in science studies. S&TS writings are consistent with the views of a handful of legal scholars working on the intersections of law and technology – for example Lawrence Lessig (1997) on the architecture of information systems, James Boyle (1996) on intellectual property, and Frederick Schauer (1998) on privacy and the Internet – although there has been little systematic conversation between these parallel strands of analysis. For the most part, legal scholarship, limited perhaps by the law's institutional commitments to resolution and finality, has been slow to incorporate the findings of scientific and technological contingency that S&TS research has elaborated over the past thirty years.

There is, accordingly, a noticeable lack of fit between legal discourse and the preoccupations of science studies. Constitutional interpretation makes do with conceptions of liberty, property, human identity and welfare that predate the industrial revolution, let alone today's dazzling developments in genetic, environmental, and information sciences and technologies (Schauer 1998). Rulings that aim specifically to take account of scientific and technological developments – for example, in cases about reproductive privacy, the rights of non-traditional parents, the prolongation of life, the ownership of human tissues, the nature of risk, or the legal status of non-humans (Stone 1974) – show little evidence of engagement with the social and cultural histories of these changes. Indeed, in seeking to defend the fundamental character of constitutional categories, courts and legal commentators often downplay the extent to which our understandings of nature, society and the self have been transformed by two centuries of scientific and technological change.

What makes this lack of reflection on science and technology more puzzling is that legal theory has been hugely influential in bringing to light some of the hidden normative assumptions that underpin supposedly neutral legal rules. Modern versions of Legal Realism, for instance, have refocused the understanding of Realism away from the indeterminacy of rules toward understanding the often-disguised substantive choices embedded in even relatively determinate rules (Fisher, Horwitz and Reed 1993; Fried 1998). Feminist jurisprudence has exposed the gender-based assumptions that undergird much legal doctrine in areas such as property and family law (Bartlett 1990). The Critical Legal Studies movement stressed the ideological contingency of legal propositions that courts often take to be natural and inevitable (Kelman 1987; Kairys 1990). With regard to economic decisions, scholars have questioned the neutrality and inevitability of the 'baselines' against which constitutional questions are considered. Cass Sunstein (1993) concluded, for example, that the legal distinction between state action and private action presupposes (and hence reinforces) a certain state-created *status quo* that established the boundaries of the *private* to start with.

Despite these turns toward self-reflexivity, legal scholars have not by and large extended their deconstructive and skeptical analysis to the ways in which legal power interacts with the authority of science and technology. An unexamined positivism still marks much writing about scientific evidence, as exemplified by a stream of work criticizing judges, juries, Congress, regulatory agencies, and not least the public for failure to heed the standards of 'good science' (Huber 1991; Breyer 1993; Foster and Huber 1997). Such critiques are often accompanied by triumphalist and historically untenable accounts of technological progress, which represent the law as an awkward impediment to the enlightened march of science. Even at its most sensitive, legal scholarship tends to treat science and law as independent sources of authority. The two domains are often seen as distinct 'cultures,' with divergent objectives, destined to clash when they occupy themselves with disputes over norms and policy (Schuck 1993; Goldberg 1994). There has been little systematic research on the ways in which modes of authorization in science and the law build upon, mimic or incorporate one another (for some exceptions, all stemming from science studies, see Wynne 1982, 1988, 1989; Smith and Wynne 1989; Jasanoff 1998a,b), even though the historical record suggests that the two cultures have supported each other for centuries in patterns of mutual construction, stabilization, and reinforcement (Shapin and Schaffer 1985; Ezrahi 1990; Shapin 1994; Porter 1995).

More generally, traditional legal inquiry generally takes for granted the boundary between nature and society or knowledge and norms. Laws and rights are held to one side as proper subjects of legal analysis; science and technology on the other side are thought to lie outside the domain of legal expertise. Rights are interpreted as preexisting technology, or at least as lying in an altogether separate normative domain, rather than as being constituted in significant part through technology. Much of the literature on science, technology and the law has thus been framed rather unproblematically in the language of technological determinism. Echoing decades-old ideas about culture lagging behind its own inventions (Ogburn 1922), law today is frequently seen as lagging behind technology, desperately trying to bridge the gaps created by rapid scientific and technological advances. Thus, genetics and genomics are thought to have leapt ahead, opening new eugenic possibilities, while the law laboriously puzzles out how to protect individual autonomy. Family law is chronically seen as trying to catch up with reproductive technology. According to one account, "genetic testing has made determining paternity simple, even routine ... But in most states, the law has not caught up with the science" (Lewin 2001). Similarly, the Internet's almost infinite capacity for copying and dissemination is seen as overwhelming the rights of creative authorship. On the whole, technology is conceived of as a threat to rights, which therefore need to be protected through vigilant enforcement of constitutional norms. The analysis of science and technology in the politics of globalization demands more deftly manipulable conceptual categories.

CONSTRUCTIVE CONSTITUTIONALISM

I have argued thus far that the constitutional formations of the era of globalization are growing not only from roots in law and politics, but also from human accommodations with science and technology. Neither science studies nor legal theory offer fully satisfying models of sustained inquiry into the ordering effects of technological change. The former has insufficiently engaged with cultural specificity and institutionalized power, while the latter has underemphasized the contingency of knowledge and the law's own role in underwriting science and technology. To find a way forward, we need to step outside the perimeters of current theory and look more closely at the ways in which actors are constructing the present world. Let us return to basic constitutional notions and ask how they are playing out in the global diffusion of science and technology.

In charting this territory, it is best to work with a flexible notion of constitutionalism that is not tied to specific institutional arrangements or codes of law. Constitutions are at bottom balance wheels between power and its abuse. They are devices for ensuring "a form of rule which both empowers a government to carry out the range of functions associated with the modern interventionist state and excludes arbitrary and despotic forms of rule" (Walker 1996, 270). In the post-national era, we may extend this notion of constitutional order by including, in the class of possible rulers, not only governments but other forms of authority that also have power to control people's lives. But the heart of constitutionalism remains the preservation of balance: between enabling and constraining power, and between individual and societal demands. Under each heading, constitutional jurisprudence clarifies and codifies certain communally sanctioned norms regarding what is worth protecting, for and against whom, by what means, to what extent, and through what processes. Laurence Tribe (1978) has referred to these principled substructures as 'models' of constitutionalism. These models, needless to say, respond to developments in science and technology by taking on board changing conceptions of such norms as 'unreasonable search and seizure,' 'privacy' or 'property rights' – all of which are affected by technologically mediated changes in human and social capability.

Yet, this account of constitutionalism leaves untouched certain fundamental ontological problems that are central to the law. What kind of entity, after all, is the state, whose powers constitutions seek to delimit, and what sorts of beings are the individuals whose rights are protected against improper state action? Other puzzles flow from these. Where do judicial beliefs about how to answer these questions stem from? How are the ideas held by courts connected to broader currents of public knowledge and understanding, or to that special branch of knowledge called science? And on what basis should courts decide when some models of constitutional decisionmaking, such as regard for settled expectations or governmental regularity (Tribe's Models III and IV), come into conflict with novel expectations arising from science?

It is here, at the level of constitutionalism's most elementary conceptual units, that we can profitably begin to inquire into the influence of science and technology. To tease out the connections, it is essential to look beyond the formal principles laid down in legal texts and elucidated by courts. We must ask instead how norms of constitutional relevance are tacitly constructed in the daily hum of technological societies: norms that are embodied in technological standards and practices, hardened into material instruments and artifacts, entrenched within professional discourses, and legitimated through public policy. Areas of rapid technoscientific change, as in the fields of genetics, informatics, and environmental science, offer specially promising sites for this kind of interpretive inquiry. Of particular interest are current debates about the self and its entitlements, the rights of the citizen-consumer against centers of private (market) power, and the legitimacy of supranational institutions of knowledge and governance.

Self, Identity, Community

When the *New York Times* criticizes the law for lagging behind science, what normative position does it seek to convey? We know that paternity can be determined today on the basis of almost foolproof, biological tests that indicate whether a given child is the progeny of the man alleged to be the father.[5] A series of cases involving such tests have come before the U.S. courts, in the form of claims by 'fathers' who discovered their lack of biological kinship to their supposed offspring, sometimes after years of living together in a trusting family relationship. In ruling on these claims, courts have juggled in different ways with the values of genetic and social kinship, support for minor children, respect for reproductive freedom, protection of economic rights, and deterrence of fraud. Not surprisingly, the outcomes look chaotic, more like *ad hoc* accommodations to the facts of the case than like principled rule-following. What principle could possibly explain why a Texas court denies a man visitation rights, but requires him to pay child support for children who are not biologically his (Lewin 2001)? Proponents of the 'law lag' theory would prefer to cut through this tangle and write the scientific 'truth' of paternity unambiguously into the law. Biology should define paternity.

Such an argument not only elevates technological (or technoscientific) determinism to a normative principle (technology *should* drive the law), but it misconceives the subtlety of the connections among science, law, and human agency in constitutional cultures. The novel meanings of selfhood, identity, and community that mark this period of social ferment owe their shape as much to the legal and political contexts in which they originated as they do to breakthroughs in science and technology. These new configurations reflect our ability to see and explain human identity and behavior in altered ways, with the aid of new scientific categories and instruments. But, equally, they are a product of our ability to imagine and enforce preferred identities through culturally sanctioned legal and political means. The Texan 'father' who wished to withdraw support from his no longer biological children – but who might have wished to keep on visiting them – was asserting a sense of kinship that was simultaneously biological and social. His rights were not already present in transcendental form, ready to be 'declared' by a court or conformed to the dictates of science. Rather, like all the newly indeterminate fathers of the genetic testing age, he was a player in a complicated ritual to redefine the meaning of paternity in a time when biological kinship can be dissociated from social kinship at any point in a family's existence. Genetic information, even in this simple context, does not determine identity so much as it enables new identity claims to unfurl.

In other cases, individuals have seized upon genetic information as an instrument of liberation or for building more complex identities and group affiliations. Their behavior confounds the theme of genetic determinism that was current in film and fiction long before the birth of the cloned sheep Dolly at a Scottish research station in February 1997. The older 'technoscientific imaginaries' of genetics were colonized by fears of state control. Aldous Huxley (1946) gave these fears their classic articulation in *Brave New World*, where people were classified and bred for characteristics esteemed by those in power. Throughout the 1980s, academic writing on the coming genetic revolution similarly dwelt on the risks of manipulation and control in connection with techniques of genetic screening, testing, and gene therapy. In the 1990s, however, the reception of genes into culture followed a more complicated script, as people actively asserted claims based on seeing themselves through newly available genetic lenses. Science and law blended into unexpected projects of social action.

Disease groups were perhaps the most prominent, though not the only, early adopters of genetic technology (Callon 1999). The discovery of genes for heritable breast cancer (BRCA1 and 2) led women activists in Britain and the United States to demand greater access to genetic tests and genetic counseling. PXE International, a citizen group committed to finding a cure for the inherited disorder *pseudoxanthoma elasticum (PXE)*, participated in the isolation of the disease-causing gene, set up its own blood and tissue bank, and in an unprecedented move, even filed for a patent on the gene (Smaglik 2000). And in August 2001, stories circulated of a privately held corporation, the San Francisco-based DNA Copyright Institute, that had urged stars and celebrities to copyright their DNA so as to prevent commercial entrepreneurs from producing genetic copies of their valuable 'original' selves. In all these cases, genetic knowledge was appropriated as an added resource for people's self-expression, not as a weapon of control by the state.

To be sure, the possible negative consequences of reading people genetically did not disappear from view, but neither did those potentially affected passively await subjugation. Genetic exploitation emerged as a new front in the longstanding liberation struggles of workers, women, and ethnic and racial minorities against oppression by dominant economic and political interests. Rhetorics of piracy, colonialism and genocide, for example, were invoked and extended in protests by indigenous peoples' organizations against researchers engaged in bioprospecting for rare, medicinally active plants or in devising protocols for sampling human genetic diversity (Reardon 2001). In these cases, social groups claiming fundamental legal rights – representation, equality of treatment, access to various state-sponsored benefits – were created, or reconfigured, through the strategic intertwining of genetics and the law.

Technoscientific activity beyond the life sciences also contributed to movements of self-expression and resistance. One of the darkest manifestations was the apparently self-willed conversion of human beings into missiles and weapons on September 11, 2001, and in the subsequent rash of Palestinian suicide bombings in Israel. These acts, conducted outside the bounds of 'normal' war, turned the narrative of technological progress into a mockery of itself, transforming people into objects, freedom into destruction, and sites of communal life into sites of carnage. That technology can kill was not the novelty here. Rather, it was the rational subject's purposeful embracing of an object-identity, and the resulting human-weapon's denial of compassion, pity or regard for its accidental victims or itself.

Other developments were less uniquely horrible. Information technology substantially lowered the barriers to worldwide communication and thereby facilitated creative processes of identity transformation. Instantaneous electronic communication and the spread of the personal computer made it possible to mobilize communal passions or communal loyalties, cutting across established lines of social identification in unpredictable ways. In cyberspace, states no longer enjoyed a monopoly on the channels of communication. The Internet diluted the power of national governments to command the 'imagined communities' of their citizens (Anderson 1991 [1983]). All manner of organizations could now control a piece of virtual territory, from disease-based groups like PXE International to the Taliban, the Islamic fundamentalists who ruled Afghanistan until after the terrorist attacks of September 11. Using electronic media, non-state actors bypassed the entrenched power of orthodox print and television media to build new group identities and affiliations, supplementing the ties of nationhood.

Some highly stable identities ceded ground, or became blurred, as new opportunities for self-identification were delineated. In the realm of gender, for example, 'transgendered' identities, neither male nor female, were said to have gained ground in the 1990s, partly assisted by the Internet:

> The movement's coalescence, which members say began over the last five years and accelerated in recent months, has gained particular momentum from the Internet, with its ability to connect far-flung people and afford them a sense of safety. On-line groups that began by swapping tips on using makeup and obtaining hormones now also spread word of the latest victims of violence and the next political protest (Goldberg 1996).

The success of various resurgent ethnic fundamentalisms – Jewish, Islamic, Hindu – can similarly be attributed, in part, to the ability of interests back home to tap into the memories, grievances, and (not least) deep pockets of the far-flung representatives of national diasporas (Anderson 1994). New hybrid categories of nationhood (e.g., Non-Resident Indian or NRI) have emerged, extending citizenship outside the borders of the nation-state.[6] Controversies over female Muslim students' right to wear head coverings in French schools or British Muslims' demand for access to state-funded, religious schools illustrate some of the more prosaically constitutional dimensions of these developments.

Corporate Rulers, Consuming Citizens

Redefinitions of individual identity and community form only one strand of quasi-constitutional change in which science and technology are centrally implicated. A second important thread is the assertion of some of the rights of citizenship by consumers against powerful corporations. There is a growing sense that today's formal constitutions do not offer adequate conceptual tools for ordering relations between individuals and private corporations, particularly when corporate operations extend across many national boundaries. Once again, action from below, by potential consumers of technoscience, has underlined the problems and focused attention on the need for solutions.

U.S. constitutional jurisprudence long since recognized that corporate power may not be exercised in ways that thwart legitimate public goals. Even property rights are not held to be sacrosanct if asserting such rights would unacceptably burden the public's lawfully sanctioned liberties. Thus, in his historic 1946 opinion in *Marsh v. Alabama*, holding that a company town could not prohibit the distribution of religious literature on its premises, Justice Hugo Black observed that

> Ownership does not always mean absolute dominion. The more an owner, for his advantage, opens up his property for use by the public in general, the more do his rights become circumscribed by the statutory and constitutional rights of those who use it. Thus, the owners of privately held bridges, ferries, turnpikes and railroads may not operate them as freely as a farmer does his farm. Since these facilities are built and operated primarily to benefit the public, and since their operation is essentially a public function, it is subject to state regulation (citations omitted).[7]

In the great burst of civil rights decisionmaking of the 1950s and 1960s, the Court held that property rights claims could not subvert overriding national goals such as racial desegregation.[8] Cases like these went some distance toward problematizing the boundary between state and corporate action – or between politics and the market. They recognized that claims of private ownership and enterprise are not alone sufficient to justify restrictions on protected public liberties, and that corporate power, when exercised in statelike fashion, needs to be curbed as much as the power of the state.

In formal constitutional jurisprudence, the sense of what constitutes an impermissible constraint on liberty is tied to structuralist notions of power and jurisdiction. The private owner of a company town, or a bridge or railroad operator, or an inn-keeper controls a piece of physical space – like a mini-state – and hence may not deprive persons using those premises or property of fundamental liberties. There is no hint in these cases that technological innovation, historically seen as a motor of progress, can function in ways that equally threaten liberty, without dominion over physical space. Upstream production and marketing decisions by corporations may constrain human behavior down the line as thoroughly as infringements of constitutional rights by the state. By the time products arrive on the market, irreversible ordering commitments have already been built into them that may deprive consumers of important freedoms. Such innovation and design decisions are not open to public questioning or other forms of accountability under existing legal regimes; if anything, they are protected against scrutiny by confidentiality rules and intellectual property rights. The market – which only comes into play when products are already on line – therefore does not function as a good surrogate for democratic control. But cracks have begun to appear in this system as users and consumers assert a more audible voice in the governance of scientific and technological production.

In the aggregate, corporate initiatives in areas like biotechnology, computing, personal communication, surveillance, tourism, and transportation – let alone weapons of mass destruction – hold the potential for a deep restructuring of human behavior. Corporations, too, have in many instances adopted the rhetoric and symbols of statehood, helped along by accidents, deregulation, and management failures that weakened the credibility of states. Thus, DuPont, a U.S.-based multinational chemical company, sought to legitimate itself to wider publics in the late-1990s in terms that straddled the line between product advertising and political campaign pledges. On its website, DuPont embraced the language of sustainability, promising to leave the world no worse off for future generations.[9] A carefully orchestrated barrage of advertisements, built on the slogan 'To Do List for the Planet,' pitched DuPont as a concerned citizen of the world, ready to use its technical know-how (the company's own term was 'knowledge intensity,' as opposed to 'capital intensity') to make life better for untold millions. Like the Hobbesian state committed to defending people against the perils of nature, DuPont promised to protect a global populace against hunger, pollution, and the vagaries of the climate; at the same time, in an appeal to well-heeled consumers, it promised self-cleaning clothes, self-sealing automobile paint, and a material that combines Lycra (a DuPont exclusive) with leather to ensure a poured-on fit. This was no offer to sell a better mousetrap, subject to the laws of supply and demand. DuPont asserted the power and claimed the privileges of an imperial state.

If corporations have taken on the symbolic, rhetorical, and behavioral attributes of states, small wonder that people have found it desirable to assert themselves as citizens against these new centers of power. The idea that major technological shifts should not be undertaken without citizen involvement steadily gained ground in the last decades of the twentieth century. The anti-nuclear protests of the 1970s and 1980s were early indicators of this change in consciousness. The incipient, large-scale commercialization of biotechnology provided further impetus for experimentation with forms of participation and governance. The term 'technology assessment' was taken off the shelf, given a dusting, and harnessed to procedural innovation, particularly in Europe, ironically at the very moment when the United States, a pioneer in this field, dismantled its congressional Office of Technology Assessment (Bimber 1996). Citizen juries, consensus conferences and public referenda were held with varying policy impacts in countries such as Britain, France, Denmark, Japan, and Switzerland (Joss and Durant 1995; Marris and Joly 1999).

By no means all of the deliberative experiments were orchestrated by governments. The politics of biotechnology in the 1990s offers a prime example of newly emancipated consumer behavior. In numerous episodes of resistance against research in plant genetics, activists tore up plots planted with genetically modified (GM) crops. Occurring in both the North (e.g., Britain) and the South (e.g., India), these demonstrations manifested growing reluctance on the part of environmentalists and farmers to accept scientists' assurances that their research was beneficial or even trustworthy. While most biologists continued to insist that agricultural biotechnology posed no threats to human health or the environment, reports of possible risk spread like wildfire through national and international media. In one case, reports by a U.K. scientist that rats experimentally fed GM potatoes showed developmental abnormalities triggered massive consumer rejection of GM products, and changes in British policy toward imports and labeling (Gavaghan 1999; Masood 1999). In another case, a U.S. researcher's finding that pollen from GM corn harmed monarch butterfly larvae received worldwide press, even though both scientists and industry dismissed the study design as too badly flawed for use in risk assessment. Nonetheless, when so-called anti-globalization forces took to the streets in Seattle, Washington and elsewhere around the millennium, monarch butterfly images and costumes provided an instantly understandable, semiotically powerful critique of biotechnology. Demonstrators in effect asserted the right to draw their own scientific inferences, overriding the credibility judgments of expert peer reviewers. Demonstrations, street theater, and consumer boycotts gave evidence that technological innovation was framed as an instrument of governance, requiring ratification by publics as well as experts. By visibly wrapping themselves in the mantle of the monarch butterfly, protesters signaled that industrial research and development could no longer be regarded as off-limits to public review and criticism.

One particular assertion of consumer-citizenship may be read in the future as emblematic of this period of constitutional ferment. This was the case of Monsanto and the so-called 'Terminator gene.' In 1998, a barely known, small cotton seed company called Delta and Pine Land (D&PL) patented a technique to switch off the reproductive mechanism of agricultural plants, thereby rendering the seed sterile (Service 1998). The company hoped that this technology would help protect the intellectual property rights of agricultural biotechnology firms by taking away from farmers the capacity to reuse seed from a previous year's genetically modified crop. Though the technology was still years away from the market, rumors leaked out of a deal by Monsanto to acquire D&PL. Such a partnership could have had enormous implications for both the speed of technology development and its worldwide distribution.

At this point, the technology's corporate sponsors lost control of the situation. The activist organization Rural Advancement Foundation International (RAFI) launched a highly effective international campaign against the technology. RAFI's executive director Pat Roy Mooney is credited with having invented the inspired label 'Terminator technology,' a name that at once translated a complex technoscientific achievement into easily accessible terms and, as in the monarch butterfly case, subverted the distinction between expert and civic technology assessment. Focusing on Monsanto's role, the RAFI campaign gathered support from a network of powerful actors, including the Consultative Group on International Agricultural Research and the Rockefeller Foundation, whose president, Gordon Conway, reportedly talked Monsanto's chief executive officer, Robert Shapiro, into publicly backing down from the company's commitment to 'Terminator technology.' In piecing together transnational politics, civil society activism, popular technology assessment, and enforced accountability from a corporate giant, this episode captured an essential moment in the transition to a global constitutional order.

In sum, these events, which uncomprehending policymakers sometimes dismissed as outbursts of a new Luddism, can be seen as trials in post-national deliberative democracy. They bypassed the electoral process, focused on technology's regulative impacts, and contested the notion that capital-intensive (or, as in the case of DuPont, knowledge-intensive) corporations should have complete leeway to determine the courses of technological innovation.

Empires of Knowledge

Historians of imperialism have pointed to the central role played by knowledge creation and appropriation in the formation of the nation-state and, later, in the extension of state power to the governance of vast, dispersed territories in the name of empire. These efforts took the form of classifying people and places, enumerating and keeping watch on them, and making histories or museums of native practices (Anderson 1991; Cohn 1996). From these efforts were born a host of new human and social sciences, such as anthropology, comparative law, geography and cartography (Foucault 1971 [1966]). Analysts of modernity have called attention to the one-sided nature of these activities. The resources and authority needed to produce imperial knowledge rested, for the most part, in the hands of the rulers, and the facts created through their scientific strivings bore, frequently, only a schematic relationship to the lived realities of those being governed (Scott 1998). Postcolonial studies and the emphasis on the subaltern perspective have offered a salutary corrective to colonial sciences, and some recent work has pointed to the complicated, mutually constitutive relationship that sometimes existed between the knowledges of the rulers and the ruled (Storey 1997).

One does not need the special context of colonialism to recognize, as many scholars in science and technology studies have done, that knowledge-making is an instrument of power, and that the scientific workplace functions as a key site for the production of social and political order. Representing the natural world is understood by philosophers and sociologists of science as a way of intervening in it (Hacking 1983; Latour 1983, 1988; also Jasanoff et al. 1995). Visual displays of natural phenomena draw on historically and culturally situated traditions of representation; in a double hermeneutic move, they also sway people's imaginations, and with this their affective selves and capacities for community-building (Fyfe and Law 1988). Methods of measuring and standardizing social or natural phenomena help create the very things they seek to characterize, while concealing the subjective judgments that enter into measurement systems (Foucault 1979; Carson 1993; Porter 1995; Bowker and Star 1999). By consolidating the means of representation in esoteric places, such as laboratories or field stations or archives, the controllers of these 'centers of calculation' form themselves into 'obligatory passage points' in high modernity's exercises of power (Latour 1988, 1990). How do these insights bear on our present era of tacit and unwritten constitutional change?

Globalization offers an obvious entry point. It is talked about in many ways, by journalists, academics, and social activists. For some, the nub of globalization is in the global extension of the free market, with the attendant transmission of capital, ideas, people, and material things around the world, whether it is intellectual property law, Islamic fundamentalism, or Microsoft and Coca Cola (Friedman 1999). For others, more sociologically inclined, it is in the formation of places and subjectivities, cities for instance, that violate or hybridize older social categories and identifications (Sassen 1991). Still others have seen it in the emergence of new kinds of politics, from the rising influence of non-state actors in international negotiations (Haas 1990; Keck and Sikkink 1998) to the production of 'human rights' as a shared, if contested, discourse (Ignatieff 1997). Hardt and Negri (2000), as we have already seen, prefer to replace the concept of globalization with that of Empire. Increasingly, too, television and the Internet have become central players in both the definition and critique of globalization. In these media, action becomes text, readable and indefinitely reproducible; the local becomes global; and the lowly can assume (if only for fifteen minutes of fame) the ancient power of the monarch to command the public gaze. Televised and multiplied, a citizen protest in Seattle or a lethal attack on New York's twin towers acquires, at least for a time, the revolutionary force of a *Communist Manifesto*.

But what of the place of scientific knowledge in all this din of making and unmaking? Does science, despite its contingent, often provisional character have the power to move beliefs, forge alliances, and underwrite norms of global application? Some have argued that this is precisely what has happened in the context of environmental decisionmaking, where knowledge of the biosphere's limited resources and of human interdependence with nature has helped build global coalitions around norms of environmental stewardship (Haas 1990). Similarly, economists would credit the spread of economic knowledge and understanding during the twentieth century with the defeat of socialism at the century's end. The completion of the Human Genome Project at the millennium produced few immediate surprises, but the project's directors used the occasion to preach a new humanism based on the now scientifically grounded observation that, under the skin, human beings are really all the same

Despite their surface plausibility, claims such as these only open the way to further inquiry for students of science and technology. If scientific knowledge is in fact traveling freely around the globe, what are the social formations and processes that enable its frictionless transfer? If a system of global governance is quietly taking shape, then what knowledge resources has it gathered to itself and where are its centers of calculation? Is there evidence of a new knowledge class that has transcended earlier political divisions and constituted itself as a transnational ruling elite? And if the politics of knowledge historically played itself out in national settings according to well-established rules of testing and credibility, what comparable processes, if any, are arising in arenas of global knowledge-making?

While extended answers to these questions are beyond the scope of this paper, there are many indications that old settlements about where knowledge ends and politics begins are everywhere being reopened and challenged. At the institutional level, we see the rise of a mass of expert bodies of global jurisdiction whose work merges the cognitive and the normative. Constituted under disparate international treaties in areas such as environment, arms control, and international trade, these bodies have the power to certify knowledge and to draw boundaries between acceptable and unacceptable knowledge claims. For the most part, these global experts have operated with traditional notions of what constitutes 'goodness' or reliability in science, accepting published, peer reviewed articles as their gold standard. But the legitimacy of judgments reached on such a basis is often highly contested. Unrest over the technical determinations of bodies like the Intergovernmental Panel on Climate Change or the World Trade Organization attest to the fact that the preconditions for credibility and expertise on a worldwide scale are still very much in flux (Miller and Edwards 2001).

Accompanying the proliferation of new expert institutions is a diversification of what we may call 'global sciences.' These include relatively new additions to the human sciences, such as various forms of risk analysis for estimating global environmental and social hazards; also observable is a cluster of global accounting systems grouped under names like 'sustainability science,' 'vulnerability science,' 'integrated assessment' or 'ecosystem services.' The rise of modeling and simulation, enabled by massive increases in computing power and by sophisticated imaging techniques, has provided a further enormous boost to the ambition of knowing the world in its entirety. A central feature of these 'sciences' is the hybridization they demand of older categories of the natural and the social – violating in this respect the line posited by Latour as modernity's foundational achievement. Self-referential, the new earth sciences constitute the very realities they purport to represent. Their credibility depends not on experimental demonstrations to peer communities, as in the 'pure' fundamental sciences, but on the construction of legitimating practices, such as 'extended peer review,' that require assent from diverse disciplinary and social groups, expert as well as lay.

From a political standpoint, then, the once distinct practices of peer review in science and participation in politics are merging at the global level into novel procedures for ensuring the accountability of scientific and industrial research (Dickson 1984; Gibbons et al. 1994; Nowotny 2001). There are increasing demands for ethics in science, particularly with rapid advances in the human capacity to manipulate the basic biological material of plants and animals. In however inarticulate a way, world polities seem to be rejecting the idea of value-free knowledge and asking for scientific inquiry to make its goals and presumptions more explicit. These and similar developments attest to the erosion of old agreements concerning the objectivity of science and its autonomy from politics. They highlight linkages between knowledge-creation and the institutionalization of power in a globalizing world.

CONCLUSION: TOWARD POST-NATIONAL DEMOCRACY

Out of the ashes of the twentieth century the phoenix of a global civil society is struggling to resurrect itself. While constitutional theorists have tied globalization largely to macro-economic and political forces, science and technology must equally be seen as linchpins of the emerging global order. Networks of new knowledge and its material embodiments are helping to frame and stabilize some of the basic elements of a global political system, such as the rights, privileges, and identities of the world's citizens and the powers of major global actors. I have argued that the totality of these changes is constitutional in scope, both enabling and constraining new political formations. Through science and technology, seen as profoundly *social* institutions, many parts of the world today are engaging in what amounts to a tacit constitutional convention. On the table are the nature of the human self, the relations of consumers and corporations, and the certification of knowledge in the conduct of global politics. In all three spheres, the initiative for generating new organizing principles lies not only with corporate and governmental actors, but also in the hands of ordinary citizens and in the proliferating networks of non-governmental associations. In this respect, the emergence of a supranational world order (the 'Empire' of Hardt and Negri) is not inconsistent with continued assertions of human agency.

Realignments of global magnitude will take generations to accomplish, and the contours of the eventual settlements remain but dimly discernible and hard to predict. Let us not forget that it is a contingent as well as an unwritten constitution whose birth we are witnessing. Nevertheless, a few generalizations can be ventured. First, science and technology have not simply deterministically constrained people's freedoms, within limits ordained by preexisting constitutional rights. Rather, science has provided resources that can expand the meanings of identity and community and help redefine the zone of individual autonomy that sits at the heart of all constitutional systems. For good or for ill, science and technology are important aids to human self-expression, not merely iron cages within which a passive humanity languishes imprisoned by forces beyond its control.

Second, without much fanfare or explicit acknowledgment, recognition is growing that technology is an instrument of governance, no less powerful in the hands of private or public actors than laws and regulations are in the hands of government. All kinds of spontaneous experiments are underway to see how far and to what extent consumer-citizens may have a say in the very earliest stages of technological innovation. These range from seemingly irrational, bottom-up acts of resistance, like destroying field trial sites planted with GM crops, to more considered, often top-down exercises in public and political consensus-building on new technologies. Whatever their merits may be, these disparate approaches suggest that the narrative equating technological progress with democracy has come under profound questioning. Technology's claims to benevolent rule must be argued and won today, not simply assumed as they were in older paradigms of development and technology transfer. Publics worldwide want a say in determining what kinds of futures they should live, and as RAFI's successful campaign against the Terminator gene demonstrated, they are sophisticated enough to see that these futures will be substantially shaped by corporate investments in technology.

Third, science's role in underwriting the global constitutional order is associated with new forms and forums of deliberation. These range from more participatory peer reviewing bodies to Internet sites at which the truth of scientific claims is exposed to public scrutiny. Less visibly perhaps, controversies such as those over agricultural biotechnology are challenging scientists' autonomy over the definition of standards of evidence and proof. The fact that findings rejected by mainstream science can exert a powerful pull on global political action should not be interpreted as a sign of public indifference to the truth. Rather, it demonstrates that a new political question has emerged on civic agendas: When is knowledge reliable enough to support collective action? The answer to that question is not seen as lying within the exclusive preserves of scientific authority.

In sum, one of the basic principles of modernity that will surely be reformulated in the course of global consolidation is the sharp disjunction between science and politics, and the separation of processes that secure the authority of scientific claims from those that safeguard the legitimacy of government. Norms of accountability that previously held only between citizens and the state are being extended to experts, with a consequent need for forums in which experts can defend their judgments to wider publics. We observe as well demands for a more 'socially robust' objectivity in science – for facts that can sustain themselves through testing by diverse social groups, rather than solely on the basis of their claimed correspondence to physical reality.

To date, all these changes remain inchoate and uncodified, lodged in diffuse and inarticulate social practices whose collective impact has yet to be felt in the citadels of organized power. To have constitutional force, they should in some sense be explicitly authorized, but by what institutions, organized according to what legitimating principles? The European constitutional debate sets both an encouraging and a cautionary example in this respect: encouraging in pointing to the prospect of an ordered supranational polity; cautionary in delineating the considerable pitfalls that lie in the path of its achievement. Yet, recognizing the very possibility of constitutionalism from below, founded on human creativity and craft, is a prerequisite for its eventual uptake into the practices of politics, law, and governance. It is a first step toward imagining a constitutional order in which the rights of *knowing* citizens – *homo sciens,* beside *homo economicus* – are explicitly acknowledged and given their place in the sun.

All these transformations, finally, demand new kinds of engagement from the social sciences. Disciplinary boundaries based on conceptual categories that are themselves in flux seem ever less appropriate to characterize, let alone analyze, the moving frontiers of global social change. Legal scholarship and political theory in particular will need to accommodate more reflexive avenues of inquiry from newer, transboundary fields such as science studies. The reward, one hopes, will be a richer scholarship of the actual – a re-theorizing of the changes happening in the world about us, and a discovery of new ways to reflect upon, and perhaps intervene in, the courses of scientific and technological change.

Harvard University, Cambridge, MA, USA

ACKNOWLEDGMENT

I am grateful to the Program in Science, Technology and Society at MIT for an opportunity to present a preliminary version of this paper. I would also like to thank my colleagues on the Yearbook committee and Stefan Sperling for helpful comments on earlier drafts.

NOTES

[1] The two murals represent Thomas Jefferson presenting the Declaration of Independence to John Hancock, President of the Continental Congress, and James Madison presenting the Constitution to George Washington, President of the Constitutional Convention.

[2] The complaint that Americans no longer trust their governing institutions grew in force through the 1990s, along with observations about the public's declining participation in national elections. According to a poll conducted in the summer of 2000, for example, 43% of registered voters said they would watch no part of the Republican convention and 38% said the same for the Democratic convention. The corresponding figures for 1996 were 23% and 21% (Source: Vanishing Voter Project, Joan Shorenstein Center, Kennedy School of Government, Harvard University).

[3] 'George Bush's Global Warming Speech', *Guardian Unlimited*, February 14, 2002, http://www.guardian.co.uk.

[4] There have been only 27 constitutional amendments since the formation of the United States, the most recent in 1992, restricting the power of Congress to raise its members' salaries. Bruce Ackerman (1998, 490–1) has questioned the validity of this enactment, which separated national assent from assent by the states by about two centuries. The failure of the Equal Rights Amendment in the 1970s, despite the growing strength of the women's movement, offers one measure of the resistance to formal constitutional change.

[5] While few question the biological validity of paternity testing using DNA markers, the reliability of the results may vary widely depending on the proficiency of the testing institution. On the standardization of genetic testing practices, see Arthur Daemmrich (1998).

[6] NRI's enjoy a number of benefits designed to capture some of their foreign earnings for the Indian state. These include a variety of tax and investment advantages and special visa privileges.

[7] *Marsh v. Alabama*, 326 U.S. 501 (1946), p. 506.

[8] In *Heart of Atlanta Motel, Inc. v. United States*, 379 U.S. 241 (1964), the Supreme Court held that a motel serving interstate travelers could not deny accommodation to African-Americans on the ground that its services were purely local in character. See, also, *Evans v. Newton*, 382 U.S. 296 (1966) ("Where private individuals or groups exercise powers or carry on functions governmental in nature, they become agencies or instrumentalities of the State and subject to the Fourteenth Amendment," p. 299).

[9] The statements have altered in interesting ways over the years. In September 2001, the company posted a statement on sustainable growth that began as follows: "DuPont is on a mission to achieve sustainable growth, which is defined as increasing shareholder and societal value while decreasing the company's environmental footprint" http://www.dupont.com/corp/overview/glance/sus_growth.html. Three years earlier, in 1998, the website offered the following DuPont commitment: "We affirm to all our stakeholders, including our employees, customers, shareholders and the public, that we will conduct our business with respect and care for the environment. We will implement those strategies that build successful businesses and achieve the greatest benefit for all our stakeholders without compromising the ability of future generations to meet their needs" http://www.dupont.com/corp/environment/commitment.html.

REFERENCES

Ackerman, Bruce A. (1983), *Reconstructing American Law*, Cambridge, MA: Harvard University Press.
Ackerman, Bruce A. (1991), *We The People: Foundations*, Cambridge, MA: Harvard University Press.
Ackerman, Bruce A. (1998), *We The People: Transformations*, Cambridge, MA: Harvard University Press.
Anderson, Benedict (1991 [1983]), *Imagined Communities: Reflections on the Origin and Spread of Nationalism*. Revised and expanded edition, London: Verso.
Anderson, Benedict (1994), 'Exodus', *Critical Inquiry* 20: 314–27.
Bartlett, Katherine T. (1990), 'Feminist legal methods', *Harvard Law Review* 103: 829–88.
Bauman, Zygmunt (1991), *Modernity and Ambivalence*, Cambridge: Polity.
Beck, Ulrich (1992 [1986]), *Risk Society: Towards a New Modernity*, London: Sage Publications.
Bijker, Wiebe, Thomas P. Hughes and Trevor Pinch (1987), *The Social Construction of Technological Systems*, Cambridge, MA: MIT Press.
Bimber, Bruce (1996), *The Politics of Expertise in Congress*, Albany, NY: SUNY Press.
Boyle, James (1996), *Shamans, Software and Spleens: Law and the Construction of the Information Society*, Cambridge, MA: Harvard University Press.
Bowker, Geoffrey C. and Susan Leigh Star (1999), *Sorting Things Out: Classification and Its Consequences*, Cambridge, MA: MIT Press.
Breyer, Stephen (1993), *Breaking the Vicious Circle: Toward Effective Risk Regulation*, Cambridge, MA: Harvard University Press.
Callon, Michel (1986), 'Some elements of a sociology of translation: Domestication of the scallops and the fishermen of St. Brieuc Bay', in J. Law (ed.), *Power, Action, and Belief: A New Sociology of Knowledge?*, London: Routledge & Keagan Paul, pp. 196–233.
Callon, Michel (1987), 'Society in the making: The study of technology as a tool for sociological analysis', in W. Bijker, T. Hughes and T. Pinch (eds.), *The Social Construction of Technological Systems*, Cambridge, MA: MIT Press, pp. 83–103.
Callon, Michel (1999), 'The role of lay people in the production and dissemination of scientific knowledge', *Science, Technology and Society* 4 (1): 81–94.
Cambrosio, Alberto, Peter Keating and Michael Mackenzie (1990), 'Scientific practice in the courtroom: The construction of sociotechnical identities in a biotechnology patent dispute', *Social Problems* 37: 275–93.
Carson, John (1993), 'Army alpha, army brass, and the search for army intelligence', *Isis* 84: 278–309.
Cohn, Bernard S. (1996), *Colonialism and Its forms of Knowledge*, Princeton, NJ: Princeton University Press.
Daemmrich, Arthur (1998), 'The evidence does not speak for itself: Expert witnesses and the organization of DNA-typing companies', *Social Studies of Science* 28: 741–72.
Darman, Richard G. (1978), The law of the sea: Rethinking U.S. interests', *Foreign Affairs* 56 (2): 381.
Dickson, David (1984), *The New Politics of Science*, New York: Pantheon Books.
Eder, Klaus and Catherine Kantner (2000), 'Nationale Resonanzstrukturen in Europa: Eine Kritik der Rede vom Öffentlichkeitsdefizit', in M. Bach (ed.), *Die Europäisierung nationaler Gesellschaften*, Wiesbaden: Westdeutscher Verlag, 306–31.
Ezrahi, Yaron (1990), *The Descent of Icarus: Science and the Transformation of Contemporary Democracy*, Cambridge, MA: Harvard University Press.
Fisher, William W. III, Morton J. Horwitz and Thomas A. Reed (eds.) (1993), *American Legal Realism*, New York: Oxford University Press.
Foster, Kenneth R. and Peter W. Huber (1997), *Judging Science: Scientific Knowledge and the Federal Courts*, Cambridge, MA: MIT Press.

Foucault, Michel (1971 [1966]), *The Order of Things: An Archaeology of the Human Sciences*, New York: Pantheon Books.
Foucault, Michel (1979), *Discipline and Punish: The Birth of the Prison*, New York: Random House.
Fried, Barbara (1998), *The Progressive Assault on Laissez Faire: Robert Hale and the First Law and Economics Movement*, Cambridge, MA: Harvard University Press.
Friedman, Thomas L. (1999), *The Lexus and the Olive Tree*, New York: Farrar, Straus & Giroux
Fukuyama, Francis (1992), *The End of History and the Last Man*, New York: Free Press.
Fyfe, Gordon and John Law (eds.) (1988), *Picturing Power: Visual Depiction and Social relations*, London: Routledge.
Gaskell, George, Martin W. Bauer, John Durant and N.C. Allum (1999), 'Worlds apart? The reception of genetically modified foods in Europe and the U.S.', *Science* 285: 384–7.
Gavaghan, Helen (1999), 'Britain struggles to turn anti-GM tide', *Science* 284: 1442–4.
Gibbons, Michael, Camille Limoges, Helga Nowotny, Simon Schwartzman, Peter Scott and Martin Trow (1994), *The New Production of Knowledge*, London: Sage Publications.
Giddens, Anthony (1991), *Modernity and Self-Identity: Self and Society in the Late Modern Age*, Cambridge: Polity.
Goldberg, Carey (1996), 'Shunning "he" and "she," they fight for respect', *New York Times*, September 8, Section 1: 24.
Goldberg, Steven (1994), *Culture Clash*, New York: New York University Press.
Goldmann, Kjell (2001), *Transforming the European Nation-State*, London: Sage Publications.
Grimm, Dieter (1995), 'Does Europe need a constitution?', *European Law Journal* 1 (3): 282–302.
Guha, Ranajit and Gayatri Spivak (ed.) (1988), *Selected Subaltern Studies*, New York: Oxford University Press.
Haas, Peter M. (1990), Saving the Mediterranean: The Politics of International Environmental Cooperation, New York: Columbia University Press.
Habermas, Juergen (1975 [1973]), *Legitimation Crisis,* Boston: Beacon Press.
Hacking, Ian (1983), *Representing and Intervening,* Cambridge, UK: Cambridge University Press.
Haraway, Donna (1991), *Simians, Cyborgs, and Women: The Reinvention of Nature*, New York, NY: Routledge, Chapman, and Hall.
Haraway, Donna (1997), *Modest-Witness@Second-Millennium.FemaleMan-Meets-Oncomouse: Feminism and Technoscience*, New York: Routledge.
Hardt, Michael and Antonio Negri (2000), *Empire*, Cambridge, MA: Harvard University Press.
Huber, Peter W. (1991), *Galileo's Revenge: Junk Science in the Courtroom*, New York: Basic Books.
Hughes, Thomas P. (1983), *Networks of Power: Electrification in Western Society, 1880–1930*, Baltimore: Johns Hopkins University Press.
Huxley, Aldous (1946), *Brave New World,* London: Harper & Brothers.
Ignatieff, Michael (1997), *The Warrior's Honor,* Toronto: Viking.
Jasanoff, Sheila (1998a), 'The eye of everyman: Witnessing DNA in the Simpson trial', *Social Studies of Science* 28 (5-6): 713–40.
Jasanoff, Sheila (1998b), 'Expert games in silicone gel breast implant litigation', in M. Freeman and H. Reece (eds.), *Science in Court,* London: Dartmouth, pp. 83–107.
Jasanoff, Sheila (1995a), *Science at the Bar: Law, Science, and Technology in America*, Cambridge, MA: Harvard University Press.
Jasanoff, Sheila (1995b), 'Product, process, or programme: Three cultures and the regulation of biotechnology', in M. Bauer (ed.), *Resistance to New Technology*, Cambridge: Cambridge University Press, pp. 311–31.
Jasanoff, Sheila (2001), 'Election 2000: Mechanical error or system failure?', *Social Studies of Science* 31 (3): 461–7.
Jasanoff, Sheila, Gerald Markle, James Petersen and Trevor Pinch (eds.) (1995), *Handbook of Science and Technology Studies,* Thousand Oaks, CA: Sage Publications.
Joss, Simon and John Durant (1995), Public Participation in Science: The Role of Consensus Conferences in Europe, London: Science Museum.
Kairys, David (ed.) (1990), *The Politics of Law*, New York: Pantheon.
Keck, Margaret E. and Kathryn Sikkink (1998), *Activists Beyond Borders: Advocacy Network in International Politics*, Ithaca, NY: Cornell University Press.
Kelman, Mark (1987), *A Guide to Critical Legal Studies*, Cambridge, MA: Harvard University Press.

Keyssar, Alexander (2000), *The Right to Vote: The Contested History of Democracy in the United States*, New York: Basic Books.
Kuhn, Thomas S. (1962), *The Structure of Scientific Revolutions*, Chicago: University of Chicago Press.
Latour, Bruno (1983), 'Give me a laboratory and I will raise the world', in K. Knorr-Cetina and M. Mulkay (eds.), *Science Observed,* London: Sage, pp. 141–70
Latour, Bruno (1987), *Science in Action*, Cambridge, MA: Harvard University Press.
Latour, Bruno (1988), *The Pasteurization of France*, Cambridge, MA: Harvard University Press.
Latour, Bruno (1990), 'Drawing things together', in M. Lynch and S. Woolgar (eds.), *Representation in Scientific Practice,* Cambridge, MA: MIT Press, pp. 19–68.
Latour, Bruno (1993), *We Have Never Been Modern*, Cambridge, MA: Harvard University Press.
Latour, Bruno and Steve Woolgar (1979), *Laboratory Life: The Construction of Scientific Facts*, Princeton, NJ: Princeton University Press.
Lessig, Lawrence (1997), 'Constitution of code: Limitations on choice-based critiques of cyberspace regulation', *CommLaw Conspectus* 5: 181–204.
Lewin, Tamar (2001), 'In genetic testing for paternity, law often lags behind science', *New York Times,* March 11, Section 1, 1.
MacKenzie, Donald (1990), *Inventing Accuracy: A Historical Sociology of Nuclear Missile* Guidance, Cambridge, MA: MIT Press.
Marris, Claire and Pierre-Benoit Joly (1999), 'Between consensus and citizens: public participation in technology assessment in France', *Science Studies* 12 (2): 3–32.
Masood, Ehsan (1999), 'Gag on food scientist is lifted as gene modification row hots up….', *Nature* 397: 547.
Miller, Clark and Paul Edwards (eds.) (2001), *Changing the Atmosphere: Expert Knowledge and Environmental Governance*, Cambridge, MA: MIT Press.
Mol, Annemarie and John Law (1994), 'Regions, networks and fluids: Anaemia and social topology', *Social Studies of Science* 24: 641–71.
Nowotny, Helga (2001), Re-thinking Science: Knowledge and the Public in an Age of Uncertainty. Cambridge: Polity.
Nye, Joseph S., Jr., Philip D. Zelikow and David C. King (eds.) (1997), *Why People Don't Trust Government*, Cambridge, MA: Harvard University Press.
Ogburn, William Fielding (1922), *Social Change with Respect to Culture and Original Nature*, New York: B.W. Huebsch.
Porter, Theodore M. (1995), *Trust in Numbers: The Pursuit of Objectivity in Science and Public Life*, Princeton, NJ: Princeton University Press.
Ratner, Sidney, James H. Soltow and Richard Sylla (1979), *The Evolution of the American Economy*, New York: Basic Books.
Reardon, Jennifer (2001), 'The human genome diversity project: A failure to coproduce natural and social order', *Social Studies of Science* 31: 357–88.
Sandel, Michael J. (1996), *Democracy's Discontent: America in Search of a Public Philosophy*, Cambridge, MA: Harvard University Press.
Sassen, Saskia (1991), *The Global City: New York, London, Tokyo*, Princeton, NJ: Princeton University Press.
Schauer, Frederick (1998), 'Internet privacy and the public-private distinction', *Jurimetrics* 38: 555–64.
Schuck, Peter (1993), 'Multi-culturalism redux: Science, law, and politics', *Yale Law and Policy Review* 11: 1–46.
Scott, James C. (1998), *Seeing Like a State,* New Haven, CT: Yale University Press.
Service, Robert F. (1998), 'Seed-sterilizing "terminator technology" sows discord', *Science* 282: 850–1.
Shapin, Steven (1994), *A Social History of Truth*, Chicago: University of Chicago Press.
Shapin, Steven and Simon Schaffer (1985), *Leviathan and the Air-Pump: Hobbes, Boyle, and the Experimental Life*, Princeton, NJ: Princeton University Press,.
Shiva, Vandana (1993), *Monocultures of the Mind: Perspectives on Biodiversity and Biotechnology*, London: Third World Network.
Shiva, Vandana (1997), *Biopiracy: The Plunder of Nature and Knowledge*, Toronto: Between The Lines.
Smaglik, Paul (2000), 'Tissue donors use their influence in deal over gene patent terms', *Nature* 408: 894–6.
Smith, Merritt Roe and Leo Marx (eds.) (1994), *Does Technology Drive History?* Cambridge, MA: MIT Press.
Smith, Roger and Brian Wynne (eds.) (1989), *Expert Evidence: Interpreting Science in the Law*, London: Routledge.
Stone, Christopher D. (1974), *Should Trees Have Standing? Toward Legal Rights for Natural Objects*, Los Altos, CA: William Kaufmann.

Storey, William (1997), *Science and Power in Colonial Mauritius*, Rochester, NY: University of Rochester Press.
Sunstein, Cass (1993), *The Partial Constitution*, Cambridge, MA: Harvard University Press.
Tribe, Laurence H. (1978), *American Constitutional Law*, Mineola, NY: Foundation Press.
Visvanathan, Shiv (1997), *Carnival for Science: Essays on Science, Technology and Development*, Delhi: Oxford University Press.
Walker, Neil (1996), 'European constitutionalism and European integration', *Public Law* Summer 1996: 266–90.
Wynne, Brian (1982), *Rationality and Ritual: The Windscale Inquiry and Nuclear Decisions in Britain*, Chalfont St. Giles, UK: British Society for the History of Science
Wynne, Brian (1988), 'Unruly technology', *Social Studies of Science* 18: 147–167.
Wynne, Brian (1989), 'Frameworks of rationality in risk management: Towards the testing of naive sociology', in J. Brown (ed.), *Environmental Threats: Social Sciences Approaches to Public Risk Perceptions*, London: Belhaven, pp. 33–45.

PART III

STS – EMERGENCE OF A FIELD

CHAPTER 9

PETER WEINGART

GROWTH, DIFFERENTIATION, EXPANSION AND CHANGE OF IDENTITY – THE FUTURE OF SCIENCE

INTRODUCTION

The beginning of a new millennium is an appropriate occasion to look at past predictions of what this transition was supposed to bring. Particularly pertinent among them is the one which, based on the extrapolation of its quantitative growth foresaw the 'end of science' (Price 1963). But the growth of science has also been considered the main factor in revolutionary social change and the transformation of industrial society to the *knowledge society* (Bell 1973). The growth of science, thus, evidently has triggered visions of doom and destiny. Price's extrapolations are, indeed, reason for concern given the centrality of the health and growth of science to the very foundation and future of modern *knowledge society*. However, extrapolations suffer from at least two kinds of simplifications: a linear concept of future developments and the assumption of self-identity of the units of analysis. Price was, of course, aware of the absurdity of extrapolated growth but he did not have the imagination to foresee what the future of science would look like. Thus, he and other observers proceeding from the same assumptions chose the pessimistic option of predicting stagnation and regression. To be sure, Price's calculations of the growth of science were by and large correct and have been corroborated many times since. But we are now in the position to see more clearly that as the growth of science has continued unabated science itself is changing. This is not surprising in itself. Growth is part of that change, but growth also creates internal structural change as well as a change of boundaries and, thus, change of identity. Identity here refers to the institution's characteristics and boundaries as seen from within and from outside. Looking at the growth of science provides a key to an understanding of its changing role in and its growing impact on society. But it cannot be assumed that this is a unilinear relationship. As science grows its growing impact on society has repercussions on itself.

The notion of *science* needs some specification. When speaking of *science* the term will be used in the encompassing meaning of *Wissenschaft*, i.e. including the social sciences and humanities and thus referring to the entire world of scholarship as it is organized in universities as the core institution. That being said, it is acknowledged that any reference to 'science' as such is obviously an unwarranted generalization hiding vast

B. Joerges and H. Nowotny (eds.), Social Studies of Science and Technology: Looking Back Ahead, 183–200

differences between the 'three cultures', even between different disciplines within each scholarly culture.

The purely quantitative growth of science, measured in the number of scientists or publications is obviously much too crude and undifferentiated to grasp the real meaning of the expansion of science. This will only become apparent if the many dimensions of growth, the causes and consequences on several levels, are explored. The look at the overall quantitative growth of science is an important first step to draw attention to potential problems. But it will be demonstrated that the quantitative analysis is only a first step pointing to the necessary sociological analysis of the interaction between the institution and its environment.

In the following section, the several dimensions of growth will be described. Since science is a peculiar communication system operating under the imperative of innovation, the analysis must focus on communication processes and their social structures. In the next section the relationship between growth and internal differentiation is analyzed, accessible primarily through the relation between knowledge production and the reception of knowledge. Here the focus is on the changing conditions of production and the resulting conditions of reception. The fourth section then looks at the repercussions of the material constraints of communication and their selective effects on the principle of publicity of scientific communication, on the general access to knowledge and on the internal mechanisms of quality control. The last section looks at the institutional identity of science whose changes are more important, it is claimed, than quantitative growth as such.

EXPONENTIAL GROWTH AND ITS END

In his 1963 *Little Science, Big Science* Derek Price had made some calculations about the growth of science which are crucial for the understanding of the special role of science as a social system:

- the growth of science exceeds the growth of the population (in the countries observed) by far. Price calculated a threefold doubling rate (Price 1971, 14).
- 80-90% of modern science is contemporaneous, i.e. 80-90% of all scientists who have ever lived, live now. Since this growth has been going on ever since the eighteenth century, the perception of temporal immediacy in science is almost as old as science itself. This means that changes in the present system of science *cannot* be explained by growth alone.
- with a doubling rate of ca. 15 years science has grown since its beginnings by five orders of magnitude. Thus, Price saw already for the turn of the millennium a transition to a slower growth rate since the consequence of continued exponential growth was absurd: that by then two scientists would count for every man, woman and dog in each respective population (Price 1971).

When growth reaches the capacity limits of a system under observation this leads inevitably to either collapse or a new definition of its constitutive variables, and thus to a new identity. But before we deal with these signs of crisis and/or change we have to look at the different patterns of growth and their structural effects.

The product of scientists are the research results which, unless they are employed in industry or military research, they publish. Thus, the growth of science can be measured approximately in number of publications, especially with reference to the volume of communication and the growth of knowledge. But the law of logistic growth does not only apply to publications but also to the number of scientists.

Probably the most comprehensive study of the development of the humanities (excluding the social sciences: sociology, psychology, economics) in the former Federal Republic of Germany showed for the period between 1954 and 1985/87 an overall growth of this area of *Wissenschaft* by a factor of seven. This exponential growth is found for professorial positions as well as for all positions in this sector. However, this growth rate levels off markedly in the 1980s (Weingart et al. 1991, 77f.). The humanities' share of the whole science system declined during this period by ca 6% (from 14,8 to 8,9) which means that their growth rate was visibly lower than that of the rest of science (ibid., 94f.) The respective development can be shown for the expenditures as well. And the same pattern prevails for publications. It corresponds roughly to manpower and also shows a growth across disciplines of a factor of seven to eight over the three decades 1954–84.[1]

There is an obvious correlation between publications and the number of scientists because in spite of individual variability there are limits within which the capacity to publish can only vary. But the connection is influenced by a number of factors. Price has estimated that of 100 publishing scientists in every year, 33 are novices. According to Lotka's law only a much smaller number continues publication activity at the research front over an extended period of time.[2] This 'demography of science' is similar to that of underdeveloped societies in which both the birth- and mortality rates are high and the majority of the population consists of children. The dynamics of the process are determined by those individuals who are qualified for research.

The obvious assumption of a linear relationship between the number of scientists and the number of publications must be also qualified for institutional reasons. Only a certain percentage of scientists publishes. While the majority of scientists is employed in industry (ca. 70%) of whom for reasons of competition related secrecy or application oriented research only a small percentage publishes (ca. 10%) the large majority of scholarly publications comes out of the universities and other research institutions (ca. 80%).

In addition to the relationship between the publication volume and the number of scientists another relationship is that between scientific manpower and *R&D expenditures* (measured in percentage of GDP). The development of financial support for an exponentially growing science is not only interesting as an indicator of growth but as an *indicator of public acceptance* of science and the *need for legitimation* on the part of science. The expenditures for universities and research institutions are public investments in the system of knowledge production and education. These investments need to be politically legitimated. The stronger the competition with other societal objectives, the more difficult it is to legitimate funds for research. This is all the more true if, as in many seemingly esoteric areas of basic research, their general utility for a broader population is difficult or impossible to demonstrate. The financial resources and the political legitimation which need to be mobilized for their allocation set limits to the exponential growth of science.

Data of R&D expenditures over longer periods of time are unreliable because they are not available and/or the units of calculation change over time (Lundgreen et al. 1986, 242). Regardless of these imprecisions they may be taken as a rough indicator of a comprehensive development.

The best comparability and estimation of the legitimation needed is given by calculations in percentage of the GDP. It rose in the Federal Republic from 1.5% in 1963 to 2.85% in 1989, i.e. a doubling within 25 years (OECD 1984; 1990). In the USA the same figure rose from 0.2% in 1929 and 0.7% in 1940 to 3.1% in 1965. The stagnation of the relative expenditure figure is evident, it fell back to 2.3% in 1977–79 and stabilized in 1985–89 at ca. 2.8% (OECD 1984, 1990; DiTrocchio 1994, 92).[3] The NSF sees a moderate decline for the R&D/GDP indicator for all G7-countries to 2.3%. 2.8% of the GDP seems to be a magic limit for modern mass democracies beyond which expenditures for science cannot rise. Vannevar Bush's slogan of *Science, the Endless Frontier* which was realized by exponential growth especially in the decades after World War II, now seems to have come to a (material) limit.

One can also assume that the *number of scientific institutions*, i.e. universities and research institutions reveal similar patterns of growth. A calculation of the number of co-existing universities reveals a logistic curve which since the seventeenth century at the latest roughly follows population growth (Wagner 1985, 83). But these numbers are too small to be taken too seriously, and comparability is limited.

Obviously, the logistic growth curve is a simplified idealization of the actual development which differs from discipline to discipline, from country to country, and with respect to historical phases. Evidently, a particular growth in many fields has occurred after World War II. R&D expenditures reached a peak between the 1960s and 1980s. This is a period of the opening of higher education to the broader population in all industrial countries and the resulting wave of new university foundations. Price calculated a doubling time of 10–15 years for published articles in all sciences. This rate differs from field to field and has declined in many areas since World War II. The geosciences show a doubling time of ca. eight years, astronomy of only 18 years. The growth of mathematics is estimated to have a doubling time of 20 years between 1870 and 1994 which after World War II had risen to 10 years. In the meantime it has fallen back to 20 years (Odlyzko 1995, 4). Chemistry may serve as a last example. Schummer has taken the number of chemical substances as an indicator, a supposedly more precise measure of actual growth of *knowledge*.[4] For this indicator he calculates an average annual growth of 5,5% since 1800 which corresponds to a doubling every 12,9 years. During the past two decades this development corresponds almost exactly to a logistic growth curve without any sign of saturation. In contrast to other disciplines this growth has not been significantly influenced by wars. The temporal replacements are caught up and do not lead to sustained replacements of the growth curve.

Schummer takes this as a reason to claim that growth is not determined by external resources but by internal methodological conditions only (Schummer 1997, 116f.). That claim is too strong, however, since in principle the science system cannot operate without resource allocations from outside. For the part of science which is not bound to the market and financed by marketing the knowledge products, this condition is inadvertable. Thus, the proportionality of the growth of publications, manpower and monetary expenditures is a given. A contraction of expenditures must inevitably lead to a deceleration or even an end of growth. This is also true for the chemists.

However, the study on chemistry is interesting nonetheless because it is an illustration (though perhaps not generalizable) that the research process is, in principle, without limits and internal stoppage rules. Contrary to the intuitive notion that research into the laws of nature must ultimately come to an end, the opposite is true: new discoveries generate new questions. Schummer shows that the exponential growth of chemistry is generated by the methodological approaches themselves. The combinatorial approach according to which known substances are combined in order to obtain new substances leads to (when limited to a pair each) exponential growth.[5] Even if this relationship is not the same in other disciplines it holds in principle: the more resources invested in science, the more scientists are educated and employed in research, the more knowledge they produce, and the more new questions are being put to research. With respect to researchable problems the *frontiers are endless*, indeed. The process is self-sustaining, it even accelerates, provided it is fed with resources from outside. It is precisely this circumstance which has driven the enormous growth of the past 350 years.

GROWTH, EXPANSION, AND DIFFERENTIATION

Exponential growth inevitably has grave consequences for science proper, i.e. for its social structures, its organizations, its communication processes and the rules which determine them, as well as for its environment, i.e. the societies in which it is institutionalized and which support it. By taking a closer look at the mechanisms which effectuate *internal differentiation,* the impact of growth can be better understood.

The focus of this analysis has to be the communication process in science, i.e. the *production* of publications, which stand for *discoveries* or *knowledge* respectively, and the *reception* of these publications, i.e. their processing, their critical assessment and their further development. The crucial question is if the *relation between producers of knowledge and recipients of knowledge* has changed in a way which is relevant for the communication process. This is at the same time the question about the relevant structures of that process. The common thesis that growth has led to an unmanageable flood of literature assumes that this relationship has changed in some fundamental way. Thus, both the products and forms of their processing as well as shifts of the balance between them have to be looked at.

The core *product* of research in the sciences is the *article* published in a scholarly journal. The article has remained largely unchanged as the typical form of scientific publication since the early nineteenth century. In the humanities for a long time the monograph was dominating. In the meantime here, too, the article gains ground as the prevalent form of publication (see Weingart et al. 1991, 284–5). Discoveries which are not published in an article or any other form do not exist for the scientific community, which means they have not been certified. The condition of this *monopoly on truth* of institutionalized science is that the discoveries contained in publications circulate within the community and are received.

The *forms* and functions of publications have changed little in the course of the history of modern science, but the *volume* of publications has, with consequences for their reception. The growth of the number of journals is exponential, as expected, but regardless of all methodological problems involved in their exact counting, the growth rate is smaller than that of publications and manpower. Price estimated a doubling time of 15 years based on new foundations only; Carpenter and Narin estimate a much longer doubling time of 57 years for the natural sciences between 1919 and 1973 (Carpenter and Narin 1980, 59). In 1830 there were approximately 300 scientific journals, by now there are roughly 40.000 in the whole world.[6] By 1830 at the latest, the point was reached that scientists could no longer absorb all the knowledge that was relevant to their work (Garfield 1987, 4). At this time still half of all journals were devoted to general themes in science. 50 years later it was less than a third. The emergence of disciplines and specialization had begun (Gascoigne 1985, 133).

The common thesis that the flood of literature inundates the solitary scientist must be qualified to some extent at this point. On the *side of the reception* the first reaction to the growth of the number of journals (and the number of articles contained in them) was the establishment of 'abstract'-journals – according to Price when roughly the first 300 scientific journals existed. 'Abstracts' increased the processing capacity of scientists considerably. The number of 'abstract'-journals follows that of journals in linear fashion, i.e. ca. one 'abstract'-journal for every 300 journals (Price 1975 (1961), 215f.; Garfield 1987, 4). Meanwhile there 'abstracts of abstracts'-journals, thus, yet another reduction of the information volume that has to be processed.

The creation of 'abstract'-journals could lead to the impression that *individual reception* of scientific information would include all available information. That has not been the case anymore since the beginning of the nineteenth century. Common talk about the 'flood of literature' also conveys the impression that a constant number of scientists is faced with a growing volume of literature they cannot master, which outruns the system of quality control, favors fraud, and distorts the diffusion of important discoveries (Broad 1988, C1, C!!, cit. in Garfield 1988, 5; Garfield 1991, 2).

A simple arithmetic relation reveals a disturbing paradox. If a scientist reads x articles and writes y articles, the average article will be read by x/y researchers, independently of the overall size of the community of scholars (Odlyzko 1994, 5). This is a simplification, of course, assuming that the number of scientists remains constant for the time. But the effect is evident: an increase in productivity would lead to a decrease of the attention for each article.

There are two mechanisms with which the communication system of science can react inwardly to growth: 1) structuration, i.e. hierarchization of attention, and 2) internal differentiation, i.e. specialization. The 'flood of literature' does not hit science like a huge wave which drowns all scientists equally.[7] A part of the entire publication volume simply remains unread. *Selective attention* is the structuring principle in scientific as in all other communication. The citation is the most obvious indicator for the fact that a scientific publication has not only been read but its information has also been processed and has become an identifiable unit of scientific communication. Publications which are not cited are de facto lost for that process. More than half of all articles are never cited. In support of the 80/20 rule it has been found that 21% of the 4400 (900) journals listed in the 1988 SCI drew 83% of 8 million citations which were processed in the same year for the Journal Citation Reports (Garfield 1989, 7). With this, SCI justifies its claim that 500 journals are enough to trace the most important scientific publications (Garfield

1990, 6). This is the best illustration of the high *selectivity of perception* which operates in scientific communication.

The consequences of this selectivity on attitudes of scientists in research and publishing are almost entirely unknown and are subject of speculation. The common assumption that the flood of literature disorients the scientist and overstresses the quality control system does not take selectivity into account and can therefore be discounted as false (Broad 1988). The real issue is the perception of selectivity itself, i.e. the motivational consequences for scientists to take part in an enterprise in which it is extremely difficult to climb into higher ranks, not only because of the competition of numbers and the differences in infrastructure but also because of the differentials in attention which can only partly be explained with science-related quality criteria. Basically two reactions are possible: an intensification of competition, and a demotivated retreat into postmodern cynicism as predicted by Forman (Forman 1997).

The second mechanism with which science reacts to the growth of literature is internal differentiation, i.e. specialization. It can be interpreted as the result of strategic action which can be observed in organizations and in markets: if the number of competitors and thus the intensity of competition increases, the incentive for differentiation by looking for 'niches' and/or by specializing and narrowing the area of competency also rises. The operation of this mechanism is, of course, also dependent on the attraction of alternatives and the availability of resources (see Whitley 1984, 109).This explains why research fields of very high practical relevance like biochemistry which obtain a high amount of public funds attract a greater number of researchers and are highly competitive.

The example of the field of English Language and Literature (*Anglistik*) in the already cited study on the humanities in Germany illustrates most clearly why and how the increase of the number of scholars and their publications must lead to *differentiation.* In 1954, 24 professors in *Anglistik* published twelve books and a smaller number of articles. This body of literature could easily be digested by all professors and their assistants and students. Thirty years later 300 professors of *Anglistik* published ca. 60 books and 600 articles, an amount which could not realistically be read by one individual. Internal specialization of the field is the inescapable result. With it the meaning of the denomination of the field also changes for the individual scholar. "The disciplinary denomination as a frame of reference for identification and orientation must have become more abstract and been replaced by internal, narrower limits. That is the supposition emerging from looking at the connection between growth of manpower and publications" (Weingart et al. 1991, 288). The field which in the mid-1950s could still be seen as a discipline has within thirty years become a mass conglomerate of specialties. The members who consider themselves as belonging to the discipline do not perceive themselves personally nor do they overlook the literature published in the field.

This *loss of unity* can also be illustrated with another indicator: the concentration of articles in journals. The greater the number of journals in which scientists of a field publish and the smaller the share of each journal of the entire volume of publications in a field, the more diversified, and thus specialized is the communication. Again the case of *Anglistik* can serve as an illustration: In 1954 professors published 10 articles in six journals, half of them in the same journal (*Anglia*). The remaining five articles appeared in five other journals. In 1984 *Anglia* contained merely 11.8% of all articles published.

314 articles appeared in 135 journals. (The same process of diversification can be demonstrated for the other disciplines analyzed. Weingart et al. 1991, 293ff.).

All this means that the *relationship* between the volume of publications and the receptive capacity of scientists, i.e. their attention is relatively stable. A growth in publications is largely due to a growth in numbers of publishing scientists. Only higher productivity, for example by writing shorter papers, leads to an imbalance, i.e. decreasing attention equal to higher selectivity. It may also lead to higher competition which is a cause of differentiation. But growth in numbers of scientists and volume of publications makes it impossible for the *individual* to maintain a command over a previously manageable amount of literature. Thus, growth inevitably leads to differentiation and specialization.

If growth is tantamount to internal differentiation and specialization, then the question arises whether there are *infrastructural* and *regulatory limits* which this growth cannot surpass and/or if in view of an unlimited differentiation, the *identity of science* as an institution (and our concept of it) must change. These two possibilities will be examined in the following sections.

GROWTH, PUBLIC COMMUNICATION AND QUALITY CONTROL

Garfield's seemingly provocative thesis that 'the flood of scientific literature is a myth' is justified insofar as selectivity and specialization counteract the effect for the individual scientist. It disregards the overall effect of this development which can bring the system as a whole very well into crisis. The obvious signs for this are the discussions on the 'crisis of the scientific communication system,' the 'crisis of the classical journal,' and on the 'future of the scientific library' as well as the reactions of scientists to the new electronic media to publish and store scientific works (Holden 1987; Grötschel and Lügger 1995; Stix 1995; Odlyzko 1997). The crises thus diagnosed have two aspects: 1) The scientific libraries run up against *economic* and *infrastructural* limits with respect to their objective to represent all literature because of the absolute growth of the number of journals; (Hyperspecialization, also called 'twigging' is manifested above all in the founding of ever new journals with small audiences and uneconomic circulation figures). This limitation may threaten the access to scientific communication and, thus, the principle of publicity; 2) The new computer- and information technology (CIT) opens alternatives with respect to both storage and speed of publication which promise a solution to growth induced problems without the need to limit communication; However, the problems related to electronic media, i.e. quality control and maintaining property rights have so far impeded the transition from classic paper to electronic publication.

In any case, if fundamental changes can be expected in science as a consequence of growth they will occur in the control of the 'appearance of truth', i.e. in the self-regulation of science and its mechanisms of demarcation. (In the next section I will discuss the analogical expectation with respect to the identity of science). The discussions on the limitations of libraries will be ignored here (see Odlyzko 1994; Grötschel and Lügger 1995; Stix 1995). Instead I will focus on the issues arising with electronic journals only.

The pressure for electronic journals comes from two sources: the economic and physical limitations of libraries (storage) and the economics of journals as well as their limitations of speed. As promising as the new technologies may be for solving the crisis of the journal, as obvious are the problems inherent to them. Their discussion points to the institutional changes that can be expected if the system adapts to the self-induced crisis. One issue is that of intellectual property rights, the other one is the securing of quality, i.e. *peer review*. Both problems refer to central mechanisms of the scientific communication system, the sustained change of which would have far reaching consequences for science as a whole. Even radical proponents of electronic journals such as Steven Harnad who sees the future in a new form of interactive publishing (*scholarly skywriting*) and who considers the communicative potential of electronic nets revolutionary admits that the net has to remain traditional in only one sector: 'in the validation of scientific ideas and discoveries through peer review' (Harnad 1996, 105). Harnad rightfully claims that there are no *essential* differences between electronic and paper media with respect to peer review. In contrast to paper journals the net would offer the possibility of a more equitable diffusion of the load of the review process by searching reviewers in electronic literature surveys, citation analyses or 'calls for reviewers' in respective 'bulletin boards'. In connection with his vision that the speed of electronic publishing would be half way between paper-mediated interaction and oral dialogue, this scenario points to yet another augmentation of the entire communication volume by several orders of magnitude (Harnad 1996, 106). In addition to increased speed of publication and augmented space, the proponents anticipate a democratization of the publishing process in view of a far more efficient covering of geographical distances and a hoped-for decentralization (OECD 1998, 23).

But the question remains whether the increase of communication opportunities will really lead to an analogical broadening of the participation in communication or if the same structuring will have to take place through the limitation of reception, the emergence of reputational hierarchies and social networks. The crisis of the traditional journal results from the connection between growth and economics, and *not* from the relation between the 'flood of literature' and the limited reception capacity of the scientists. The same applies to the selectivity of quality control. The higher processing capacity of the electronic media will possibly draw a larger number of scientists into the communication process, it may soften the structure of center and periphery, and it may broaden or at least stabilize the accessibility of communication in science. Thus, the electronic future of scientific publishing may push the infrastructural limits of the system 'forward' yet another time and thereby allow it to grow even more. But that is precisely the reason why in all probability it will not prevent but rather contribute to a change of the institutional borderlines and thus the identity of science.

GROWTH AND INSTITUTIONAL IDENTITY OF SCIENCE

The previously determined connection between quantitative growth and differentiation already signals that the concept of an institution of science remaining self-identical over time is misleading. The global measures of quantitative growth hide in more than one respect the differential developments between disciplines, research fields and organizations, cultures and countries (Ziman 1994, 81). Above all they suggest that *science* always means the same. Quantitative growth does have obvious consequences both on the level of intellectual contents of the knowledge produced (e.g. the subject matters of research, the methods of their analysis) and also on the level of social organization (e.g. the demarcation of disciplines and specialties). Price's diagnosis that exponential growth must come to a limit remains correct in principle. But it does not say anything about the categorical adaptations which determine the moment in time when that is to happen.

The transition of the system of classifications of sciences of the 16th to the 18th centuries to that of disciplines of the 19th and 20th centuries is a first historical example that the system reacted to communicative overload with a change of epistemic objectives and their social organization. The classification techniques of descriptive natural history reached their limits at the latest at the end of the 18th and the beginning of the nineteenth century. This crisis of growth, triggered by the pressure of empirical data on the classification systems forced about a shift to techniques of temporalization, i.e. evolutionary thinking as a new form of systematization (Lepenies 1976, 16ff., 52f.).[8] Science of the eighteenth century did not know disciplines in today's sense of the term. The demarcations of areas of knowledge were ordering schemes, places to 'store secured knowledge' and not yet the "focus of actual efforts of a disciplinary community held together by common problems" (Stichweh 1984, 12). The important ordering principle, in contrast to the unimportant disciplines, was that of hierarchies of areas of knowledge. In German science this was Christian Wolff's distinction between history, philosophy, and mathematics, based on the methods of gaining knowledge. The dissolution of this order occurred with the constitution of a particular empirical experience through experimental physics, the specification of a subject matter and a problem definition as well as the renunciation of comprehensive description. *Growth, differentiation* of science as an institution from a non-scientific environment, and *internal differentiation* into a multitude of specialized disciplines go hand in hand. The emergence of disciplines in the modern sense is the reaction of the scientific communication system to its growth-related overburdening (Stichweh 1984, 28f., 40).

One direction of internal differentiation is greater abstraction, i.e. mathematization. The result of this process is that science is less and less concerned with immediate factual and experience-based knowledge and its categorization but increasingly reorganizes and reconstructs this knowledge under its own perspective. Concepts, theories, instruments and models mediate experience and render it to a growing degree *constructed.*

The second direction of internal differentiation is directly related to the first. It is the expansion of scientific investigation to ever new subject matters. The 'concreteness of the subject matter' which characterizes the beginning of every disciplinary development, is replaced by problem perspectives specific and constitutive to a discipline which can then be applied to ever-new subjects (Stichweh 1984, 47f., 49). This self-referential multiplication of subject matters and their scientific analysis is the mechanism of *specialization* which constitutes a centrifugal force within science.

The more abstract and generalizable the methods and instruments of knowledge production are, the more effective their application to new subjects, and the more numerous the possible linkages to already existing disciplines and their subject matters.[9] This growth-driven process of specialization of science can also be seen as the motor of a ubiquitous *scientification.* As long as science grows the abstraction of modes of gaining knowledge and the expansion of subject matters will continue to operate as interdependent mechanisms.

Internal differentiation of science and scientification of society as the 'external' impact of science on its societal environment continually change the identity of science compared to the form it had found in the nineteenth century. Differentiation leads to the *loss of the concept of the unity of science*. The process of scientification is the result of an *expansion of the institutional demarcations* of science and leads to *new forms of coupling* with other social subsystems.

In a brief comparison of the ideologies of science held by the academies in France, England and Germany at the end of the 18th century Daston states that nowhere else was the striving for the unity of science stronger than in Germany, and nowhere was it associated more strongly with the unity of the academy than in Berlin. During the second half of the 19th century Leibniz was hailed as the 'one-man-embodiment of the unity of science and at the same time as the warden of pure as opposed to applied science.' In contrast, Theodor Mommsen deplored the loss of the 'one master' of science who could rejoice over the work of many 'journeymen' (Daston 1997, 5, 18). In similar fashion his fellow academician Emil Du Bois Reymond diagnosed the division of labor among scientists and the speed of development which had been reached as a consequence: "Renouncing high glory thousands of busy workers produce countless details daily, carefree of inner or outer perfection, only concerned to attract attention for a moment and to obtain the best price for their commodity". The consequence of this development: "The stream of discovery bifurcates into ever more numerous, ever more irrelevant creeks" (Du Bois Reymond 1886, 450). Twenty years earlier Herrmann von Helmholtz had made the same observations: "Who is to overlook the whole, to hold the threads of connectedness in hand and to find one's way? The natural consequence becomes evident in that every individual researcher is forced to choose an increasingly smaller field his own workplace, and can maintain only incomplete knowledge of neighboring fields" (Helmholtz 1896, 162). The dynamics of specialization is well expressed by the philologist August Boeckh who declared in 1885 that "no problem was too small not to be worthy of serious scientific analysis" (cited in Daston 1997, 20).

This specialization into disciplinary research fields had its impact on the role of the scientist. In the context of the classificatory, hierarchically structured science of the 18th century the professor was a 'scholar' (*Gelehrter*) who wrote textbooks and compendia and strove to master several areas of knowledge during his career. With the newly emerged disciplines at the beginning of the 19th century, *originality* and the *discovery of the novel* had been established as primary values. The respective 'publics' of the disciplines to whom publications were addressed were defined in much narrower terms: the small community of specialists replaced students and *Bildungsbürgertum* (Stichweh 1984, 11; Daston 1997, 30). In view of the division of labor in research, the multitude of highly specialized 'journeymen' and the deplored loss of an encompassing communication linking all of science as it was attempted by the academies, the role of a universal genius such as Leibniz had become a thing of the past. This role was not only rendered impossible for communicative but above all for social structural reasons. The small number of outstanding heroic figures in the history of science had been replaced by an army of workers. This was an expression of the leveling of status hierarchies and of differentials of societal visibility, i.e. of *democratization*.

The academy as the institution representing the unity of science was organizationally overcharged by this development. Thus, specialization also expressed itself in the founding of a multitude of disciplinary and subdisciplinary scholarly societies (by 1887 there were 892 such groups in Germany alone) which undermined the preeminence of the academies, not least because they operated on the model of the academies. Research emigrated *de facto* from the academies into the universities where it found a new home. With it the specialized and faster journals of the scientific societies severely constrained the importance of the academies' publications (in 1900 there already existed 1258 journals for mathematics and the natural sciences in Germany) (Stichweh 1984, 69, 73; Daston 1997, 19).

This change of the identity of science – from the unified science of the academy and its universally erudite scholars to the disciplinary, university based research of specialists organized in a massive division of labor – as it came to be expressed in the nostalgic formulations in the last third of the 19th century echoes into the present. The attempts to find a unity in the fractured landscape of disciplines and to harness its integrative powers have been continued ever since. Du Bois Reymond had identified synthesizing principles which, in his view, connected the diversified specialties such as the conservation of energy or evolution (Daston 1997, 22) At the end of the 1930s the 'unity of science' – movement formed itself around Rudolf Carnap, Otto Neurath, Betrand Russell, Niels Bohr and John Dewey. It was explicitly defined in line with the tradition of the French *Encyclopédie* and operated on the expectation that the 'mosaic of empirical science' could be assembled with the help of logical empiricism Neurath et al. 1938 (1970)). This attempt at unification has been the implicit basis of the discourse on interdisciplinarity that reached its first peak at the end of the 1960s. Here again the complaint was directed against departmentalization and specialization of science, against a structure, which supposedly did not correspond to the practical problems of society and created more open questions than it answered (Jantsch 1972). Since the mid-1980s the reality of disciplinary specialization seems to be regarded as irreversible. At least the authors of the tome *Interdisciplinarity Revisited* come to the conclusion that the concept of interdisciplinarity had lost its momentum. Departments and faculties of universities as their most important organizational units not only had survived but gained in strength (Levin and Lind 1985, 9).

Since then the concept of interdisciplinarity as a reductionist promise of a (re-)unification of the disciplines in a distant future has been replaced by the programmatic conjuration of interdisciplinarity as an innovative form of research organization. The promise is that of innovation by linking the variety of disciplines. The paradox of the increased intensity of the call for interdisciplinarity in light of continued specialization can be explained.

> At the same time that the heterogeneity of science becomes the primary experience within science and the prevailing perception of science from the outside, the unavoidable transformation of the traditional order of knowledge is processed in particular fashion: with the (futile) attempt at mediating the old with a new order not yet found (Weingart 1997b, 596).[10]

The potential long-term impact of this development is evident. The transition from the integrative concept of the unity of science to the weaker concept that views the unity of the heterogeneous disciplines only in their cooperative links certainly takes the diversification of disciplinary cultures into account.[11] For the time being an orientation to university science still encompasses all disciplines because the university still is the core institution of science. But once the universities no longer represent all disciplines and begin to specialize themselves, a change of the internal identity of science similarly fundamental and far-reaching as that at the beginning of the 19th century can be expected.[12] Helmholtz raised a similar question already in view of the differentiation into the four 'faculties': "Is the unification of the four faculties in one university only a remainder of the middle ages?" (Helmholtz 1896, 166).

One scenario of the future development of science, thus, is that different areas of knowledge, e.g. the humanities, the social sciences, bio- and medical sciences become institutionally independent and separate so that there will be no longer a common orientation to science as a common institutional framework. Such a differentiation would probably explode the epistemological concepts on which the fields are based and which already differ considerably.

The change of the *external* identity of science occurs in parallel to internal differentiation and is propelled by it. Internal differentiation is tantamount to its *expansion*, i.e. the capture of ever more areas of experience by the analytical power of science. The differentiation into disciplines, subdisciplines and highly specialized research fields is not only the partition of an already demarcated territory of knowledge into ever smaller claims. It is also the expansion of the scientific perspective to areas of experience which hitherto lay outside its reach. In part this development is connected to attempts at professionalization of certain occupations, i.e. their academization. But even independently from this and preceding it, realms of experience and action are *scientized.* A particularly pertinent example is the entire complex of individual behavioral control, control of feeding habits through dietary information, control of reproductive behavior through hereditary information, and control of social behavior and individual psychic disposition through psychological information and scientized therapeutic techniques (Weingart 1983; Maasen 1998). In the wake of this development, a plethora of research fields has sprung up – science of nutrition, human genetics, research and therapy of human sexuality, epidemiology, to name but the most important. As these fields are all directly related to issues of human health and ethical values the line between 'pure' science and societal values can no longer be drawn unambiguously.

A second illustrative area are the environmental sciences which have expanded from ecology to social science based risk research and to environmental psychology. The common characteristic of this far-reaching research area is also the inescapable political, economic, and societal evaluation which enters into the conception of subject matters and the interpretation of results (Funtowicz and Ravetz 1993; Gibbons et al. 1994). Both areas, environment and energy, are considered paradigmatic for a new type of science christened with fashionable terms such as 'post-normal science' which is characterized by proximity to politics, sensitivity to legitimation and dependence on knowledge. Wholly aside the question as to which degree of generalizability is granted these diagnoses, it can be concluded that all the new research areas are indicators of a progressive *scientification* of society (Weingart 1997a).

Scientification, like specialization, also has repercussions on the identity of science. Scientific knowledge becomes an increasingly indispensable instrument of problem solving in practically all areas of social activity. Among other things this is expressed in the characterization of science as the "direct means of production" and the resulting treatment of "science as a commodity" (Richta et al. 1968; Gibbons and Wittrock 1985;

Stehr 1994, 214ff.). Still more current is the emergence of concepts like 'information-societies' or 'knowledge societies' (Stehr 1994). These descriptions reflect the stronger interrelations or *closer couplings* between science and other social systems. As scientific knowledge gains a ubiquitous instrumentality and a correspondingly wide use, the expectations regarding this instrumentality retroact 'internally': knowledge production orients itself to the market, to the media, to politics. In the context of this mutuality of expectations, *scientification* always means *economization, politicization* and *medialization of science.*

SUMMARY

The *growth* of science, and thus its described development can be summarized: it must be differentiated into several processes which change its structure in order to come to an adequate understanding of the *repercussions of growth on the change of identity of science as an institution.* Continued *specialization* and concomitant differentiation of subject matters, methods, specialized languages and research cultures leads to a *loss of internal unity of science.* This loss of unity has its critical threshold in the university as that organization which represents and comprises the institutional clamp of disciplinary variety. If the universities begin to specialize on a relevant scale, that will be the end of a concept of the unity of science.

The *expansion* of science driven by specialization leads to a shift of the institutional borderlines. The demarcations of science are pushed 'outward' into areas which used to lie outside the realm of legitimate concern: politics and the economy are scientized, ethics, moral and social values become objects of scientific discourses. With this the line between 'pure' and 'applied' research is blurred.

Derek Price's partly scary, partly absurd prognoses according to which shortly after the turn of the millennium there would be two scientists for every man, woman and dog did turn attention to the obvious limits to exponential growth of science. But they did not take the flexible identity of science as an institution into account. Rather, like almost all prognoses of the growth of scientific literature and the limits to processing it, they rest on monolithic and a-historical definitions of science not to mention the problematic nature of data) (Ziman 1994, 67ff.). Nonetheless, there can be no doubt that the changes of identity of science which were analyzed above are the outcome of a growth process which cannot be unlimited. Growth and change of identity go hand in hand.

There is a lesson from this for science studies and the sociology of the sciences. In a sense Price's analysis, notwithstanding its oversimplification, had a predictive value that is probably better than that of many other perspectives on the future. It predicted an end to a particular pattern and left it open to speculation what the consequences would be for the system of science. In another way it revealed the typical flaws of extrapolation. It also testifies to the power of quantitative analysis as much as it reveals its limitations, and, thus, it demonstrates once again that there is little wisdom in playing out one against the other. During the last decade of the past century a host of analyses of the fundamental changes of the institution of science have appeared. Some filled in the gaps left by Price with informative detail. Others dared making new predictions, sometimes crossing the boundaries into the world of fiction and ideology. The former may have a chance to be remembered as sound scholarship. The latter will be shelved as innovative and provocative *Zeitgeist*-commentary that has stimulated discourse and research. Apparently both are essential elements for the field of science studies to progress and leave lasting traces at the same time.

Meanwhile, the evolution of science as a social institution continues. If we are not deluded by presentness we bear witness to an epoch in which fundamental changes take place in the order of knowledge. Science studies will have to focus not only on the institutional changes but the greatest challenge may lie in making sense of the diffusing identity of what used to be clearly delineated as 'science'. Ultimately, this may force the field to commit a shift of category: Instead of focusing on 'science' it may be necessary to take the interaction of 'knowledge systems' into view. But that will be the theme for another Yearbook!

University of Bielefeld, Germany

NOTES

1 The numbers are given in the study cited. They vary between the fields analyzed, i.e. German Language and Literature (Germanistik), English Literature (Anglistik), and history depending on the data bases used. Truly exact numbers do not exist for any of the three indicators.

2 Lotka's law says that for every 100 authors who publish one article in a given time period, 25 publish two, 11 three etc.

3 The American NSF gives slightly different figures. According to it, R&D expenditures have never reached 3.1% of the GDP but only 2.7% in 1991. For the Federal Republic the highest number was 2.9% at the end of the 1980s. Numbers vary depending on whether expenditures for defense research are added or not. All numbers are R&D total expenditures, i.e. including industry for which the legitimation argument does not apply (NSF 1998, 4-37).

4 This procedure is only possible for sciences which have similar codified, cumulative classification systems as a research task. It is also questionable if the data collection can reach the same degree of exactness as in the case of publications (Schummer 1997).

5 Schummer mentions two more approaches, the analogical and the feedback approach which do not reveal the same simplicity as the combinatorial approach that is exemplary for the mechanism (see Schummer 1997, 119ff.).

6 Calculations about the number of journals are inexact and contradictory. Price's estimate in the 1960s of 100.000 for that time did not take into account the number of discontinued journals. Bernal gives the same number (Bernal 1961, 860, and 1970, 1168) which remains the same in both editions even though the number of scientists has doubled during the same period. For journals appearing between 1900 and 1960 a "mortality" of 33% is assumed. Only 40% of all entries represent actual titles, the rest are double publications or "world list ghosts" (Barr 1967, 111). For the years 1950-1960, Gottschalk and Desmond (1963) do not see a growth in the number of journals. New foundations and discontinued ones neutralize each other. The growth in the number of articles is absorbed by growth in volume and higher frequency of publication. Kronick (1976) analyzed the beginnings of scientific journals for 1665-1790. Both Barr and Carpenter and Narin estimate for 1965 that ca. 25.000 journals were published in the sciences including psychology but excluding the social sciences and humanities (Carpenter and Narin 1980). Gottschalk and Desmond's estimate of 35.000 for 1961 is deemed too high because it included marginal areas and discontinued journals (ibid., 61, 62).

7 The only organization in science which is, indeed, hit by the wave is the library with its claim to cover all fields comprehensively. On the growing impossibility of this objective see the next section.

[8] Lepenies' *Ende der Naturgeschichte* (1976) contains a host of detailed descriptions of this process in different areas of knowledge and thus the illustrations of what is meant here.

[9] A particularly impressive example is provided with the history of molecular biology which emerged from the systematic application of methods of physics to the problem and explanation of life and inheritance. See Cairns et al. 1966 as well as Mullins (1972) who analyzes the relationship between intellectual and social processes in the emergence of specialties.

[10] For a more detailed description of the many facets of the discourse on interdisciplinarity, its metaphors and organizational analogues see Weingart (1997a). A comprehensive account with affirmative objectives is provided by Klein (1996).

[11] A systematic analysis of different disciplinary 'cultures' and the resulting forms of organization is given by Whitley (1984).

[12] If one disregards the foundation of the *Technische Hochschulen* at the beginning of the twentieth century because they were mostly turned into full universities afterwards, specialization has already begun with the wave of new foundations in the 1960s and 70s and will most likely continue under the pressure of budget cuts. On the different "transformations" of universities in the eighteenth, nineteenth, and twentieth centuries see Wittrock (1993).

REFERENCES

Barr, K D. (1967), 'Estimates of the number of currently available scientific and technical periodicals', *Journal of Documentation* 23 (2): 110–16.

Bell, Daniel (1973), *The Coming of Post-Industrial Society: A Venture in Social Forecasting*, New York: Basic Books.

Bernal, John D. (1961), *Die Wissenschaft in der Geschichte*, Darmstadt: Progress Verlag Fladung.

Bernal, John D. (1970), *Wissenschaft*, Reinbeck: Rowohlt.

Broad, William J. (1988), 'Science can't keep up with flood of new journals', *New York Times,* 16 February: C1, C11.

Cairns, John, Gunther S. Stent and James D. Watson (ed.) (1966), *Phage and the Origins of Molecular Biology*, Cold Spring Harbor: Cold Spring Harbor Laboratory of Molecular Quantitative Biology.

Carpenter, M.P. and F. Narin (1980), 'The subject composition of the world's scientific journals', *Scientometrics,* 2 (1): 53–63.

Daston, Lorraine (1997), *Die Akademien und die Klassifizierung der Wissenschaften: Die Disziplinierung der Disziplinen*, Ms., Berlin: Berlin-Brandenburgische Akademie der Wissenschaften.

De Solla Price, Derek J. (1963/1971), *Little Science, Big Science* (pb edition 1971), New York and London: Columbia University Press.

De Solla Price, Derek J. (1975), *Science Since Babylon*, New Haven, Conn.: Yale University Pres.

De Solla Price, Derek J. (1978), 'Toward a model for science indicators', in Y. Elkana, J. Lederberg, R.K. Merton, A. Thackray and H. Zuckermann (eds.), *Toward a Metric of Science*, New York: Wiley, pp. 69–95.

DiTrocchio, Federico (1994), *Der große Schwindel: Betrug und Fälschung in der Wissenschaft,* Frankfurt a.M.: Campus.

Du Bois Reymond, Emil (1887), 'Über die wissenschaftlichen Zustände der Gegenwart', in *Reden, Zweite Folge*, Leipzig: Verlag von Veit & Co., pp. 448–64.

Forman, Paul (1997), 'Recent science: Late-modern and post-modern', in T. Söderquist (ed.), *The Historiography of Contemporary Science and Technology*, Amsterdam: Harwood Academic Publishers, pp. 179–213.

Funtowicz, Silvio O. and Jerry R. Ravetz (1993), 'The emergence of post-normal science', in R. von Schomberg (ed.), *Science, Politics, and Morality. Scientific Uncertainty and Decision Making*, Dordrecht, Boston, London: Kluwer Academic Publishers, pp. 85–123.

Garfield, Eugene (1987), 'The 170 surviving journals that CC would have covered 100 years ago', *Current Contents* 26: 3–12.

Garfield, Eugene (1988). 'Too many journals? Nonsense!', *Current Contents* 45: 5–6. (Reprint from *The Scientist* 2: 5, 7 March 1988).

Garfield, Eugene (1989), 'Citation behavior – An aid or a hindrance to information retrieval?', *Current Contents*, 18: 3–8.
Garfield, Eugene (1990), 'How ISI selects journals for coverage: Quantitative and qualitative considerations', *Current Contents* 22: 5–13.
Garfield, Eugene (1991), 'In truth, the 'flood' of scientific literature is only a myth', *The Scientist* 5: 17 (2. September: 11).
Gascoigne, Robert M. (1985), *A Historical Catalogue of Scientific Periodicals, 1665-1900*, New York and London: Garland Publishing Inc.
Gibbons, Michael and Björn Wittrock (eds.) (1985), *Science as a Commodity. Threats to the Open Community of Scholars*, Harlow: Longman.
Gibbons, Michael, Camille Limoges, Helga Nowotny, Simon Schwartzman, Peter Scott and Martin Trow (1994), *The New Production of Knowledge*, London: Sage.
Gottschalk, C.M. and W.F. Desmond (1963), 'Worldwide census of scientific and technical serials', *American Documentation* July 1963: 188–94.
Grötschel, Martin and Joachim Lügger (1995), *Wissenschaftliche Kommunikation am Wendepunkt – Bibliotheken im Zeitalter globaler elektronischer Netze*, Technical Report TR 95-1, Konrad-Zuse-Zentrum für Informationstechnik Berlin.
Harnad, Stevan (1996), 'Implementing peer review on the net: Scientific quality control in scholarly electronic journals', in R. Peek and G. Newby (eds.), *Scholarly Publication: The Electronic Frontier*, Cambridge, MA: MIT Press, pp. 103–8.
Helmholtz von, Hermann (1896), 'Über das Verhältniss der Naturwissenschaften zur Gesammtheit der Wissenschaften, Akademische Festrede (1862)', in *Vorträge und Reden*, 1. Bd., 4. Aufl., Brauschweig: Vieweg und Sohn, pp. 157–86.
Holden, Constance (1987), 'Libraries stunned by journal price increases', *Science* 236: 908–9.
Jantsch, Erich (1972), 'Towards interdisciplinarity and transdisciplinarity in education and innovation', in CERI, *Interdisciplinarity. Problems of Teaching and Research in Universities*, Paris: OECD, pp. 97–120.
Klein, Julie (1996), *Crossing Boundaries. Knowledge, Disciplinarities and Interdisciplinarities*, Charlottesville, VA: University Press of Virginia.
Kronick, David A. (1976), *A History of Scientific and Technical Periodicals: The Origins and Development of the Scientific and Technical Press 1655–1790*, Metuchen, NJ: Scarecrow Press.
Lepenies, Wolf (1976), *Das Ende der Naturgeschichte. Wandel kultureller Selbstverständlichkeiten in den Wissenschaften des 18. und 19. Jahrhunderts*, München: Hanser Verlag.
Levin, Lennart and Ingemar Lind (ed.) (1985), *Interdisciplinarity Revisited*, OECD/ CERI, Swedish National Board of Universities and Colleges, Linköping University.
Lundgreen, Peter, Bernd Horn, Wolfgang Krohn, Günter Küppers and Rainer Paslack (1986), *Staatliche Forschung in Deutschland 1870–1980*, Frankfurt a.M.: Campus.
Maasen, Sabine (1998), *Genealogie der Unmoral: Zur Therapeutisierung sexueller Selbst*, Frankfurt: Suhrkamp.
Mullins, Nicolas C. (1972), 'The development of a scientific speciality: The Phage Group and the origins of molecular biology', *Minerva* X (1): 51–82.
Neurath, Otto, Rudolf Carnap and Charles Morris (1938/1970), 'Encyclopedia and unified science', *Foundations of the Unity of Science* 1,1, Chicago: University of Chicago Press.
Odlyzko, Andrew M. (1994), *Tragic Loss or Good Riddance? The Impending Demise of Traditional Scholarly Journals*, November 6, 1994: ftp://elib.zib-berlin.de/netlib/att/math/odlyzko/index.
Odlyzko, Andrew M. (1994/1995), 'Tragic loss or good riddance? The impending demise of traditional scholarly journals', *Intern. J. Human-Computer Studies* (formerly *Intern. J. Man-Machine Studies*) 42 (1995): 71–122; and in the *Electronic J. Univ.Comp.Sci.*, pilot issue 1994; Kurzfassung in: *Notices Amer. Math Soc.* 42, Januar 1995: 49–53; deutsch in: *Deutsche Math. Ver. Mitteilungen* Nr. 1, 1995: 19–24: http:www.research.att.com/~amo/doc/tragic.loss.lomg.pdf.
Odlyzko, Andrew M. (1997), *Silicon Dreams and Silicon Bricks: The Continuing Evolution of Libraries*, AT&T Labs-Research, rev. draft, Ms.
OECD (1984), *Science and Technology Indicators*, Paris: OECD.
OECD (1998), *The Global Village: How Information and Communication Technologies Affect the Science System*, Background Report to the international conference *Global Research Village II: Maximising the Benefits of Information Technology for Science*. Sintra, Portugal, 17–18 September 1998. Ms.
OECD (1990), *Main Science and Technology Indicators; Statistical Publications on Diskette*, Paris: OECD.
Richta, R. et al. (1968), *Ziviliation am Scheideweg (Richta Report)*, Freiburg: Verlag an der Basis.

Schummer, James (1997), 'Scientometric studies on chemistry I: The exponential growth of chemical substances, 1800–1995', *Scientometrics* 39 (1): 107–23.

Stehr, Nico (1994), *Arbeit, Eigentum und Wissen. Zur Theorie von Wissensgesellschaften*, Frankfurt a.M.: Suhrkamp.

Stichweh, Rudolf (1984), *Zur Entstehung des modernen Systems wissenschaftlicher Disziplinen. Physik in Deutschland 1740–1890*, Frankfurt a.M.: Suhrkamp.

Stix, Gary (1995), 'Publizieren mit Lichtgeschwindigkeit', *Spektrum der Wissenschaft* March 1995: 34–9.

Wagner, S. (1985), 'Die Entwicklung der exakten Naturwissenschaften von der Antike bis zur Gegenwart. Eine Quantifizierung ihrer Geschichte', *Science Studies Report* 27, Bd. II. Bielefeld: Kleine Verlag.

Weingart, Peter (1983), ‚Verwissenschaftlichung der Gesellschaft – Politisierung der Wissenschaft', *Zeitschrift für Soziologie* 12 (3): 225–41.

Weingart, Peter (1997a), 'From 'Finalization' to 'Mode 2': Old wine in new bottles?', *Social Science Information* 36 (4): 591–614.

Weingart, Peter (1997b), 'Interdisziplinarität – Der paradoxe Diskurs', *Ethik und Sozialwissenschaften* 8 (4): 521–9.

Weingart, Peter (1997c), 'Interdisziplinarität im Kreuzfeuer: Aus dem Paradox in die Konfusion und zurück', *Ethik und Sozialwissenschaften* 8 (4): 589–97.

Weingart, Peter, Wolfgang Prinz, Maria Kastner, Sabine Maasen and Wolfgang Walter (1991), *Die sog. Geisteswissenschaften: Außenansichten*, Frankfurt a.M.: Suhrkamp.

Whitley, Richard (1984), *The Intellectual and Social Organization of the Sciences*, Oxford: Clarendon Press.

Wissenschaftsrat (1992), *Zur Förderung von Wissenschaft und Forschung durch wissenschaftliche Fachgesellschaften*, Bremen.

Wittrock, Björn (1993), 'The modern university: The three transformations', in S. Rothblatt and B. Wittrock (eds.), *The European and American University Since 1800. Historical and Sociological Essays*, Cambridge, UK: Cambridge University Press, pp. 303–62.

Ziman, John (1994), *Prometheus Bound. Science in a Dynamic Steady State*, Cambridge, UK: Cambridge University Press.

CHAPTER 10

STEPHEN HILGARTNER

INSTITUTIONALIZING SCIENCE & TECHNOLOGY STUDIES IN THE ACADEMY

In many ways, science and technology studies (STS) seems poised to become one of the most dynamic and influential areas of social scientific and humanistic inquiry during the twenty-first century. The intellectual sophistication and empirical strength of much work in the field is impressive, and its subject matter is of undeniable importance. The field is providing rigorous interpretations of the place of science and technology (S&T) in social orders; it is exerting growing intellectual influence in the academic world and in public affairs; and it is illuminating some of the fundamental issues of our technological age. But at the same time, the intellectual strength of STS is not matched by a corresponding institutional robustness. One can therefore confidently predict that the field will fail to realize its potential unless it takes steps to consolidate its position in the academy, to become more fully institutionalized, and to expand.

To be sure, STS has already made impressive strides toward institutionalization. STS scholars have built outposts in many universities, especially in the United States and Europe but also in Asia and elsewhere. In addition to free-standing departments of STS and interdisciplinary programs, the field is represented in a wide range of academic departments, including anthropology, communication, environmental studies, government, history, history of science, philosophy, sociology, as well as in schools of engineering, health sciences, public policy, information studies, and so on. Its professional societies have become vibrant international organizations. A number of graduate programs offering Ph.D. degrees in STS have been founded, and their alumni are, by and large, doing well on the tight academic job market. Undergraduate courses in STS have become increasingly common. The institutionalization of the field is also reflected in the strength of its journals, in the publication of handbooks, readers, and textbooks, and in the devotion of an entire volume of the International Encyclopedia of the Social Sciences to science and technology studies. At the same time, many STSers advise governments and NGOs, or are active in political arenas. The uptake of STS ideas by governments and political bodies varies cross-nationally; for example, right now STS seems to be more influential in Brussels than in Washington. But there is growing recognition that STS is capable of shedding useful light on important problems of governance.

B. Joerges and H. Nowotny (eds.), Social Studies of Science and Technology: Looking Back Ahead, 201–210.

Nevertheless, one should not mistake these achievements for a solid base in the academy. Too few universities have made long-term commitments to STS, and the field remains institutionally fragile. In many universities, STS depends on an unstable mixture of ad hoc arrangements, temporary solutions, and the dedication of a few individuals. Even some of the strongest STS departments and programs many not be as institutionally robust as they appear at first glance. For these reasons, STS cannot achieve its potential unless the field takes steps to improve its standing in the university.

Under these circumstances, STS urgently needs to consider how it can improve its institutional strength. Strategic planning, difficult in any organizational context, is especially hard for a heterogeneous and highly-international academic field, populated by independent thinkers and home to national, local, and disciplinary differences. Moreover, because the salient considerations vary among universities, STS cannot develop a one-size-fits-all 'master plan' applicable to all environments. What is needed is not a narrow step-by-step plan, but an ongoing discussion aimed at enhancing the field's collective capacity for strategic action. This chapter is intended to contribute to such a discussion, arguing that the field should place several points at the top of its strategic agenda. The strategic points developed here are most immediately relevant to universities in the United States, although in many cases they may also be pertinent elsewhere.

PRESENT THE INTELLECTUAL FOCUS OF THE FIELD CLEARLY

To begin with, STS must do more to present its distinctive intellectual focus clearly. Any undergraduate can provide the standard one-word summary of the subject matter of most established academic disciplines: for sociology, 'society'; for biology, 'life'; for history, 'the past;' for economics, 'choice.' The generality of these terms is the key to their utility. They position fields on a crude academic map; they stake out an area of fundamental importance; and they provide assurance that something profound unites the diverse work that takes place under their rubric.

STS needs to promote such a one-word rubric actively, and, without a doubt, 'knowledge' is the leading contender. The term *knowledge* stresses the distinctive concern of the field, the thing that unifies STS work on such superficially disparate topics as, say, seventeenth century mathematics and twenty-first century biotechnology. The term is general enough to apply to all societies and historical periods, and it easily encompasses both 'science' and 'technology,' even as it reaches beyond them. Most important, knowledge is truly fundamental; it pervades all social space and is among the most distinctive characteristics of human beings. Indeed, knowledge is every bit as fundamental as the corresponding core concepts, such as 'society,' 'culture,' and 'power,' of traditional scholarly disciplines. (The plural form, *knowledges*, which emphasizes diversity and historical location, is more accurate, but it smacks of jargon, making it unsuitable for general use.)

STS's one-word rubric should be supported by a succinct mission statement of the field. An effective mission statement must be vague enough that it does not become a straightjacket, yet concrete enough to show the connections between different flavors of STS work. Thus, a good mission statement must stake out the common ground shared by technology studies and science studies, archival research and ethnography, descriptive approaches and normative ones. Something along the following lines captures the essential points:

> STS is an academic field concerned with the study of knowledge as a social phenomenon. It considers knowledge (especially, but not only, scientific and technical knowledge) in all its forms – whether embodied in technologies, texts, persons, competencies, practices, etc. In particular, the field examines: (a) how knowledge is produced, spread, contested, contained, challenged, stabilized, displaced, destroyed, evaluated, and prevented; and (b) the consequences and normative dimensions of knowledge and the means of producing it.

This mission statement presents STS as an autonomous area of scholarship, defined with respect to a set of intellectual concerns rather than in terms of traditional scholarly disciplines, such as history, sociology, philosophy, and anthropology.[1] Such a vision recognizes that the academic study of S&T initially developed within traditional disciplines, where it was typically understood as consisting of 'subfields' (such as history of science, philosophy of science, or sociology of science). Over time, much of the scholarship in these subfields grew increasingly connected by shared intellectual concerns and a common literature. Ultimately, as STS matured, it began to transcend its original disciplinary roots, creating the new intellectual domain that the above mission statement describes.[2] Pointing out the existence of this distinctive intellectual domain is a key step toward building the institutional strength of the field.

ELABORATE A VISION OF WHAT THE FIELD CAN ACHIEVE

Beyond presenting its intellectual focus, the field should provide a clear picture of why STS is important and what it intends to accomplish in the decades ahead. STS has reached a stage in its development where it should promise, demand, and deliver more. STS scholars should frequently remind people of three key points:

First, there is an urgent need for scholarship on the social dimensions of science and technology. The expansion of S&T was among the most salient features of the last century, and the centrality of S&T in the contemporary world is, if anything, increasing. Developments in information technology, environmental sciences, nanotechnology, genomics, medicine, biotechnology, neurosciences, and engineering are presenting contemporary societies with unprecedented opportunities, challenges, and dilemmas. Those universities that invest early in STS research will be better prepared to contribute to societal debates about many of the most important issues of our time.

Second, the university that fails to prepare its students to grapple with the complex choices of our technological age is doing them a profound disservice. The citizens and professionals of the future need to understand science and technology in social context. They need to understand the history of S&T to inspire their vision with the lessons the past. They need to comprehend the forces that shape science and technology and to gaze down roads not taken. And, in a shrinking world increasingly dominated by global issues and North-South politics, they need to understand how science and technology articulate with different cultural contexts.

Finally, meeting the educational and research needs of twenty-first century societies will require STS to attain a much stronger position in the academy. This goal can in principle be achieved, and STS should work to ensure that in the next fifteen years, the field will come to be represented at every research university worthy of the name. STS will also be increasingly represented in colleges of engineering, medicine, and business. As the field grows, STS research – on both historical and contemporary topics – will form a larger, deeper, and more integrated body of scholarship, becoming an essential part of the curriculum of a wide variety of academic programs. Exposure to STS will be considered an essential part of the basic education of scientists, engineers, physicians, managers, information technology specialists, and policy professionals, and it will aim to equip them with a capacity for critical thinking and self-reflection about S&T. STS will also become a standard undergraduate major in the liberal arts, and strong graduate programs in STS will exist at a number of research universities. As a result, STS research and education will have tremendous impact beyond the academy, enhancing the ability of contemporary societies to reflect on, and critically evaluate, the problems and opportunities that confront them during a period of unparalleled technological and scientific change.

IDENTIFY THE STRUCTURAL BARRIERS TO SUCCESS

The full potential of STS cannot be realized without a period of sustained growth in the field. For STS capacity to reach a level commensurate with the research and educational needs of twenty-first century societies, the number of Ph.D.-level STS scholars working in the U.S. universities must eventually double. To be sure, most of these positions will not be housed in departments or programs devoted exclusively to science and technology studies: on the contrary, the strong societal interest in the topics that STS addresses ensures that the field will continue to be represented in a wide variety of academic departments and professional schools. But there is little doubt that existing STS departments and programs should be strengthened, new programs and departments must be created, and graduate training in STS should be expanded.

Achieving these goals, however, is a tall order, and a hard-nosed assessment of structural barriers is needed. New initiatives in the academy face a number of well-known obstacles, such as institutional inertia, lack of imagination, conflicting priorities, entrenched interests, budgetary constraints, complacency, and weak incentives for working for change. But beyond these familiar problems, which affect many substantive areas, the growth of STS is constrained by simple fact that the prevailing division of academic labor does not offer STS a suitable institutional home.

Despite the undeniable importance of S&T in the contemporary world, studies of their social aspects occupy a marginal position in the traditional social science and humanities disciplines. In the United States, the situation is best in history, for the history of science is relatively well-institutionalized and a handful of universities have excellent programs in this area. However, the prospects for expanding history of science and technology significantly within history departments is limited. Even the largest history departments will have room for only a few historians of S&T after all the obligatory temporal periods and geographic regions are covered. History departments are also unlikely to offer a suitable home for much work on contemporary S&T, although (in a testament to the diminishing significance of the boundary between historical and contemporary studies of S&T) a few scholars who study very recent S&T are now found in some of the leading history of science programs.[3] The sociology of science and technology is less thoroughly institutionalized. Most sociologists regard it as a minor subfield, far removed from the central specialties, such as organizational sociology or social stratification. Similarly, the philosophy of science is a well-known academic area, but it is numerically small and unlikely to grow appreciably. In anthropology departments, studies of science and technology, virtually unheard of a generation ago, are beginning to make their mark, but the prospects for continued growth remain unclear. Among departments of government and political science, the politics of science and technology is barely recognized as an academic domain, although a handful of programs on science policy exist.

The marginality of research on S&T in the traditional academic division of labor imposes severe demographic constraints on the growth of STS: unless they are loosened, they will prevent STS from meeting the needs of twenty-first century societies. But these constraints have much more than a purely numerical effect; they also reduce the intellectual coherence of the field by making it difficult for most universities to build groups of STS faculty that achieve a critical mass. Collegial interaction, graduate and undergraduate education, and collaborative initiatives all suffer. Moreover, in most cases, those clusters of like-minded faculty that form serendipitously remain utterly dependent on a few key personnel, leaving them perpetually unstable and prone to spontaneous disintegration. Uncertainty about the stability of the field also chills the enthusiasm of many young scholars, whose advisors urge them to pursue safer paths.

In short, the disciplinary system leaves the sustained, systematic study of science and technology without a homeland. STS scholars, scattered in a diaspora across far-flung departments, have often found it hard to organize collectively to rectify this situation. To make matters worse, some traditional social science and humanities departments – fearing a loss of resources – have opposed efforts to open some space for STS. In many environments, these factors have produced administrative inaction. For these reasons, the inflexibility of the system of academic disciplines is clearly the greatest impediment to the success of the field.

DEVELOP SOME CORE STRATEGIC PRINCIPLES

Despite these serious obstacles, the case for building STS is so compelling that faculty and administrators have a responsibility to find ways to strengthen the field. Already, some notable successes have been achieved. However, I believe that STS has reached a point where it needs to articulate some strategic principles for creating space for the field and solidifying its institutional position. A first step is to consider three common models of the institutional future of STS:

- *The subfield model.* This model frames STS as set of subfields housed mainly within traditional disciplinary departments (e.g., sociology, history, anthropology, etc.), The subfield model leaves the existence of STS at any given university dependent on the hiring decisions of departments with no particular mandate to study science and technology.
- *The interdisciplinary model.* This model envisions a future in which STS is institutionalized primarily within interdisciplinary programs. Such programs sometimes achieve considerable institutional strength, even if they are less than fully autonomous in that they do not completely control their own appointments. However, the

degree of institutional robustness varies, depending on precisely how such interdisciplinary programs are organized.

- *The new discipline model.* This model imagines a future in which STS ultimately becomes institutionalized as a new discipline, with all of the autonomy, resources, and stability that disciplinary status conveys, including independent departments, secure tenure lines, undergraduate majors, and, in some universities, graduate programs. STS clearly has not yet achieved full disciplinary status, but in some places it appears to be moving in this direction.

STS can only achieve its long-term potential if it takes steps that will ultimately enable it to be recognized as a distinct academic discipline. But disciplines are not built overnight, and in American universities today, STS clearly exists mainly in the subfield and interdisciplinary forms. Accordingly, I suggest that the field self-consciously adopt three strategic principles for improving the institutional status of the field:

Principle 1: STS should seek to move the field up the gradient of institutional strength.

The three main visions of the future of STS – the subfield, interdisciplinary area, and new discipline models – represent increasingly robust positions on a gradient of institutional strength. STS should self-consciously aim to advance up this gradient.

Principle 2: Local goals and strategies should be matched to the specific organizational context.

Given that STS takes a variety of institutional forms in different universities, strategies for strengthening its overall status must respond to these variations. The goal at all times should be to move up the gradient of institutionalization, but local conditions will shape appropriate strategy. In universities and colleges where STS does not exist, pursuing the subfield model (e.g., by encouraging traditional disciplines to hire in STS) may offer the best way to promote the field. In settings where STS exists as a collection of subfields, a key strategic objective should be to build connections among them, so as to pave the way toward future integration. In universities where strong interdisciplinary programs exist, the goal should be to achieve strong control over faculty appointments, curricula, and other resources, with an eye toward ultimately achieving departmental status. In places where true STS departments exist, the goal should be to build their intellectual and institutional strength, making them indispensable. In pursuing these goals, it is extremely important that STS scholars not make the mistake of assuming that the parochial politics of their own universities are mirrored everywhere. Members of review groups and visiting committees have a particular responsibility in this regard.

Principle 3: The field should be framed as an 'emerging discipline.'

Given the ambiguous position of STS on the gradient of institutionalization, the question of how to refer to the field is a salient one. The choice of terms has a performative dimension. On the one hand, equating STS with a group of subfields, or an interdisciplinary 'meeting point' among them, undercuts the notion that STS represents a distinct intellectual domain that properly belongs at a higher position on the institutionalization gradient. On the other hand, pretending that STS is a well-established 'new discipline' is an exercise in wishful thinking, since the field is still far from having achieved standing as a full-fledged discipline. In these circumstances, what should one call STS?

In my view, STS should refer to itself as an 'emerging discipline.' This phrase accurately describes the ambiguous position of STS on the gradient of institutionalization. At the same time, the phrase lays claim to disciplinary status by asserting that this status can and should be achieved in the future. Finally, the 'emerging discipline' concept is consistent with the history of a field: STS began as a set of subfields that over time grew increasingly linked, forming an interdisciplinary domain. As this interdisciplinary field matured, it became increasingly clear that STS had become a discipline in-the-making. Thus, the emerging discipline concept does not directly contradict the subfield or interdisciplinary model so much as treat them as stages in a process of institutional change that the field should work to advance.

ADDRESS THE OBJECTIONS OF SKEPTICS

STS should not expect that its efforts to develop over time into an influential scholarly discipline in the twenty-first century will be unopposed. Experience teaches that the most vocal opposition to expanding STS often comes from social science and humanities departments that hope to avoid competition. But beyond such attempts to defend professional turf, many thoughtful administrators and faculty remain skeptical about the wisdom of making a significant commitment to STS. Accordingly, the field should devote considerable attention to responding to their concerns. To stimulate discussion, some of the most common objections and answers to them are presented below:

Is STS really a coherent field? It seems to lack the intellectual coherence of the traditional disciplines.

STS is diverse, but its focus on knowledge gives it unity. But let us consider the premises of this question: Can we assume that the traditional scholarly disciplines are highly 'coherent'? Consider sociology as an example. Sociologists share an interest in 'society,' but they pursue extremely diverse research agendas from a variety of often incompatible theoretical, methodological, and epistemological perspectives. The American Sociological Association, for example, has some 40 sections on topics as diverse as aging and the life course, international migration, Marxist sociology, and the sociology of emotions. Sociology is not unique in this respect. The fields of history, philosophy, anthropology, and political science are also extremely diverse. Should one expect STS to exhibit kinds of coherence not found in established scholarly disciplines?

STS seems analogous to an interdisciplinary 'area studies' field, such as Latin American Studies or European Studies, in which diverse specialists in a geographic region interact. In area studies programs, each discipline brings its own perspective to the study of the region, and all benefit from the exchange. Why can't the future of STS follow such a model?

The creation of interdisciplinary STS programs is a salient symptom of the limitations of the subfields model. In many interdisciplinary fields, the area studies metaphor has been an influential one; indeed, STS programs themselves have drawn on it at times, especially during the early years. But before uncritically accepting this metaphor as a guide for the future of STS, we should consider its implications. In particular, the area studies metaphor casts traditional disciplines, not perspectives emerging from scholarship with a distinct focus on knowledge systems, as the sources of intellectual vision. The established disciplines offer theories, paradigms, and methods; that is, 'lenses' for examining of the world. Science, technology, and knowledge become 'topics,' the raw materials for traditional modes of investigation.

There are two serious problems with this formulation. First, as an academic specialty, the study of knowledge is really not analogous to the study of a geographic region. The production, spread, evaluation, and effects of knowledge are fundamental processes in all societies; understanding them poses challenges for social theory and empirical research that little resemble the challenges of understanding a region, such as Europe or East Asia. Second, the area studies metaphor privileges existing perspectives at the expense of new ones. It suggests that to comprehend S&T, one needs merely to add together traditional perspectives – as if one could use a 'disciplinary stereoscope' to combine the viewpoints of sociology, history, economics, political science, anthropology, and philosophy to form a complete picture. But each disciplinary 'lens' has blind spots, and these do not neatly compensate for one another. Moreover, STS research has advanced significantly not only by applying existing approaches to new terrain but also by developing new theories and perspectives specifically aimed at grappling with the unique challenges of studying knowledge.

But couldn't the existing disciplines make room for the new perspectives arising from STS? Do we really need to take the radical step of establishing a new discipline?

As a purely theoretical matter, the existing disciplines certainly could make room for STS; for the boundaries of disciplines are flexible. But will they? Experience teaches that the largest and best traditional departments will often find room for some coverage of the most strongly institutionalized STS-related subfields, such as history of science. This sort of permeation of traditional departments with STS scholars and ideas should be encouraged. Nevertheless, as a practical matter, the structural barriers discussed above make it extremely unlikely that existing disciplines will make a major commitment to STS. That is why, ultimately, STS will need to emerge as a new discipline. Its focus – knowledge – is every bit as fundamental, consequential, and in need of sustained study as the foci of today's recognized disciplines.

Establishing a new discipline may seem like a radical step, but from time to time such adjustments in the division of academic labor are necessary. Today's disciplinary system is not an optimal design for organizing intellectual work but the product of historical contingencies. There is no reason to assume that established disciplines completely cover all phenomena of intellectual interest, with no gaps or uncharted territories. Nor can we assume that today's disciplines will naturally stretch to embrace all issues of intellectual import, expecting that those problems ignored by one field will automatically be taken up by others. Indeed, science and technology provide an excellent example of something that has largely fallen through the cracks: for no established discipline has accorded the study of science and technology a place in its scholarly agenda that is commensurate with their significance in the contemporary world. It is high time we stopped relying on a century-old map of intellectual domains to guide scholarship on science and technology. The radical changes associated with their rapid growth surely justify a minor remapping of the academic division of labor.

The standards of established disciplines provide a means of maintaining quality and protecting against shallow scholarship when making academic appointments, developing curricula, and building programs. How will STS uphold standards?

This is an important concern: the vision of academic fields lurching off on ill-defined intellectual missions, without the guidance of scholarly standards and methods, is certainly not a pretty one. University administrators and faculty are wise to worry about maintaining quality. But in the social sciences, humanities, and natural sciences alike, research programs that are truly innovative, such as much work in STS, do not tend to become bastions of mediocrity. To be sure, novel areas often exhibit a frontier culture; valuing the new and the bold, they inspire experimental approaches and attract young, untested talent. Mistakes are sometimes made, but overall such high-risk, high-reward areas tend to be far more productive than are safer, tried-and-true domains, where intellectual complacency often reigns. The proper way to address the administrative challenges of building innovative fields, therefore, is not to fail to develop them – especially when they concern matters of pressing societal importance. On the contrary, the solution is to provide important, innovative areas with the resources that they need to succeed. Quality can be assured only by hiring faculty with high standards and enabling them to create strong departments and programs, make sound appointments, enhance research capacity, and strengthen curricula. Forcing innovative areas to remain within established disciplines does not somehow automatically ensure quality. Excellence cannot be borrowed, second hand, from other disciplines; it must be built into the intellectual and institutional architecture of STS.

CONCLUSION

There is an undeniable need for enhancing the capacity of contemporary societies to understand the social dimensions of science and technology. STS has an important role to play in this endeavor, but its potential cannot be achieved without substantially improving its position in the university. The field has reached a stage in its development where a more concerted, collective effort to institutionalize the field is necessary. This essay is intended to stimulate discussion about how to proceed as STS continues to work to advance its intellectual and institutional agendas.

Department of Science & Technology Studies, Cornell University, Ithaca, NY, USA

NOTES

[1] Peter Dear's notion of epistemography describes a similar focus in different terms (Dear 2001, 130): Epistemography investigates "science 'in the field,' as it were, asking such questions as these: What counts as scientific knowledge? How is that knowledge made or certified? In what ways is it used and valued?"

[2] See Edge (1995) for an account of the history of the field.

[3] On the limits of the past-present dichotomy as a principle for organizing the study of S&T, see Jasanoff (2000).

REFERENCES

Dear, Peter (2001), 'Science studies as epistemography', in J. Labinger and H. Collins (eds), *The One Culture? A Conversation About Science*, Chicago: University of Chicago Press. pp. 128–41.

Edge, David (1995), 'Reinventing the wheel', in S. Jasanoff, G.E. Markle, J.C. Petersen and T. Pinch (eds.), *Handbook of Science and Technology Studies,* London, Thousand Oaks and New Delhi: Sage Publications, pp. 3–23.

Jasanoff, Sheila (2000), 'Reconstructing the past, constructing the present: Can science studies and history of science live happily ever after', *Social Studies of Science* 30 (4): 621–31.

CHAPTER 11

MICHAEL HAGNER AND HANS-JÖRG RHEINBERGER

PROLEPSIS

Considerations for Histories of Science After 2000

Anticipation, presumption, and promise are what the term prolepsis conjures up. There is, however, no anticipation without retrospection, no presumption without prior consumption, and no promise without stake. And just as there is generally more than one way of looking at what happened in the past, proleptic excursions also contain an element of choice. Several grand perspectives for the historiography of science have followed each other in the past decades. In the 1960s and 1970s, it was fashionable to talk about revolutions. In the 1980s, revolutions were passé and 'turns' became the order of the day – thus we could have taken the so-called 'practical turn' as our point of departure. We would then have asked how far and where that turn has taken us since and what we might expect from it in the near future. In the 1990s it was fashionable to take issue with that 'relativistic turn' in philosophy and history of science which in the second half of the decade provoked a battle for truth, objectivity, and reason. This was a battle fiercely fought by those who saw the grand scientific project of the occident endangered, against those accused of abandoning these values altogether. We could have asked whether the ensuing 'science wars' did or did not bring about a climatic change in talking about science that will prove lasting and whether, as a consequence, the claims and terms of inter- and transdisciplinarity will be redefined. And indeed, we will come back to all these issues of the 1980s and the 1990s, but we will tackle them from a different perspective.

STRUCTURALISM VERSUS CONSTRUCTIVISM

Structuralism and constructivism have become key labels for a number of intellectual movements and developments in the humanities and social sciences in the last fifty years. There is perhaps no general agreement about the meaning of the terms structure and construction and some might prefer to use the notions only in the plural. But there is certainly little doubt that the two concepts have been regarded as more or less incompatible terms of opposition. When Pierre Bourdieu characterized himself as a 'constructivist structuralist' some years ago, he could do so only by deliberately abstracting from

B. Joerges and H. Nowotny (eds.), Social Studies of Science and Technology: Looking Back Ahead, 211–228.

the historical context in which the concepts had been shaped and had taken on radically different connotations. In an effort to explain this apparent paradox, Bourdieu defined structuralism as assuming 'objective structures' in the social world, objectivity meaning a certain independence from the consciousness and will of actors. These structures are assumed to function as possibility horizons but also as barriers, thus setting the margins for shaping practices and conceiving ideas. Constructivism, on the other hand, was defined by Bourdieu as emphasizing the 'social genesis' and thus a certain subjectivity and flexibility of such structures, patterns of perception, practice, and thinking. These structures and patterns are, in turn, taken to be responsible for the constitution of habitus, social relations, group identities, and social classes in particular (Bourdieu 1987, 147).

Bourdieu argued for a reciprocality between the two – and he referred to Hegelian categories when speaking of a 'dialectic relation' between structure and construction. The charming impudence of Bourdieu's approach lies in the fact that his use of both terms is decontextualized and ahistorical. When he talks about structuralism, he is far from positioning himself with respect to, say, structural linguistics in the tradition of Ferdinand de Saussure and the Prague circle or to structural anthropology in the tradition of Claude Lévi-Strauss, let alone Michel Foucault's structural discourse analysis. When he talks about constructivism, he certainly does not have in mind things like radical constructivism à la Heinz von Förster, Ernst Glasersfeld, and Humberto Maturana, or the social constructivism of the Edinburgh school. Rather, and generally speaking, it is Bourdieu's aim to find a way out of the dichotomy between a philosophy *without* subjects and a philosophy *of* the subject.

This is not the place to answer the question of whether Bourdieu succeeded in offering the clues to a deeper understanding of the intricate relation between facts and acts, between objective structures and subjective constructs in society. Nevertheless, it is heuristically tempting to play around with a decontextualization of the sort just mentioned, because doing so might shed light on some methodological and historiographical conundrums in recent history of science. We do not pretend to do full justice to Bourdieu's intentions, but rather take the liberty of decontextualizing Bourdieu himself. Starting from his paradox allows us to draw together various things which are usually kept in different boxes. For example, different brands of social historians, marxists, and discourse analysts may have taken issue with one another in many respects, but they have one thing in common: They tend to avoid basing their accounts on a philosophy of the subject and argue instead that agency, creative dimensions and human actions in general are epiphenomena of social and economic structures, of discourses or even of self-organizing machines; this can justifiably be called structuralism. Against this tendency, there is a movement – by no means a homogeneous one – arguing for the irreducible importance of human agency in circumscribed and local contexts. Many positions in cultural history, those of *histoire des mentalités*, *microstoria*, and *Alltagsgeschichte*, can conveniently be mapped along the line of human agency: constructivism. Social versus cultural, macroscopic overviews versus microscopic insights, structures and laws versus contingencies – these are the frontlines in an ongoing debate in contemporary historiography (Hardtwig and Wehler 1996; Evans 1997; Chartier 1998; Wehler 1998; Daniel 2001).[1]

In history of science, these polarities have only recently become important. For a long time, the main categorizations were quite different. First of all, there was the classic divide between an internalist and an externalist history of science. These two accounts were in perfect harmony, and they practiced a clear-cut division of labour. For the externalist, political, economic and ethical frames constituted the guiding principles and limitations for scientific activity; for the internalist, the lawfulness of nature itself and the universal and historically invariant categories of objectivity, rationality, and truth were seen as the driving forces of scientific progress. This divide, however, is not matched by the opposition between structuralism and constructivism. Yet, both views shared a reliance upon a deliberately de-individualized understanding of scientific activity.

Besides these two accounts, there was yet another, very different approach that had its origins in the heroic understanding of science. One might, with Nietzsche, call it the 'monumental,' or memorial, history of science. This attitude was often promoted by scientists themselves, who incorporated categories like individual mental power, genius, creativity, and the relationship between the teacher and his pupils in their accounts. Wilhelm Ostwald's early-twentieth-century project on 'Great Men' of science is a telling example of this kind of history (Ostwald 1910). Hard-boiled structuralist philosophers or sociologists of science would at best admit that such aspects may, if at all, complete the picture of the scientific enterprise.

The cold world of rational structures and systems and the passionate world of freedom, playful creativity and power-play, anxiety and joy – the science of the day and the science of the night, in François Jacob's words (Jacob 1987; Jacob 1997): one would expect these two extremes to stand in staunch opposition, permanent debate, and competition. Quite surprisingly, that is not the case. This is due to the fact that the two approaches serve different purposes and are often directed toward different audiences and identified with different genres of writing about science. Rationality and objectivity occupy center stage in metascientific reflection, in the hands and minds either of professional philosophers of science, or of scientists themselves as self-made, spontaneous philosophers. The personal dimension, on the other hand, is rarely found in textbooks on philosophy of science or in the public pronouncements of scientists. Here it is the 'context of justification' that counts. The 'context of discovery,' in contrast, is more likely to be cultivated in personal genres of scientific prose such as (auto)biographies, obituaries, or popular histories. This division of labour has effectively precluded the question as to whether and how far these two aspects of scientific activity might complement each other, or, more radically, the extent to which the division might be seen as a contingent historical phenomenon. Are we confronted with two incommensurable conceptions, or is there a way to set aside the categorial differences between structural necessity and contingent construction, the objective and the subjective? What might it mean to be an intelligent 'constructivist structuralist?'

Bourdieu's answer to this question is to be found in his *Méditations pascaliennes*, where he talks about the double face of scientific reason:

> A realistic vision of history forbids a fictive transgression of the intranscendable limits of that history and prompts to examine how, and under what historical conditions, truths can be squeezed out of history which are not reducible to that history. One has to admit that reason did not fall from heaven like a mysterious gift destined to remain inexplicable, and that it is, therefore, historical through and through. But one is not at all forced to conclude, as is usually done, that it is reducible to history. It is in history, and in history alone, that one must look for the principle of the relative independence of reason from history whose product it

> nevertheless is; or more precisely, one has to look at the genuine, but very specific, historical logic according to which those universes of exception have been instituted in which the singular history of reason accomplishes itself (Bourdieu 1997, 130–1).[2]

It is only recently that history of science has even begun to take notice of the historicity of categorial choices such as rationality and objectivity. Historians of science have started to examine the circumstances in which these choices were shaped, reshaped, and gained their historically changing meanings (Daston 1992; 2001; Daston and Galison 1992). In his classic book on the history of the life sciences in eighteenth century France, Jacques Roger did not yet feel compelled, as we do today, to justify a statement such as: "Scientific truth itself is a product of history" (Roger 1993, xxvi). For today's new, self-empowered defenders of scientific objectivity, this statement would certainly count as an example of the fiercely combated "flight from reason" (Gross, Levitt and Lewis 1996; Sokal and Bricmont 1997). To this end, the grail watchers of science have constructed one Enemy to the scientific enterprise, an enemy intent on deconstructing the triple myth inherited from the nineteenth century: that of – again in the words of Roger – the one Scientist, the one Science, and the one Nature. It is an act of intellectual honesty rather than of revolutionary radicalism to maintain that a historicizing analysis of the role of the scientist, of basic epistemological categories, and even of the meaning of science itself is much more an attempt to liberate the concept of reason from its self-chosen chains rather than a flight from reason. We want to remain interested in how these categories become historically embodied in different scientific practices and how they are shaped by the various materialities in which they are embedded. This is by no means tantamount to claiming that these categories are mere epiphenomena of the deterministic power of socially constructed hardware (whether economic conditions, technological systems, or media). We will return to this point.

PHYSIOGNOMICS OF EXPERIMENTATION

Over the past ten years, our efforts have focused on the role of scientific experiment and experience, of epistemic objects and of spaces of representation in the historical development of the sciences. One of the guiding ideas was that analyzing the historical forms and configurations of experimental systems may provide important clues to an understanding of the dynamics of modern science. We envisaged that such a bottom-up analysis would serve as an alternative to top-down approaches, generally in the structuralist tradition, be they framed in terms of a sociology of disciplines or a philosophy of scientific reason. Instead, we chose to examine the materials and material forms in which scientific activity is enacted and through which it represents its results. In history of science, the exploration of certain objects and phenomena has for a long time been described as a process revolving around core concepts assumed to be historically invariant. Although this does not mean that these concepts were supposed to have been present throughout history, it was taken for granted that once they had been *discovered* or had proven to be successful, they could be regarded as integral parts of the science at issue. In the life sciences, concepts such as organism, species, race, gene, or neuron have been placed in this category. Undoubtedly, they were of tremendous importance for nineteenth and twentieth-century life sciences. Yet none of them had any significance for dealing with the world of living beings in the early modern period. And we do not have an answer to the question with which François Jacob ends his account on the history of heredity:

> Today the world is messages, codes and information. Tomorrow what analysis will break down our objects to reconstitute them in a new space? What new Russian doll will emerge? (Jacob 1973, 324).

Since the seventeenth century, the successive scientific accounts of living beings have gone from a classificatory *historia naturalis*, an anatomy of visible structures, and a physiology of obvious body functions to the multidisciplinary investigative enterprises of the late nineteenth and early twentieth century, with their research objects located on multiple levels.[3] The dimensions of these objects reach from biodiversity, through the social behavior of organisms, right down to macromolecules. To each of these levels of organization and investigation there corresponds a phenomenology, along with a set of analytical approaches and of ways in which the objects are determined. In a history of science from the top down, one has become accustomed to seeing these developments as a social process in which scientific disciplines become more and more differentiated, or as a succession – evolutionary or revolutionary – of theoretical paradigms in a series of ever more refined representations of nature. Both accounts – history of disciplines, history of ideas – are typically structuralist: They identify and lay bare (in the anatomical sense) key elements and structures with an ultimate appeal to universal recognition. In this view, disciplines develop according to a general pattern of differentiation, although they are a relatively modern phenomenon starting only in the early nineteenth century; and ideas develop according to an infinite approximation toward truth/reality.

While the exclusive interest in ideas always risks degenerating into a matter of academic genealogy that searches for a true origin, disciplines are certainly important categories to understand professional developments and social and economic structures of modern science. The disciplinary framework allows particular actors or groups of actors to raise an authoritative voice. It seems, however, that disciplinarists have a tendency to lose touch with scientific contents. In trying to understand the epistemic activity of science, we find it inevitable to ask how research fields, and ultimately disciplines, aggregate around ensembles of practices and technologies, including methods, concepts, and theoretical conjectures. Such aggregations appear to be socially as well as conceptually underdetermined. If, on the one hand, they lead to relatively stable configurations of epistemic power and explanatory relevance, these configurations are, on the other hand, as a rule also historically transient and prone to disaggregation. It is probably no coincidence that we take our bottom-up approach at a time when a thoroughly molecularized and technologized biology is in the process of subverting and blurring all the disciplinary boundaries that, for the short hundred years between 1859 and 1953, between Darwin's *Origin of Species* and Watson and Crick's DNA model, enabled a friendly coexistence between academic specialists in the diverse field of the life sciences under the roof of an overarching evolutionary perspective.

If a pragmatogonic view of the history of the life sciences is taken seriously, historical assessment cannot begin with the reconstruction of a perceived immanent logic of the development of biological concepts. Rather, it will start from a reconstruction of the practical contexts in which particular concepts and associated theories gained momentum and eventually came to dominate. Here, the design of experimental arrangements for the generation of concepts, generalizations, and explanatory patterns in the modern biological sciences becomes crucial. Over the past ten years, we have assembled a number of studies focusing on the role of experiment in the biomedical sciences (Rheinberger and Hagner 1993; Hagner, Rheinberger and Wahrig-Schmidt 1994; Rheinberger, Hagner and Wahrig-Schmidt 1997). Our case studies of experimental systems have always been cautious about the question of when these systems historically came into being and whether they are of comparable importance in other areas of science (Rheinberger 1997). Whatever the future answers to these questions may be, it remains one of the big tasks for history of science to provide a comprehensive analysis of how experimentation itself became a – historically – variable scientific form of life that has undergone decisive shifts of structure and content from the early modern period to the present. These shifting and sifting boundaries reveal themselves best through broadly conceived comparative examinations, which eventually may lead to a classification of the historical settings, structures and meanings of experiment.

As early as in the seventeenth century, natural philosophers can be said to have 'experimented' with living things. Concerning England, Steven Shapin has argued for a sharp distinction between demonstration experiments, displayed for others and characterized by being successful, and more private trials, in which the outcome remained unclear (Shapin 1988, 400–3). In a more diachronic view, the demonstrative experiment of the seventeenth century differs starkly from the experiment as a more or less systematic extension of the art of observation in eighteenth-century French life sciences (Roger 1993). In contrast, in early nineteenth-century Germany, particularly in what one might call Goethean science – though not necessarily that practiced by Goethe himself, who saw the experiment as a 'mediator' of object and subject (Goethe 1962, 305–15) – observation and experiment were regarded as being radically opposed. The former was viewed as the natural attitude of the scientist, the latter as artificial and at least suspect, if not fallacious. This opposition included ethical and aesthetic aspects which formed an integral part of the question of how properly to investigate nature. Criticism against useless, superfluous and badly organized experiments began at the same time, and has developed its own history to the present.

Later in the nineteenth century, in many areas observation itself became dependent on experimental exploration. In the course of these changes, the practice of experimentation took on different forms and became greatly diversified. Concomitantly with this diversification, methodological reflexivity grew. A brief example from mid nineteenth-century experimental physiology may illustrate the point. In his private notebooks, Claude Bernard repeatedly thematized the difference between his approach and the *école allemande*:

> Science is enriched in two ways: either by adding new facts or by simplifying what already exists. [...] I have been reported as finding what I did not look for, whereas Helmholtz only finds what he is after (Bernard 1965, 145).[4]

Experimentation is offered here as a systematic mode of verification and refinement within a certain horizon of knowledge on the one hand, and on the other as an inventive practice in which the unexpected and the unforeseeable is brought into existence. Bernard's plea for the role of contingency and surprise might be read in the light of another note, in which he criticizes the practice of excessive quantitative measurements in physiology:

> Ludwig, du Bois-Reymond, etc., are wrong because they believe in certain precisions they cannot attain (Bernard 1965, 174).

In contrast to the German physiologists mentioned, Bernard was convinced that physiology was not merely applied physics or chemistry (Bernard 1965, 188; Mendelsohn 1965; Coleman 1985). However, this does not explain sufficiently Bernard's own views on his strategy of experimentation. This caveat does not amount to saying that experimental practice is entirely disconnected from metascientific statements, but that their relation has to be investigated – instead of assuming that experimental practice is the direct result of a metascientific, that is conceptual, methodological, epistemological or cultural, position. There is no doubt that a number of mid nineteenth-century physiologists discovered precision as a value and attempted to organize their experimental research around this value, among others.[5] Nevertheless, Bernard expressed strong reservations against precision and quantitative measurement in general as a value. How should we historically interpret such an account?

Even if Bernard did not write his notes for the public, we should not take his statements as an immediate reflection of the experimental process itself. It is quite likely that the passages quoted above do not have an immediate addressee, and that they are nurtured by simplifying dichotomies such as German-French rivalry, competing schools, etc., which were part of the physiological reality of the time. It is thus imperative to compare these confessions with published statements as well as with published and unpublished documents that allow some insight into the experimental work and its conduct itself.

More generally speaking, these rudimentary remarks call for something that could be called a historical physiognomics of experimentation. A comparative study of the forms of experimentation is part of a comprehensive historical epistemology of the sciences and their culture. It is a truism that the epistemological and cultural gaze on experimentation leads far beyond the narrow focus of a philosophy of science in the hypothetico-deductive and in the later analytical tradition that have accorded to the experiment the mere, trivial function of testing hypotheses. But various attempts over the last fifteen years to open 'Pandora's box' will only be successful if the full assembly of new methodologies is mobilized (Gilbert and Mulkay 1984). This includes the comparative analysis of different genres of scientific writing such as research notes, lectures, and published texts of various formats such as handbooks, textbooks, and research articles. It will also require a semiotic approach to visualizations from unpublished sketches or drawings to copperplates, photographs, recordings of traces and inscriptions of various sorts. If these are the media within which the scientific discourse organizes itself, the study of the macro-material culture will address historical laboratory structures including their architecture, the technological functioning of laboratories, the role of laboratory objects, their impact on structuring behavior, and the trajectories of instruments.

EPISTEMIC OBJECTS

In calling for comprehensive comparative studies, one must not overlook the danger of all classificatory enterprises: becoming an endless and fruitless collection of cases. Historical studies of the laboratory do not necessarily have to remain in the realm of microstudies that speak only for themselves. On the contrary, the focus on the materialities of science, on objects and phenomena, opens the perspective for long-range macrostudies of experimentation. A decisive aspect of scientific innovation lies in choosing, shaping, reshaping, and sometimes in abandoning such objects or phenomena. Studying the history of these entities presupposes that they take on variable forms within the different research contexts where they acquire their epistemic virtues and powers. Embryos, brains, mice, or bacteria, for example, are not epistemic objects by themselves; they are things which become epistemically laden insofar as they locate monsters, memory, oncogenes, or messengers in an ensemble of practical and theoretical relations. In contrast to an abstract analysis of pure concepts, we propose an analysis of these impure epistemic entities in their impure settings, half-way between crude pieces of matter and rarefied ideal types: material enough to exert the characteristic resistance and resilience encountered in the exercise of science, and symbolic enough to serve the vested interests of a quest for knowing.

One example of the historical complexity of an object is the transformation in brain research from the 'organ of the soul' to the 'brain' in the years around 1800 (Hagner 1997a). In the early modern period, the organ of the soul played the role of a material medium between the body and the soul. It provided an understanding of man which combined metaphysical and religious aspects with medical and natural historical approaches. The reconceptualization of the brain as an organ in which various mental qualities resided in proximity with each other marked a profound epistemic rupture. Before, there existed a categorical difference between the body and the soul, and between reason and the passions. This accorded with an enlightened, yet strictly hierarchical, view of man. The new development led to brain research as a forceful field within the sciences of man. The previous radical difference now became localized in the brain itself. The brain came to represent an epistemic field of combat for various attributions and oppositions that were fundamental and foundational for the coordinate system of modernity – man/woman, madness/genius, wildness/civilization, and sense/sensibility.

In the life sciences, the choice of organism has always played and continues to play a decisive role in the process of research (Clarke and Fujimura 1992; Lederman and Burian 1993; Kohler 1994). Since the end of the eighteenth century, the meaning of key biological concepts has been shaped by the availability of particular model organisms. Late eighteenth-century vitalism, for example, was by no means a purely speculative endeavor; it is unthinkable without the frog as a model and emblem of bioelectricity. The same animal, however, became crucial in other historical contexts in a quite different manner. It was instrumental in the establishment of physiology as a discipline in Germany around 1830; two decades later, it became the material carrier for the success of the Berlin organic physicists; at the turn of the nineteenth to the twentieth century, it helped to establish *Entwicklungsmechanik*; and it became one of the first higher organisms to enter the realm of molecular biology in the second half of the twentieth century (Rostand 1964a; Rothschuh 1969; Rothschuh 1973; Pera 1991; Holmes 1993).

The eighteenth-century conception of epigenesis accreted around the freshwater polyp *Hydra viridissima* as a model of organic regeneration and, thus, of the vital force of *Bildungstrieb*. We can hardly imagine today the repercussions that were provoked throughout the scientific community by this little animal which Abraham Trembley had retrieved from the ponds of Count Bentinck near Den Haag. For example, the annalist of the Académie Royale des Sciences exclaims in 1741:

> The story of the Phoenix who is reborn from its ashes, fabulous as it is, offers nothing more marvellous than the discovery of which we are going to speak. The chimerical ideas of Palingenesis or regeneration of Plants & Animals, which some Alchemists believed possible by the bringing together and the reunion of their essential parts, only leads to restoring a Plant or an Animal after its destruction; the serpent cut in half, & which is said to be rejoined, gives but one & the same serpent; but here is Nature which goes farther than our chimeras. From one piece of the same animal cut in 2, 3, 4, 10, 20, 30, 40 pieces, & so to speak, chopped up, there are reborn as many complete animals similar to the first (Dawson 1987, 7).

Perhaps to a lesser extent, but significantly nonetheless, the emerging discourse of epigenesis accreted around the incommensurability between the existence of monsters and the doctrine of preexistence. Pierre Louis Moreau de Maupertuis, for example, hinged his arguments against preexistence on this point when he stated:

> Certain monstrosities, either by excess or by default, perpetuate themselves ordinarily from one generation to the other, and through several generations. [...] This phenomenon [...] is inexplicable from the point of view of today's systems of generation [...] supposing that the child is completely prefigured either in the father or in the egg of the mother before the copulation of the two sexes (Maupertuis 1965, 159-60).

Nowhere, however, were monsters used more excessively as an epistemic "instrument of science" (Canguilhem 1975, 179)[6] than in the epigenetic approach of Caspar Friedrich Wolff. He worked for almost twenty years on an encompassing theory of monsters with which he closed his scientific career in St. Petersburg (Hagner 1999). A few decades later, Etienne Geoffroy Saint-Hilaire inserted teratology into the experimental landscape of the early nineteenth century (Rostand 1964b).

The history of genetics, to take another example, could almost in its entirety be written as a succession of model organisms. It started from eighteenth-century observations on human pedigrees of deviant characteristics such as surnumerary fingers. Peas and other plants such as tobacco became favorite objects of breeding experiments as well as for horticulture in the later eighteenth and the nineteenth century. The fruit fly *Drosophila melanogaster* played a crucial role in the establishment of classical formal genetics at the beginning of the twentieth century. The mold *Neurospora crassa* mediated the transition from formal to physiological or biochemical genetics. Without bacteria such as *Escherichia coli* and viruses such as *phage* and the tobacco mosaic virus, molecular biology would have been unthinkable. The molecularized developmental biology of today leads back to fish, mice, and humans. Each of these organisms, in its day, exhibited features that fitted with the means of manipulation available, and they usually resonated with broader concerns of the contemporary life sciences, nutrition regimes, and medicine.

To mention yet another perspective on model organisms, experimental medicine has been epitomized since its beginnings by a debate on what counts as a good animal model for studying human diseases and testing pharmacological substances. A closer analysis of such examples shows that the uses and scientific 'careers' of certain organisms are bounded by a complex set of conditions. These include technicalities such as utility, disposability, aptness for laboratory work, and ease of manipulation. However, they also extend to ethical and economic considerations; forms of everyday life; theoretical presuppositions; and, not least, contingency. These conditions may even include the natural and cultural histories of organisms, diseases, and entire environments. Cultural, social and epistemological aspects of the investigative enterprise are here tightly interwoven and should accordingly be studied within this texture, as Jean-Paul Gaudillière has shown for the example of mice as model organisms for breast cancer in the United States (Gaudillière 1999).

SPACES OF KNOWLEDGE

A natural object or phenomenon such as an organism or an associated structure or function can become attractive for many reasons. Its epistemic relevance, however, depends on its fitting into historically specified real and symbolic spaces of knowledge acquisition. According to Michel de Certeau, space is a dynamic network of mobile elements. It depends upon different speeds and the variability of time.

> Space occurs as the effect produced by the operations that orient it, situate it, temporalize it, and make it function in a polyvalent unity of conflictual programs or contractual proximities (Certeau 1984, 117).

This is a rather unusual understanding of space. What Certeau has in mind is conceptualizing space as an 'anthropological space' as opposed to a 'geometric space.' Following Maurice Merleau-Ponty in this respect, Certeau understands anthropological spaces as being created by human experiences, perceptions, and practices. What can such a conceptualization mean for an understanding of spaces of knowledge like natural cabinets, botanical gardens, laboratories, the camp and field of the anthropologist or the traveling naturalist, or – today – the vast and ever growing spaces of simulation and virtual reality? At first glance, one might be tempted to say that, for example, a laboratory is a laboratory is a laboratory. It consists of stable structures serving the process of research including architecture, electricity, instruments, apparatuses, etc. But such an account says nothing about the intrinsic variability of laboratory equipment. Laboratories are as variable as the experiments performed in the confines of their walls. It is not sufficient to ask whether or not frogs, as objects of scientific inquiry, are proper model organisms. The question must also be posed how the spaces are shaped within which the objects are located and relocated, established and dismissed. In short, the creation of a space is linked to a history, and that space remains a space of knowledge production only as long as it is variable and has open boundaries. If this is no longer the case, the space degenerates to a junk room, filled with objects no longer needed for a specific research enterprise. Undoubtedly, the history of science is full of spaces like this, late witnesses of past successful endeavors or failures.

In contrast to these abandoned or even extinct spaces, there is another kind of space that is best described with Foucault's concept of *heterotopia*, which he contrasted to *utopia*. The latter term signifies 'unreal spaces,' related to the 'real space of society' by 'immediate or inverted analogy.' According to Foucault, utopias 'present society itself in a perfected form, or else society turned upside down.' Heterotopias are characterized as

> real places – places that do exist and that are formed in the very founding of society - which are something like counter-sites, a kind of effectively enacted utopia in which the real sites, all the other real sites that can be found within the culture, are simultaneously represented, contested and inverted. Places of this kind are outside of all places, even though it may be possible to indicate their location in reality. Because these places are absolutely different from all sites they reflect and speak about, I shall call them, by way of contrast to utopias, heterotopias (Foucault 1986, 24).[7]

The representational power of these spaces is enormous, although they do not necessarily invent, create, order or systematize the objects and phenomena they are filled with, independently and splendidly isolated from their entourage. Foucault illustrates the historicity of *heterotopia* with reference to libraries and museums. Until the late eighteenth century, both institutions were the 'expression of an individual choice.' Very much in accordance with his genealogy in *Les mots et les choses*, Foucault locates the rupture in the nineteenth century:

> By contrast, the idea of accumulating everything, of establishing a sort of general archive, the will to enclose all times, all epochs, all forms, all tastes, the idea of constituting a place of all times that is itself out of time and inaccessible to its ravages, the project of organizing in this way a sort of perpetual and indefinite accumulation of time in an immobile place, this whole idea belongs to our modernity (Foucault 1986, 26).

Of course, one could add here the archive, which in its modern form is also an invention of the nineteenth century, standing for virtual completeness and timelessness, maximized order and preservation (Nora 1984–1992). If one thinks of the many thousands of specimens in nineteenth-century pathological museums, the hundreds of monstrous deformations, or the thousands of ethnic skulls on the archival shelves of a Natural History Museum, the physiognomy of 'unlimited amassing' becomes plainly visible. Usually these spaces were not accessible to the public, but were reserved for the anatomist, the naturalist, or the anthropologist. There are, however, also examples of public exhibitions in late nineteenth-century science museums, in which the showcases and glass cabinets were overloaded with objects. The idea behind such amassments was the 'total impression' that was supposed to speak for itself in its overwhelming abundance.[8] Only a few years later, however, this type of exhibition began to be replaced by more didactically, structured and clearly delineated exhibitions, of which the Hygiene exhibition in Dresden in 1911 became an epitome (Offizieller Katalog 1911). The tendency of today's science museums to become places of entertainment and surprise events is only the most recent step in a long history of the visual disposition of scientific knowledge. The point to be made here is that different shelves, cabinets, visual arrangements and designs tell more about perceptions and ways of characterizing scientific objects than the objects themselves can display in their supposedly natural, naked and de-contextualized form.[9]

These few examples call for a more refined historicization of spaces of knowledge in general, and in particular an examination of the strategies of inclusion and exclusion associated with them. Following Foucault in this respect, one can regard the years around 1800 as a watershed, which is not to say that this divide is sufficient to grasp the full range of spaces of knowledge. Here Certeau's dynamic concept of space enters the stage again. Take the spaces opened up by experimental physiology in the second half of the nineteenth century: The first institute of physiology was built in Leipzig in 1869 and was headed by Carl Ludwig (Lenoir 1998, esp. chapter 5). What was special about this place? In an analysis of Emil du Bois-Reymond's physiological work in Berlin, Sven Dierig has argued that the space into which du Bois-Reymond's mode of doing science became inserted developed from a 'parlor physiology' in the 1840s to a 'workshop physiology' in the 1850s and 60s, and to large-scale 'industrial physiology' from the 1870s on (Dierig, forthcoming). This development was accompanied by the industrial and urban development in nineteenth-century Berlin, including economic and infrastructural systems such as water supply and electricity, as well as aspects related to education, art and mentality. Such detailed analyses will have to be carried out for other cities and disciplines, and for other protagonists and time periods. The historical development of the spaces of twentieth-century molecular biology, for example, may provide further insight into the dynamics of contemporary science and broaden the existing perspectives on experimental systems, epistemic objects, and the instruments and technologies that knot them into the fabric of experimental practice.

MENTALITIES

Finally, to a particular material culture there corresponds the mental equipment of the scientist.[10] In fact, *the scientist* is itself a figure displaying historically contingent configurations of explicit and implicit skills, of bodily and mental discipline, and of gestural repertoires. As already mentioned, for a long period in the history of science the creativity and the ingenuity of the scientific genius were regarded as the personal, individually unique complement to structural rationality and objectivity, which were taken to be located beyond the scope of the scientist's personality. Apart from psychoanalytically motivated attempts to develop a psycho-history of scientific creativity, the personal factor was hardly ever a subject of closer historical examination. Only with the advent of the history of the body and the application of Foucault's concept of discipline to the scientist was the body also thematized in history of science (Kutschmann 1986; Feher et al. 1989; Lawrence and Shapin 1998).

These accounts mainly focus on the body's contribution in the engagement with science, whereas they give little credit to cognitive and psychological aspects. The predominance of the body seems to be an exaggerated reaction against the long philosophical tradition of separating mind and body and almost ignoring the latter, or else reducing it to a machine. The history of the body has corrected this image, but at times the focus on the body and on 'science incarnate' seems to dwell uncritically in these dualisms, as if they were primordial distinctions. Our guess is that the instantiation of the scientist's self in modern science does not primarily work along the lines of a mind-body dualism. It is rather the scientist's mind-body as a *mixtum compositum* that flexibly and variably becomes incorporated into and in turn embodies different scientific activities. Foucault's rigorous claim, that the mind is a mere result of the practices of body discipline, fails in explaining explorative activities, which do not follow strict rules and necessities. To be sure, there is no mental activity or expression without embodiment or in-corporation. This, however, does not merely mean discipline, control, and restriction, but also excess, transgression, and 'imagination', to take an expression used by the physiologist Johannes Müller (Müller 1834, 3–4).

Certeau proposed an alternative model in *The Practice of Everyday Life*, in which he conceived of the everyday behavior of consumers as an active mode of practice and not mere passive conformity. Although he accepted Foucault's notion of microphysics of power and tacit technologies behind the scenes, by which institutional procedures are shaped and regulated, Certeau observed that individuals and social groups can be quite successful in escaping these structures. What are the strategies and practices capable of building a counterweight to those silent procedures that organize the socio-political order? Transferring Certeau's account to scientists in action, one might arrive at the following scenario: Discipline does not make very much sense without subversion; rigid frameworks are not very effective without spontaneous bricolage, but they do prevent bricolage from leading to mere noise; practitioners mobilize artfulness, skill and imagination in order to redistribute the spaces in which they work. It is important to note that Certeau does not retreat into a traditional theory of the individual subject. Individuality is, rather,

> a locus in which an incoherent (and often contradictory) plurality of [...] relational determinations interact.

In order to examine this phenomenon more precisely, Certeau suggests examining

> readers' practices, practices related to urban spaces, utilizations of everyday rituals, re-uses and functions of the memory through the 'authorities' that make possible (or permit) everyday practices, etc. (Certeau 1984, xi, xv).

Certeau was interested in consumers, and thus did not turn his attention to mind-bodily practices occurring in the scientific arena. There is, however, no reason why these patterns of analysis should not be extended to the minds and bodies of scientific actors.

Before we give an example to illustrate what this could mean, we must consider a problem encountered by any cultural historian dealing with the composite mind-body (mentality, perception, ideas, imagination, experience, etc.) of an actor: the very explanatory power and reliability of these categories themselves. An influential attempt of highlighting individual forces is Michael Polanyi's category of 'tacit knowledge' or 'personal knowledge' (Polanyi 1958). More or less ignored in the 1960s and 70s, when structures, theories and paradigms were dominant, Polanyi's approach has recently been taken up again and has come to include craft and skill, competence and commitment, bodily discipline and gestures, perception and passion; in short, all aspects of scientific activity that resist explicit formalization. Though being mainly a challenge for traditional philosophy of science, Polanyi's account has also been criticized as inaccessible to historical analysis. In a paper on tacit knowledge and school formation, Kathryn Olesko critically remarked that Polanyi's concept of the inarticulable and untractable nature of tacit knowledge had serious historiographical consequences: If tacit knowledge was unspecifiable, then

> the actual mechanisms for acquiring skills and values central to the formation of a school fell outside the domain of direct historical investigation (Olesko 1993, 20).

In contrast to Olesko's, our problem is not so much whether or not scientists in their environment are to be perceived as conscious and self-reflexive actors; following de Certeau, we would rather contend that skills, gestures, preferences, perceptions, etc., can be historically analyzed independently of whether they are explicitly reflected by the historical actors or not.

We are aware, however, that such a claim bears risks and problems. In his study on the history and anthropology of sensory perception, Alain Corbin pointed out that historians often fall into the trap

> of confusing the reality of the employment of the senses and the picture of this employment decreed by observers.

For example, experts in nautical medicine noted about the sensory perception of sailors that their senses of smell, taste and touch, in particular, were dramatically reduced. Except with respect to vision, sailors had degenerated to 'senseless creatures.' Descriptions of this kind, Corbin argues, display the perspective of the authors and clichés about sailors and social hierarchies (Corbin 1995, 187). A similar problem arises in history of medicine when the world of patients is reconstructed through the world of doctors' records and hospital files.

In the case of the scientist, the historiographical situation is easier, since in most cases scientists do not report about other scientists; they report about themselves. Yet even here one has to be careful, because the documents in question – be they textbooks or public speeches, letters or diaries – may very well represent the rhetoric of received and canonized views instead of undistortedly documenting the uses and abuses of the mind-body. Historians are always to some extent imprisoned within the confines of written documents and the languages these documents speak. However, the point here is less to display a purified and undistorted self than to examine the interplay between the given material culture and the mental equipment. That includes a close look at the apparatuses, instruments, and locales which were part of the experimental activities. This is not to say that experiments in physiology, in particular, would have necessarily to be reworked as has effectively been shown to be possible in physics (Sibum 1995). For many of them, this is not feasible for ethical reasons, but at least in certain areas of self-experimentation such an approach could help to gain a closer understanding of the mastering of instruments in a confined experimental situation, as well as the training of the muscles, the senses, or other cognitive and/or bodily functions. Besides the senses, an analysis of the art of practice will have to encompass the ideas, conceptual tools, virtues, passions, mental dispositions, and idiosyncrasies of the researcher. All that belongs to their mental equipment which is characterized by an assembly of heterogeneous elements. Their configurations and reconfigurations range from the prominence of sensibility in late eighteenth-century life sciences and the role of heroism in romantic self-experimentation to attempts to either include or exclude the participating observer in different versions of twentieth-century anthropology.

One striking example is the history of vertigo and attention as an integral part of self-experimentation. Scientists' understanding of attention underwent fundamental alterations in the course of the eighteenth and nineteenth centuries. These transformations were linked to the role of the scientist as subject and as object in the self-experiment, as well as to the establishment of attention as a fundamental virtue of observation. The ambiguous and reflexive character of attention lay in the dual role which it came to assume as an experimental tool and object of investigation. Until the eighteenth century vertigo was regarded as a merely pathological phenomenon and thus a subject fit only for physicians. In the late eighteenth century, however, empirical psychologists (*Erfahrungsseelenkundler*) came to see vertigo as a disturbance of the regular activity of the mind and started to investigate it more closely. From this time on, the phenomena of vertigo and attention became polarized. While attention was defined as a state of orientation, vertigo was understood as its opposite – disorientation, distraction. As a consequence, the history of these two mental qualities became inseparably intertwined. Moreover, this story shows that the formation and the transformation of categories such as attention and vertigo played a crucial role in establishing the criteria of scientific experience (Hagner 1998, 2001).

CONCLUSION

Looking back at the intellectual projects and projections of the twentieth century that have been put forward as ways of understanding the development of the sciences, it is apparent that dichotomies played a central role in most of them. These dichotomies include the difference between theory and practice, internal and external factors of scientific development, context of discovery and context of justification, basic and applied research, experiment and observation, representation and intervention, engineering and bricolage. We do not deny the historical importance of these categories, but we propose that they be treated deliberately as an *explanandum* and not as an *explanans*. Solutions for overcoming these dichotomies like the universalization of representation in the 1980s lead into a dead end, because ultimately an overly generalized notion of representation explains everything and nothing (Hagner 1997b; Lynch 1994). Bourdieu's solution, a constructivist structuralism, risks leading to a rather ahistorical Hegelianism that does not seem very promising for an understanding of the dynamics of historical processes. Bourdieu tackles the problem of dichotomies without seriously asking how they came into being and why they became so powerful. When we take up the oxymoronic motif of a constructivist structuralism, we do not attempt a neutralizing synthesis of approaches to the history of scientific cultures which have passed their climax and need not be rescued from oblivion. Instead, the conception of artful practice relying on difference, not contradiction, in combination with an encompassing notion of material culture relying on repetition, not identity (Deleuze 1994), can contribute to an understanding of the relation between scientific activity with its heterogeneous arrangement of spaces, objects, and mentalities, and more general lifeworldly and social ideas, practices, and values. Notions such as experimental systems, epistemic objects, spaces of knowledge, and mental equipment are tools to develop a framework for replacing the traditional dichotomies mentioned above. Such a pragmatogonic view aims to salvage what Jacques Derrida once declared to be the very point of the 'structuralist invasion':

> an adventure of vision, a conversion of the way of putting questions before us (Derrida 1978, 3).

And it aims to retain from the constructivist legacy that sense of an irreducible historicity which must accompany all our reflection about the processes of gaining knowledge in their multiplicity of expressions. There is certainly a danger that talking about historicity might become as well a truism as talking about representation, power networks, disciplines, etc. Not a single epistemological leitmotif in science studies is immune against fading away after a while. The traces that it might leave behind are certain unspectacular standards, beyond which no one might want to return. But at the same time, standards are not the driving forces for scientific activity. Science studies and history of science have pointed out again and again that the notions of progress and linear knowledge accumulation in the sciences are problematic. Why should we be surprised if that problem would turn out to be also one of our own field?

ACKNOWLEDGMENTS

We are grateful to Bernward Joerges and Jens Lachmund for their comments and criticisms.

Max Planck Institute for the History of Science, Berlin, Germany

NOTES

[1] The three books by Chartier, Evans and Wehler are subject to an illuminating interdisciplinary debate. See Kiesow and Simon (2000).

[2] Translations are our own, if not mentioned otherwise.

[3] A detailed and stimulating attempt to chart twentieth-century biological kinship categories is undertaken by Haraway (1997, 219–29).

[4] See also a similar note on p. 128.

[5] On the German precisionists see Holmes and Olesko (1994).

[6] This term is taken from George Canguilhem, who has argued that monsters were used as an instrument in the debates between preformationists and epigeneticists.

[7] See also the insightful collection by Smith and Agar (1998). In contrast to our perspective, however, Smith and Agar are mainly interested in the "relations of knowledge to authority, power and control." Emphasizing local cultural contexts, they are interested in "issues of scientific and technological authority, embodied spatially in sites" (Smith and Agar 1998, 3–4, 9). Without underestimating the importance of discursive power for an understanding of modern science, we do not think that it is possible to explain the specificity of the dynamics of science and the production of new knowledge from a historiographical perspective that mainly focuses on authority, power and control.

[8] This strategy of display was developed in the museums of ethnology and anthropology in, for example, Berlin or Dresden. See Zimmerman (2001).

[9] The state of the art and possible future of displaying the sciences in the museum is discussed in Lindquist (2000). On the historicity of cases in natural history cabinets see (te Heesen 2001).

[10] The term 'mental equipment' plays an important role in the tradition of Annales-historians. It attempts to incorporate ideas and conceptual tools into historical analysis. See e.g. Febvre (1985), Le Goff (1988, 100).

REFERENCES

Bernard, Claude (1965), *Cahier de Notes 1850–1860,* Paris: Gallimard.

Bourdieu, Pierre (1987), 'Espace social et pouvoir symbolique', in P. Bourdieu (ed.), *Choses dites,* Paris: Éditions de Minuit, pp. 147–66.

Bourdieu, Pierre (1997), *Méditations pascaliennes,* Paris: Seuil.

Canguilhem, Georges (1975), 'La monstruosité et le monstreux', in G. Canguilhem (ed.), *La connaissance de la vie*, Paris: Vrin, pp. 171–84.

Certeau de, Michel (1984), *The Practice of Everyday Life* (translated by Steven F. Rendall), Berkeley, CA: University of California Press.

Chartier, Roger (1998), *Au bord de la falaise. L'historie entre certitudes et inquiétude*, Paris: Albin Michel.
Clarke, Adele and Joan Fujimura (eds.) (1992), *The Right Tools for the Job. At Work in Twentieth-Century Life Sciences*, Princeton, NJ: Princeton University Press.
Coleman, William (1985), 'The cognitive basis of the discipline: Claude Bernard on physiology', *Isis* 76: 49–70.
Corbin, Alain (1995) 'A history and anthropology of the senses', in A. Corbin (ed.), *Time, Desire and Horror. Towards a History of the Senses* (translated by Jean Birrell), Cambridge: Polity Press, pp. 181–95.
Daniel, Ute (2001), *Kompendium Kulturgeschichte*, Frankfurt a.M.: Suhrkamp.
Daston, Lorraine (1992), 'Objectivity and the escape from perspective', *Social Studies of Science* 22: 597–618.
Daston, Lorraine (2001), *Wunder, Beweise und Tatsachen. Zur Geschichte der Rationalität*, Frankfurt a.M.: S. Fischer.
Daston, Lorraine and Peter Galison (1992), 'The image of objectivity', *Representations* 40: 81–128.
Dawson, Virginia P. (1987), *Nature's Enigma: The Problem of the Polyp in the Letters of Bonnet, Trembley, and Reaumur*, Philadelphia, PA: American Philosophical Society.
Deleuze, Gilles (1994), *Difference and Repetition* (translated by Paul Patton), New York: Columbia University Press.
Derrida, Jacques (1978), *Writing and Difference* (translated by Alan Bass), Chicago: The University of Chicago Press.
Dierig, Sven (forthcoming), *Experimentierplatz in der Moderne – Physiologie in Berlin, 1840–1914*.
Evans, Richard (1997), *In Defence of History*, London: Granta.
Febvre, Lucien (1985), *The Problem of Unbelief in the Sixteenth Century, the Religion of Rabelais*, Cambridge, MA: Harvard University Press.
Feher, Michel, Ramona Naddaff and Nadia Tazú (eds.) (1989), *Fragments for a History of the Human Body*, 3 vols. New York: Zone.
Foucault, Michel (1986), 'Of other spaces', *Diacritics* 16: 22–7.
Gaudillière, Jean-Paul (1999), 'Circulating mice and viruses. The Jackson Memorial Laboratory, the National Cancer Institute, and the genetics of breast cancer', in M. Fortun and E. Mendelsohn (eds.), *The Practices of Human Genetics*, Sociology of the Sciences Yearbook 1997, Dordrecht: Kluwer, pp. 89–124.
Gilbert, Nigel and Michael Mulkay (eds.) (1984), *Opening Pandora's Box. A Sociological Analysis of Scientists' Discourse*, Cambridge: Cambridge University Press.
Goethe von, Johann Wolfgang (1962), 'Der Versuch als Vermittler von Objekt und Subjekt', in Dorothea Kuhn (ed.), *Die Schriften zur Naturwissenschaft*. Erste Abteilung: Texte, Bd. 8, Weimar: Hermann Böhlaus Nachfolger, pp. 305–15.
Gross, Paul R., Norman Levitt and Martin W. Lewis (eds.) (1996), *The Flight from Science and Reason*, New York: The New York Academy of Sciences.
Hagner, Michael (1997a), *Homo cerebralis. Der Wandel vom Seelenorgan zum Gehirn*, Berlin: Berlin Verlag.
Hagner, Michael (1997b), 'Zwei Anmerkungen zur Repräsentation in der Wissenschaftsgeschichte', in H.-J. Rheinberger, M. Hagner and B. Wahrig-Schmidt (eds.), *Räume des Wissens*, Berlin: Akademi Verlag, pp. 339–55.
Hagner, Michael (1998), 'Aufmerksamkeit als Ausnahmezustand', in N. Haas, R. Nägele and H.-J. Rheinberger (eds.), *Aufmerksamkeit*, Eggingen: Edition Isele, pp. 273–94.
Hagner, Michael (1999), 'Enlightened monsters', in W. Clark, J. Golinski and S. Schaffer (eds.), *The Sciences in Enlightened Europe*, Chicago: University of Chicago Press, pp. 175–217.
Hagner, Michael (2001), 'Psychophysiologie und Selbsterfahrung. Metamorphosen des Schwindels und der Aufmerksamkeit im 19. Jahrhundert', in A. and J. Assmann (eds.), *Aufmerksamkeiten*, München: Fink, pp. 241–63.
Hagner, Michael, Hans-Jörg Rheinberger and Bettina Wahrig-Schmidt (eds.) (1994), *Objekte, Differenzen und Konjunkturen. Experimentalsysteme im historischen Kontext*, Berlin: Akademie Verlag.
Haraway, Donna (1997), *Modest_Witness@Second_Millennium. FemaleMan©MeetsOncoMouse™*, New York: Routledge.
Hardtwig, Wolfgang and Hans-Ulrich Wehler (eds.) (1996), *Kulturgeschichte heute*, Göttingen: Vandenhoeck & Ruprecht.
Holmes, Frederic L. (1993), 'The old martyr of science: The frog in experimental physiology', *Journal of the History of Biology* 26: 311–28.
Holmes, Frederic L. and Kathryn M. Olesko (1994), 'The images of precision: Helmholtz and the graphical method in physiology', in M. Norton Wise (ed.), *The Values of Precision*, Princeton, NJ: Princeton University Press, pp. 198–221.
Jacob, François (1973), *The Logic of Life. A History of Heredity*, Princeton, NJ: Princeton University Press.

Jacob, François (1987), *La statue intérieure,* Paris: Seuil.
Jacob, François (1997), *La souris, la mouche et l'homme,* Paris: Editions Odile Jacob.
Kiesow, Rainer Maria and Dieter Simon (eds.) (2000), *Auf der Suche nach der verlorenen Wahrheit. Zum Grundlagenstreit in der Geschichtswissenschaft,* Frankfurt a.M.: Campus.
Kohler, Robert (1994), *Lord of the Fly. Drosophila Genetics and the Experimental Life,* Chicago: University of Chicago Press.
Kutschmann, Werner (1986), *Der Naturwissenschaftler und sein Körper,* Frankfurt a.M.: Suhrkamp.
Lawrence, Christopher and Steven Shapin (eds.) (1998), *Science Incarnate. Historical Embodiments of Natural Knowledge*, Chicago: The University of Chicago Press.
Le Goff, Jacques (1988), *Histoire et mémoire,* Paris: Gallimard.
Lederman, Muriel and Richard M. Burian (eds.) (1993), 'The right organism for the job', *Journal of the History of Biology* 26: 235–367, Special Section.
Lenoir, Timothy (1998), *Instituting Science: The Cultural Production of Scientific Disciplines,* Stanford, CA: Stanford University Press.
Lindquist, Svante (2000), *Museums of Modern Science,* Canton, MA: Science History Publ.
Lynch, Michael (1994), 'Representation is overrated: Some critical remarks about the use of the concept of representation in Science Studies', *Configurations* 2: 137–49.
Maupertuis de, Pierre Louis Moreau (1965), *Système de la nature, Oeuvres II,* Hildesheim: Olms.
Mendelsohn, Everett (1965), 'Physical models and physiological concepts: Explanation in nineteenth century biology', *The British Journal for the History of Science* 2: 202–19.
Müller, Johannes (1834), 'Jahresbericht über die Fortschritte der anatomisch-physiologischen Wissenschaften im Jahre 1833', *Archiv für Anatomie, Physiologie und wissenschaftliche Medicin,* 1–201.
Nora, Pierre (ed.) (1984–1992), *Les lieux de mémoire,* 3 vols, Paris: Gallimard.
Offizieller Katalog der Internationalen Hygiene-Ausstellung Dresden, Mai bis Oktober 1911, Berlin 1911.
Olesko, Kathryn M. (1993), 'Tacit knowledge and school formation', *Osiris* n. s. 8: 16–29.
Ostwald, Wilhelm (1910), *Grosse Männer,* Leipzig: Akademische Verlagsgesellschaft.
Pera, Marcello (1991), *The Ambiguous Frog. The Galvani-Volta Controversy on Animal Electricity,* Princeton, NJ: Princeton University Press.
Polanyi, Michael (1958), *Personal Knowledge. Toward a Post-Critical Philosophy*, London: Routledge and Kegan Paul.
Rheinberger, Hans-Jörg (1997), *Toward a History of Epistemic Things. Synthesizing Proteins in the Test Tube,* Stanford, CA: Stanford University Press.
Rheinberger, Hans-Jörg and Michael Hagner (eds.) (1993), *Die Experimentalisierung des Lebens. Experimentalsysteme in den biologischen Wissenschaften 1850/1950,* Berlin: Akademie Verlag.
Rheinberger, Hans-Jörg, Michael Hagner and Bettina Wahrig-Schmidt (eds.) (1997), *Räume des Wissens,* Berlin: Akademie Verlag.
Roger, Jacques (1993), 'L'histoire des sciences: Problèmes et pratiques. Histoire des sciences, histoire des mentalités, micro-histoire', in J. Roger (ed.), *Les sciences de la vie dans la pensée française au xviiie siècle,* Paris: Albin Michel, pp. xix–xxxix.
Rostand, Jean (1964a), 'Des grenouilles et des hommes', in J. Rostand (ed.), *Biologie et humanisme,* Paris: Gallimard, pp. 271–301.
Rostand, Jean (1964b), 'Etienne Geoffroy Saint-Hilaire et la tératogenèse expérimentale', in J. Rostand (ed.), *Biologie et humanisme,* Paris: Gallimard, pp. 135–50.
Rothschuh, Karl Eduard (1969), 'Von der Idee bis zum Nachweis der tierischen Elektrizität', in K.E. Rothschuh (ed.), *Physiologie im Werden,* Stuttgart: Fischer, pp. 139–54.
Rothschuh, Karl Eduard (1973), 'Laudatio ranae exploratae', *Sudhoffs Archiv* 57: 231–44.
Shapin, Steven (1988), 'The house of experiment in seventeenth-century England', *Isis* 79: 373–404.
Sibum, Heinz-Otto (1995), 'Reworking the mechanical value of heat: Instruments of precision and gestures of accuracy in early Victorian Britain', *Studies in the History and Philosophy of Science* 26: 73–106.
Smith, Crosbie and Jon Agar (eds.) (1998), *Making Space for Science. Territorial Themes in the Shaping of Knowledge,* London: MacMillan.
Sokal, Alan and Jean Bricmont (1997), *Impostures intellectuelles,* Paris: Editions Odile Jacob.
te Heesen, Anke (2001), 'Geschlossene und transparente Ordnungen. Sammlungsmöbel und ihre Wahrnehmung in der Aufklärungszeit', in Gabriele Dürbeck et al. (eds.), *Wahrnehmung der Natur, Natur der Wahrnehmung. Sehen und Sichtweisen um 1800,* Dresden: Verlag der Kunst, pp. 19–34.
Wehler, Hans-Ulrich (1998), *Die Herausforderung der Kulturgeschichte,* München: Beck.
Zimmerman, Andrew (2001), *Anthropology and Antihumanism in Imperial Germany*, Chicago: The University of Chicago Press.

CHAPTER 12

MICHAEL GUGGENHEIM AND HELGA NOWOTNY

JOY IN REPETITION MAKES THE FUTURE DISAPPEAR

A Critical Assessment of the Present State of STS

THE CURRENT JOYS OF STS

Every scientific field has to set up its own notions of progress, even though these may turn out to be time-dependent.[1] Although it might be argued that STS as a field is still too young for a comparison of its initial goals and visions with its actual achievements, such an argument ignores the pressure on any new field to build up an intellectual, disciplinary and institutional identity.[2] There are various ways of constructing such an identity: breaking with a tradition, and hence opposing it; inventing a past that serves to provide normative reference points for present directions; or continuously readjusting history to legitimize present and future courses of action. It is our contention that STS as a field has largely failed to set up criteria for its progress. Its development has proceeded for the most part in an unplanned manner and without a real center. The intellectual and social organization of STS was built around single books (and their authors) rather than with an attempt to establish a tradition in textbooks.[3] The vagaries of the institutionalization of STS, although different in Europe and in the United States, have undoubtedly contributed to a process whose outcome resembles what Richard Whitley has called 'fragmented adhocracies' (which occasionally shade into 'polycentric oligarchy'). The dominant feature of these sciences is their intellectual variety and fluidity. They lack a stable configuration of specialized tasks or problem areas, nor do they have strong co-coordinating mechanisms which systematically interrelate results and strategies. The underlying matrix is a combination of high task uncertainty with a low degree of mutual dependence (Whitley 2000).

As long as STS was producing new literature clearly distinct from the 'old' in sufficient quantities, this alone sufficed to mark progress, and no additional criteria had to be supplied. Due to circumstances mentioned also by Richard Whitley in his retrospective (contained in this volume), the field began to flourish with a multitude of strands developing side by side. Efforts to build a shared identity with recourse to a few founding fathers (and no mothers, as usual) remained either isolated or were simply ignored.

B. Joerges and H. Nowotny (eds.), Social Studies of Science and Technology: Looking Back Ahead, 229–258.

There was a widespread but illusory assumption that STS could imitate what was believed to be one of the ingredients of success in the sciences, namely the inherent 'opportunism' attributed to natural scientists. What was overlooked was that for opportunism to work, the 'right' opportunities have to be present. STS could also – and, unfortunately, rightly so – claim that questions and criticisms raised from outside the field were either based on blissful ignorance or willful misrepresentation, as evidenced in the recent 'science wars'. Collective and individual responses which arose in defense of the 'correct core' of STS literature and its theoretical and empirical foundations hardly went beyond the *status quo* to question the direction in which the field was moving. Under these circumstances, no vision of the future is needed, since the present delivers all that is seemingly wanted: more of the same.

The history of STS enabled it to conceive itself as a modern Sisyphus. One of the implicit assumptions of its early programme was that the prevailing understanding of what science was and how it worked was deeply flawed, and that it was the calling of STS to correct these distortions. Deconstruction and de-mystification were means of achieving this. But this task did not come to an end, since additional misunderstandings and distorted views arose inevitably from different quarters. No longer were the 'old' philosophers of science the main culprits. They simply ceased to be interlocutors. The epistemological questions they had raised no longer were intellectual issues over which academic battles could be fought. Once the philosophers were no longer the main contenders, another spate of misunderstandings emerged, many now internal to the field itself. In the general eagerness to eradicate heresies a number of vociferous campaigns in journals and in conferences were fought in the name of STS. Reflexivists were pitted against empiricists, the 'social' became opposed to the 'cultural', SSK confronted ethnomethodology and ANT had to stand up to social constructivism.[4] While these lively debates were initially conducive to an atmosphere of intellectual dynamism, over time their argumentation either tended to exhaust themselves or led to a narrowing of focus. The debates essentially became a self-serving matter for insiders. Meanwhile, in numerous empirical case studies, most of them situated on the micro-level, a fascination, if not an infatuation, with the artifacts under study set in. The style of presentation began to resemble in its performativity the practice of 'show and tell', a characteristic of the U.S. school system, but also of certain media styles (Wagner 1996).

There remained, however, two powerful sources that provided a constant flow of distorted or misunderstood notions about science for STS to correct. One was latent and difficult to detect, the other manifest and seemingly easy to deal with. The latent but immensely powerful source of 'false' images, self-interpretations and public representations was science itself. This source remained hidden for a long time, since its only public manifestation was found in the work of philosophers acting as the self-appointed spokespersons of the scientists. The moment of revelation came when the different 'versions' of science exploded onto the public scene with Sokal's hoax. The virulence, utter hostility and aggression that followed took STS by surprise, even if it can be explained in retrospect.[5] What was at stake now became a political issue, namely who is allowed to speak 'in the name of science': STS or the scientists?[6] The other, more manifest source of public images, representations and the proper 'understanding' of science in need of continuous adjustment and correction is 'society' itself. As shown, for instance, by Pestre in a programmatic article in 'Annales', 'society' has a false understanding of science, and STS, together with the historians of science, is here to correct it (Pestre 1995). This is the reason why STS has carved out for itself an honourable role as authentic representative and spokesperson for a 'society' whose images and understanding of science were either wrong or too powerless to make its voice heard. Hence also the celebration of laypersons, whose knowledge and understanding are alleged at times to be superior to those of the scientific experts.

The convenience of having chosen the role of Sisyphus is not to be underestimated. His fate is that of muted rebellion, of having assumed the literally unending and unfinished task of responding in a repetitive and unchanging way to circumstances not of his making which continue to re-generate themselves. Sisyphus does not have to ask many questions about himself and his destiny, since all is given. *Joy in repetition*[7] makes the future disappear and therefore renders it harmless. An unwanted side-effect not publicized by the gods, who wanted to punish him, is the intellectual blindness that befell Sisyphus. It never occurred to him that he himself might question what he did and how he got there, questions that could lead him to alter his agenda.

STS likewise has never established its own research agenda of unresolved problems. Yet, if it does not want to become trapped in an unending present of joyful repetition, albeit in the guise of rebellion, it will have to create a future that is neither Sisyphus-like, and therefore closed, nor utopian or dystopian, as so many future expectations of other actors in society are when assessing the benefits and burdens of science and technology. In a caricature of its own theoretical stance that scientific problem choice is not the outcome of an inherently rational selection process within science, nor that Nature whispers into the ears of scientists which problems to choose, STS hardly ever showed interest in the functioning of its own processes of problem selection or which issues, originating in wider societal developments, it did take up or not. Nor did it delve into the processes through which contrasting theoretical positions within STS are negotiated or eliminated. What, if anything, did Society whisper into STS ears? If problem choice within STS is at least partly seen to be the outcome of larger societal processes, then a vision of the future would be needed which takes into account the processes of its own professionalisation and its likely effects. This implies the capability to select problems for their own – scientific and intellectual – sake and to transform them into research priorities which are seen as a collective, and not simply as an individual task. It also implies the willingness and capability to transmit the relevance of scientific activities to a wider audience and perhaps even to one's 'clients'.

We perceive STS to have reached a critical point in its development. The change is one from exploring new intellectual ground and joyfully inventing stimulating concepts and novel theoretical directions in blissful ignorance of where it might lead, to having to confront its own possible futures. But even if the road towards professionalisation and consolidation is chosen, it is still full of pitfalls, since different interpretations are possible. One is a more narrow, academic interpretation, while the other takes into account the larger set of societal, economic and political forces that impinge on every academic endeavour. Our task in this paper will be to critically examine the state of the art and, with occasionally polemical remarks, look for new future directions.

We proceed as follows: In the first section we analyse the historiography of STS. We claim that the historiography of STS is still not STS-like and we will outline what a more STS-like approach to the historiography of STS would have to take into account.

In the second section we reconsider the link between STS and its subject of inquiry. We will argue that STS has never attempted to define its subject matter consistently. We attempt to offer such a definition by moving away from the terms 'Science' and 'Technology' replacing them with the terms of 'the production of new knowledge and things' as the defining focus of STS. The next two sections start with the observation that STS on its way towards becoming a 'real' discipline has lost some of its previous links to the other social sciences. We will inquire how STS could strengthen again its grounding in social theory. We do this, first, by showing what STS has to offer to social theory and how it could lead in reformulating social theory based on the assumptions of STS. Second, we will present some illustrative examples from social theories which are neglected inside STS, but could be very fruitful if taken up by STS. In the last section we present a polemical outlook on a utopia (some might say: a dystopia) of STS restructured along our theoretical assumptions and developed organisationally towards becoming a profession.

Before developing these arguments further a precautionary remark is in order. What is termed STS in the following pages is necessarily an abstraction which can never encompass the entire, very broad and richly textured field of science and technology studies as it exists today. When we speak of STS, it stands for what we perceive to be the current mainstream of a scientific field in the sense used by Pierre Bourdieu. Mainstream therefore stands for structuring positions in a structured field. Mainstream is most visible and 'shown and told' in a performative mood. When making this abstraction we follow the 'logic' of the field and its specificity in accumulating scientific capital. We try to identify the strategies through which strands of knowledge are turned into scientific capital, who owns it and which kinds of strategies, methods and rhetorics are used in order to increase its accumulation. The diversity of the field STS may be an indicator of a relatively low degree of hierarchization inside the field, because it is relatively easy (and occasionally very useful) to accumulate scientific capital in neighbouring fields and import it. Nevertheless this does not mean that a hierarchy of themes, methods, theories and strategies does not exist inside the field. They form part of the tacit assumptions of all actors in the field. Although it is impossible here to attempt an even brief empirical study of the scientific field of STS we firmly base our analysis on the assumption that STS, just as any other scientific field, must be conceptualised in a sociological way to understand its functioning. To speak with Bourdieu, such an analysis of one's own field can even form a prerequisite for its functioning:

> The particular difficulty which sociology has in conceiving science scientifically is related to the fact that sociology is situated at the very bottom of the social hierarchy of the sciences. Whether it rises to conceive other more scientific sciences better than they conceive themselves, or descends to record the triumphant image produced and propagated by scientific hagiography, sociology always encounters the same difficulty in conceiving itself, i.e. conceiving its own position in the social hierarchy of the sciences (Bourdieu 2000, 44).

And we should add: STS encounters the same difficulties as does sociology in conceiving the positions inside its own field. Such a conception of mainstream (a notion that is needed for the claim that there are heretics) obviously ignores important differences among national, disciplinary and political strands as well as a-synchronic developments occurring in different parts of the world. There also exists a built-in tension between what we perceive to be the mainstream and this Yearbook which needs to be explicitly acknowledged. The Yearbook, as described by Whitley (this volume), was established under historical circumstances, partly in confrontation with the mainstream of the time. It is certainly too simple to claim that one mainstream has been replaced by another and that the Yearbook's role has been, and continues to be, a kind of monitoring and steering against the tide. Hence, when we plead for a critical re-examination of what the term 'science' in science studies stands for and urge a re-thinking of the role of social theory for the future development of STS, we do not suggest that others have not also raised similar criticisms or opened up new directions. And yet, the tension between what we call the mainstream and what the contributions in this volume represent, is real and needs to be spelled out.

A further remark is necessary here. Our arguments against the mainstream require clarification as to where we stand in relation to it and therefore to position ourselves among the meanderings of other streams that, together with the mainstream, make up the dynamics of the field. We do not consider ourselves to be outsiders of STS. On the contrary, we see ourselves as partaking in the successes and failures of STS upon which the empirical content of the subsequent analysis rests. One of the authors (Helga Nowotny) shares with the other first generation members of the Editorial Board of the Yearbook a common background. In fact, she had been a student of Robert K. Merton in her Columbia University days and, like others, found it necessary to depart from what Mertonian sociology of science had come to embody. The common intellectual interests that brought together the members of the editorial board have been recorded elsewhere in this volume. Her biography led her back to countries in Europe in which it would take decades (and sufficient self-reliance in her own institution-building efforts) to set up an institutionalized basis in which STS could take root. Both Austria and Switzerland share the situation of somewhat peripheral countries that, due to size and geography as well as due to the (different) legacy of their national histories, have remained outside the forces that have shaped the social sciences in countries like France, Germany, the UK and USA. This creates a tension between being a member of an international scientific community in which scientific reputations and scientific careers are made and a 'career' in a more narrow sense as belonging to a quite different, and 'deviant', local community. Helga Nowotny's visibility and career were built on continuously interweaving opposites, combining also science with science policy analysis and advice. In the end came the (successful) attempt to institutionalize STS in places where it had not existed before. Having first had to build up an infrastructure confers perhaps immunity against hyper-reflexivity and certainly induces impatience with any move to give up too fast what had taken long to acquire.

Michael Guggenheim, the other author, found himself to be a student of the local *Volkskunde* at the University of Zurich, an academic sub-speciality of cultural anthropology, which exists only in German-speaking countries. His intellectual curiosity and other, more contingent, factors led him to discover science studies while still a student. What he found was an already prefigured and well-established field, which had its mandarins, its 'great books' and exemplary models of how to conduct research on topics, which were also given. Like others in his generation, he had no choice but to try to identify gaps in research and to articulate theoretical lacunae that would allow to position himself at a time where the field was already structured with no 'revolution' in sight. STS presented itself to him as any other academic discipline – with the difference that, unfortunately, it could not yet be studied in Switzerland. The local vacuum was in stark contrast to international developments, but it was easily filled by the passing through Zurich of the jet-set of STS luminaries. In any case, STS appeared to him much more coherent than the local variations of either sociology or *Volkskunde*. Compared to their inferiority complex, he could not understand what the proclaimed identity problems of STS should be all about. This highly localized and somewhat idiosyncratic view of STS offered a habitus and a strong sense of identity. It could even be turned into a glamorous 'view from somewhere, where the view is best' kind of attitude (remember that Switzerland is located in the Alps and part of the Swiss unconscious is to strive for places where the view is best). The authors joined efforts, curious about the convergence of their ideas and outlooks despite – or because – their different departure points and in search of establishing an intergenerational dialogue. But the present contribution can also be read as a prosaic attempt to pass on scientific capital from the older to the younger author.

HOW TO GET THERE: THE MISSING HISTORY OF STS

The missing future of STS is paralleled by its missing past. This does not mean that there are not many stories in circulation claiming to constitute its history. Rather, and curiously enough, these stories resemble the kind of history being told by any other discipline, or found in other introductory textbooks. It consists of standardised *topoi*, many of which are transferred uncritically from the founding generation to the next, without attempt to analyze their making. Where historians of science normally tread very carefully in unveiling the constructions of scientists as active participants in establishing their history, and scrupulously attempt to spell out the alternating functions served by historiography, there appears to be a collective blind spot within STS. It is perhaps indicative in this respect to note that the history of STS, as written by David Edge and published in the *Handbook of Science and Technology Studies*, is rendered as a first person account.[8] It begins with Edge, as the lonely hero, arriving at "... Waverley Station in Edinburgh, on the night train from London, to start the Science Studies Unit at Edinburgh University." His office was empty, "...no phone, no books, no bibliographical resources, no files, no staff – indeed, it was tempting to think, *no subject!*" (Edge 1995, 2). Although this narrative of the history of STS is very thoughtful, its beginning is more reminiscent of the account of an explorer or an ethnographer in the classic era of colonialism than of a historian or sociologist of science. The reader, in this case the uninitiated younger generation of STS scholars, are invited to take part in the adventures of a founding father and to delight in the times when STS was still more of a pioneering venture than a discipline. STS in those days appears to have resembled finding Colonel Kurtz in the jungle: nothing, not even a subject. Now we have phones, books, bibliographical resources, files, staff and much more, such as computers and search engines. We even have a subject or – to be honest – some hundreds of subjects, programmes, theories. We have material cultures and experimental cultures, bodies and artefacts, factishes, actants and much more – but still not the history we deserve.

In this respect, STS is becoming like any other discipline, too normal, perhaps. The standard history of scientific disciplines is usually a history of progress, of emancipation. Mertonian sociology of science, the older sociology of knowledge and analytic philosophy of science were, with different degrees of emphasis, all caught in the unwarranted belief that the substantive content of science – scientific knowledge – was impenetrable, if not taboo. It was considered either too difficult to pry open this forbidden zone, or the zone was thought immune to study by sociological eyes, since it contained the heart of Western science, namely rationality. Enter the courageous STS tribe!

Having acquired telephones, hired some staff, read some books and found a subject, they now fearlessly enter the laboratories whose threshold no social scientist had ever dared to cross before. They tunnel their way into historical archives and, in alliance with historians of science, who quickly become infected by their enthusiasm, search there for 'hard data', which turn out to exist in abundance. They reveal that science is not what it was thought to be, nor what it publicly claimed to be. It is a conglomerate of (cultural) practices, just like other fields of activities or other kinds of work.[9]

This is not intended to be a lament about the state of historical consciousness, or rather the lack of it, in STS. But the parallels with any other disciplinary history or with accounts of the pre-STS history of science are nevertheless striking. Irrational belief is overturned by 'real facts', brought to light by the researcher, who replaces mere speculation with data, texts and figures. One is reminded of the anecdote of the three soccer referees who explain to each other how they know when the rules have been broken. I whistle because a foul has been committed, says the first. I whistle because I have observed that a foul has been committed, says the second. For the third it is also clear. When I whistle, the foul occurs, he says. It may well be argued that a discipline, even if it takes pride in having discovered that fouls occur when referees whistle, also has to resort to the kind of history which shows the progress from a naïve realism to a sophisticated constructivism. It may well be necessary for STS, too, to prove itself capable of providing students with a sense of disciplinary identity and of continuity over time linking them to their founding ancestors. STS also may be in need of its distinct founding myths as a form of social bonding. But reflexivity entered the social sciences precisely by critically examining how social functions mingle with substantive content, and it spoke out against naïve dichotomies that tried to separate the two. It had argued in other cases that such unexamined constructions of history may lead to fateful misconceptions, uncritically favouring the actions of heroic individuals and promoting a notion of progress as inherent in the field. But in the end, STS turned a blind eye to the risk of a far too optimistic image of a discipline, the result of ignoring how changes in the broader institutional and political environment affect its own historical development.

We propose therefore an assessment of STS in terms of its own standards and criteria. This means that a socio-cultural history of STS must be written to take into account questions like the following: what is the connection between the development and institutionalization of STS and broader economic and political developments?[10] How did the specific cultural and geographic patterns of STS and its diversity in orientation and research style develop as they did? Why are there relatively many women in STS as compared to, say, economics, history, or even sociology and anthropology? What were the factors leading to the establishment of STS units in the universities of some countries but not in others? What were the relations between STS and science policy and what has changed, if anything, with the growing importance of EU research policies? Why did the political strand inside STS undergo a relative decline? Such a history, were it to be written, would probably show how STS people fought for institutionalization inside their university and how those who succeeded were also fortunate enough to have used university reforms or other academic restructuring to their advantage. It would probably show how STS people have entered specific 'trading-zones' with other academic tribes to exchange concepts and methods, and how they have fared when encountering the numerically superior cultural studies tribe. It would tell how STS, despite its initial claims of overcoming the restrictions of Western science, remained caught in its hegemonic and patriarchal tradition. It would show how the 'feminist focus' within STS, despite its considerable breadth and depth in conceptual and empirical studies, has failed to lead STS as a whole towards collective action against widespread gendered practices. It would show how Bruno Latour, as a 'wild capitalist' of his own making, was to further his career as an entrepreneur in STS, mirroring the development of his rich oeuvre.[11] We could continue with more detailed chapters of this imaginary future history of STS, but need to content ourselves here with one further speculative hint arising from the criticism voiced above.

If we return to the history of STS as told by David Edge, an important but today often neglected strand of its origins is its use of the Marxist critique of science and its links with radical science and other protest movements. STS in this view grew at least partly out of the broad stream of other political and social movements of the late sixties and early seventies, as did feminism and the environmental movement.[12] Such a co-evolutionary development between social movements and their subsequent academic integration or appropriation is nothing new in itself. The constitution and establishment of the social sciences in parts of Europe in the late 19th century also grew out of a strong reform movement and for a while remained intellectually linked to its political agenda (Nowotny 1990). In contrast to social movements, however, whose *raison d'être* does not cease to exist as long as the grievances and conditions persist against which its protest is directed, the translation of social problems into scientific problems has to follow a different kind of logic. Science is under the obligation to bring forth continuous novelty, be it the discovery of novel phenomena or novel explanations or be it the novel recombination of older problem formulation. STS, therefore, no more than any other scientific field, cannot retain the same politically or socially inspired research agenda over time.

Given the origin of STS at least in part in the protest movements of the seventies, it is interesting to compare the subsequent development of STS with that of the environmental movement. In our view the environmental movement succeeded in transforming its initial moral protest into the refined structures of academic environmental studies programmes. These turned out a new generation of environmental *professionals* who found work and employment in NGOs with their world-wide networks, in the offices of national and international law-makers, in political parties and their think-tanks as well as by setting up their own small but thriving consultancies.[13] Each of these heterogeneous sites of knowledge production contributes in turn towards further enhancing the dynamics of interaction between itself and the generation of new knowledge in other sites. In contrast, STS has succeeded merely in establishing its own academic base, which remains unsatisfactory and deficient in some countries and vulnerable to the calculatory whims of the new academic entrepreneurs in deregulated universities in others. STS has gradually lost contact to the social movements it was originally affiliated with, without building up other sorts of ties.

The point here is not to describe what happened as a form of depoliticisation or to lament the loss of moral outrage. Moral outrage needs to transform itself into another kind of sustainable programme that fits into (and may even subvert) already existing structures, be they scientific, political or juridical.[14] In contrast to the environmental movement, STS has simply failed to establish its own professional structures that would have linked its academic institutionalization to the wider political, economic and cultural world and the issues that dominate there. Was it, we might ask, because the habitus of STS people has become too divergent from that prevalent in social movements, making it difficult to find a pragmatic common ground for common work?[15] Or is there something inside STS's mode of knowledge production which focuses its intellectual attention and analytic strength on the production of texts? Whereas each environmental studies programme is still grappling with some 'real' issues of environmental change and degradation in addition to a host of other political, institutional, legal and cultural problems, the political impetus within STS seems to have become confined to professing its good intentions.[16] As with cultural studies, whose origin, at least in the U.S., also lies in part in a wide-spread mood of unease and political protest against the white-male canon, STS followed the move and retreated into its academic niche.

It seems that we will have to 'reinvent the wheel', to use David Edge's phrase. He argues that "the original aims of the STS pioneers [sic!] are still very much alive and relevant...but that, for all the progress of the past quarter century...the insights we have gained still seem, too often, to be almost wantonly disregarded. The potential audiences for our 'messages' are stubbornly unreceptive" (Edge 1995, 4). Edge sadly sees others, mainly outside the STS community, finding anew the old sense of practical urgency and unease at the social status and role of science and technology. In attempting to end this unease, they then 'reinvent the wheel' and, in so doing, largely ignore STS's achievements. But reinventing the wheel turns out to be a tricky business, whoever the reinventors may be. Why the wheel, one might ask, and why blame the 'potential audiences' for not having heard the message? And, while we share David Edge's amazement at the puzzling tenacity with which scientists still cling to the 'old', positivistic image of science as an abstract, timeless search for irrefutable truths, we cannot help wondering about STS's irresponsibility in having failed to see the interdependencies between the authority ascribed to Nature and the social authority acquired by those who act in its name. Was it only when 'society started to speak back to science' that STS was forcefully reminded of its own analytical capabilities, which were far superior to those of the various 'Public Understanding of Science' campaigns? Will it reinvent the wheel or is it ready to move towards the invention of a more advanced technology?

WHAT IS THE 'SCIENCE' IN SCIENCE STUDIES?

The word 'science' in science studies still seems largely to mean physics, although the weight is shifting in the direction of the life sciences. This statement does not imply that other fields and subjects are not studied, but that whatever is within the scope of STS is still held against some kind of standard image of science which is definitely not geology or psychology. We claim that science studies in their actual practices carry a hidden assumption of what science is. Our intention here is to highlight this assumption and to question it.

Despite the vast and continuously expanding range of topics, objects and fields of study of STS, a certain hierarchy among them persists. First and foremost, the 'science' in science studies refers to the natural sciences. While this may be self-evident in English, it is far less clear in German where STS, for instance, is called *Wissenschaftsforschung* and where the term *Wissenschaft* is known to comprise the social sciences and humanities as well. Among the natural sciences the most rewarding fields of study for STS are those having a long, established tradition and those that are considered to be 'hard'. This confers prestige and renders attractive those scientific activities involving scientific instruments, experimental set-ups and visualising practices, or which at least focus on something tangible, like graphic representations capable of being 'shown' and 'told'. Scientific work carried out mainly in symbols or conducted in thinking and writing is only marginally represented in science studies.[17] While this might partly be due to methodological considerations and relative ease of access, it is also the fascination with 'things' that accounts for their privileged disposition in becoming favourite STS objects of study. The inclusion of the social sciences or of law has hardly begun to attract STS attention.[18] It seems that STS has never been really interested in what it, as a social science, is most familiar with, thereby unwittingly reproducing the history of anthropology: one starts with the most exotic societies and their strangest customs before eventually being allowed to return home. The problem is that this exoticism veils the reason for an object's having being chosen for study in the first place. It eclipses the criteria according to which research objects are selected, because a choice based upon their unfamiliarity and strangeness obliterates the necessity to come up with theoretically based arguments.[19]

'Strangeness', unfamiliarity or distance as criteria for choosing one's field of study are problematic terms, however. If the hierarchy of the sciences and the differences in reputation which result from dealing with the more or the less prestigious sciences were openly acknowledged, the reasons for the actual choices made would be clearer. The STS community consists of those who have a background in the natural sciences and those who come from a variety of social sciences or from the humanities. For the latter, the strangeness criterion applies more than to those whose disciplinary origins lie in the natural sciences, but there is also more prestige for them to gain from turning to the natural sciences. This is also the reason why the social sciences never attracted the attention of the former natural scientists. There is also a certain gap remaining between what is considered science in the sense of original research and what is considered further 'downstream' in the sense of applied knowledge. Engineering and medicine are seen as being somewhat different and peripheral to 'science'. This is partly so for historical reasons, because these disciplines were already thought to be open to social influences long before STS came into existence. Research on engineering and medicine continues to engage research communities not identical with STS, who gather in conferences of their own and publish in different journals.[20] Last but not least, 'science' refers – with notable exceptions – to science in Western countries and STS accordingly seems to perceive itself as producing knowledge of relevance only to the Western world.[21]

Whatever the origins and explanations of these persistent hierarchies and how they differ in detail, they go almost unnoticed within STS and remain uncontested. Nor is it easy to locate attempts to delineate the field of STS and to analyze how particular boundaries became established. Explicit definitions of what 'science' is are extremely rare. For example, David Hess, in his introductory book *Science Studies*, is very careful to delineate disciplinary boundaries *inside* STS but remains silent on what constitutes science (Hess 1997). Dominique Pestre argues that the history of science (for him co-terminous with STS) should not attempt to define what science is, since it should only take as 'science' the notions that were current in certain historical circumstances (Pestre 1995, 494). What might be the reasons for such self-imposed abstinence?

Mario Biagoli provides an answer in the introduction to his *Science Studies Reader* (Biagoli 2000). He argues that science studies, unlike other academic fields, do not "have to define [their] subject matter in relation to neighboring disciplines; over the years, the scientists have done much of that work" (Biagoli 2000, XI). He develops his argument with two comparisons. First, he compares STS to classical ethnography. As does ethnography, STS identifies 'tribes' inhabiting a circumscribed territory as its chosen objects of study. But unlike ethnography, STS never was a colonizing force; rather, it was colonized by its subject. The second comparison is with literary studies. Biagoli maintains that the object of literary studies, i.e. literature, retains a passive role and that therefore literary studies must always debate their proper object – the literary canon – anew. STS, again unlike literary studies, does not face such a problem, since the scientists themselves delineate what counts as science. What remains for the student of science studies is therefore to investigate "how science works" (Biagoli 2000, XII).

Interestingly, such a self-imposed division of labour runs counter to the usual arguments of scholars in the humanities and the social sciences – from Foucault to Garfinkel, from Bourdieu to Giddens. They believe that the social sciences and humanities can never aspire to create a 'real theory', simply because their subjects never remain immutable and separate, but react in unforeseen ways to the interpretations of social scientists.[22] There is one further problem with Biagoli's position: it is blatantly anti-STS. Why should the otherwise widely accepted notion of a co-construction between research and subject not hold for STS? Even if we grant that scientists as primary producers have a certain power to define and delineate what counts as science, why should the STS student be obliged to feel bound by such a view, especially since we know that such delineations are always subject to cultural differences and will be contested and altered over time? There is also a curiously a-historical tinge to Biagoli's argument. It may be true that STS in its present state takes a rather passive stance in defining its subject matter. Ironically, the science wars, which are for Biagoli evidence of the colonizing fervour of the natural sciences, came at a period when the delineatory ambitions of STS were low. Some decades earlier, demarcation of science from non-science was a hotly contested issue, especially among natural scientists and philosophers, who in the end had to withdraw. As Michael Aaron Dennis has shown in one of the rare articles on the history of the (social) history of science, the definition of science over time was closely linked to vested interests of U.S. research policies. Especially the debates about internalism and externalism and about the distinction between fundamental and applied science/technology were closely linked to the necessities of legitimating the U.S. (military) research budget when facing an ignorant public (Dennis 1997).

How did the curious renunciation of its own definition of science come about, leading surreptitiously to others' definitions seeping into STS? In the founding era of STS, one salient driving force was dissatisfaction with the Mertonian programme, since it left out the substantive content of science. Conventional wisdom holds that the reason for this was that even sociologists mistook science as existing outside society (that science is part of society should be a truism for any sociologist, since no human activity can exist outside society). Before Merton, the sociology of science was yet another sub-field of sociology concerned with a societal sub-system, such as law or religion, which had arisen in the course of functional differentiation. SSK, the sociology of scientific knowledge, arose as a response to this grave deficiency, but subsequently parted company with sociology. It started with inquiries into the epistemic core of science and gave itself a name that reflected this task in a programmatic way. It focused on scientific *knowledge,* rather than on things, artefacts or technologies. Obviously, in order to show it could achieve its goals, it had to crack the 'hardest' of all possible cases. In principle, SSK could have returned to general sociology/anthropology/history after having accomplished its mission and could subsequently have enriched the sociology of science with its newly gained insights and methodological tools for analysis. If that had happened, we could have today an equally vibrant and dynamic field of science studies, but one where 'science' would still refer to the peculiar characteristics and processes of a societal sub-system, thereby making it also easier to link the sub-system of science to other societal sub-systems. Science would have been kept as an object of study, which thrives *within* society.

Unfortunately, STS took another route. *First,* it added technology to its name. While this may have been simply due to a pragmatic-strategic move through which its initially small territory could be expanded, it had unanticipated consequences. The technology strand in STS never gained the stature or weight that the science strand has, thereby creating an unresolved tension. Technology is definitely not a societal sub-system. Depending on the definition used, it is considered a strict coupling of cause and effect, whatever is taken to be the coupling mechanism (Luhmann), an ensemble of human-made artefacts, from household gadgetry to large technological systems, or a simple means-ends relation.[23] Since technology cannot be confined to any societal sub-system, but is diffused throughout society, its only link to science is at the point at which science forms the basis for the emergence or development of *new* technologies. This is in fact what the widely used term 'techno-science' is meant to convey. But it hardly accounts for the multiform plurality of what Joerges and Shinn call the 'regimes' of science and technology, which are defined by their thematic axes and markets of diffusion. The 'transversal regime', for instance, is engaged in producing generic instrumentation, such as the ultracentrifuge, lasers or spectrometers. The members of this community move between science and technology and its markets are widely diffused (Joerges and Shinn 2001). The study of scientific instrumentation, embodying science but also giving rise to new science, seems for the most part to be left for historians to study. Technology studies in STS deal mainly with glitzy high-tech and with the emergence, impact and use of these new technologies and not of those already known and in use – a fascination it seems to share with the rest of society. But to limit the social study of technology to those areas where science gives rise to them (and the relatively short periods of techno-genesis in the strict sense) cuts out the vast terrain on which technologies actually emerge, evolve, are shaped by and shape in turn the societal fabric. Much of it occurs outside science strictly speaking or on the porous boundaries with the science system. Edgar Zilsel, one of the forgotten founders of STS, remarked that until the end of the 16th century new technologies emerged among craftsmen and not at universities at all (Zilsel 1942). Even today, most technologies such as telecommunication technologies and those of the entertainment industry are developed outside of the system of science or are byproducts of scientific research. One prominent example for the latter is the World Wide Web, which originated within CERN, where it had 'merely' been a means to process the huge amount of data generated by and for the use of an otherwise totally different scientific research agenda.

Second, STS usually argues explicitly or implicitly against functional differentiation (Knorr Cetina 1992): implicitly, because so many of its studies demonstrate the manifold influences of power, legal or cultural constraints or how economic factors and similar considerations intervene in the actual working of science. The explicit refutation of functional differentiation occurs when it is argued that science cannot be compartmentalised or when the 'seamless web' characteristics of networks constituted by humans and non-humans are emphasised. But when STS is confronted with the question of what 'science' is, if it is not a societal sub-system with its own distinct mode of functioning, the answers become vague, as shown above. The point here is not to argue that a specific theory of functional differentiation, such as Luhmann's, has to be espoused. The point is rather that *either* there exist different sub-systems or fields, which must necessarily have their distinctive characteristics and definable modes of operation, *or* no such distinctions can be made. And if this is the case – why should one still speak about 'science'?

If STS is right in claiming that we cannot define science as something distinct and stable and if, indeed, negotiation is at the heart of science in the making, how else can science be defined? If overflows and the 'framing' between different sub-systems, as claimed by Callon, are as essential as is their control, does this not also hold for 'science' (Callon 1998a)? And what would be the substance of its overflows? As far as we can see, there are no commonly accepted or explicit assumptions about the macro-evolution of societies or about the dynamics through which science is part of these developments, with the possible exception of Latour's idea that societies become increasingly more complex by creating ever longer networks made up of humans and non-humans alike (Latour 1993).

If we were to accept the arguments made in the two points above, then we could reach an agreement on a new definition of what STS is all about: STS is not about observing, describing and analysing the societal sub-system of science. Rather, it is the study of *the production of new knowledge and the production of new technologies* (or things). This would imply that there is not one special and privileged site where to localize 'science', such as within universities, but that the sites of knowledge production are distributed throughout society. Scientists would remain the most important producers of scientific knowledge, but their great diversity would become more visible, as would their multiple interdependencies with other co-producers in many different and heterogeneous sites. Above all, we would come to see clearly what distinguishes STS from other social sciences and what it has to offer as a kind of vanguard: the basic, underlying assumption *that modern societies are predominantly shaped by knowledge and technology.* This assumption can then no longer simply be taken for granted by the social sciences, only to be ignored afterwards. Instead, it would have to become an utmost research priority and the basis of all further social scientific research, including the knowledge and technologies that underlie the functioning of the social sciences themselves. It would also greatly facilitate to link STS back to social theory and highlight at the same time the potential contributions of STS to general social theory.

THE PLACE (AND A PLEA) FOR STS IN THE SOCIAL SCIENCES

On its way towards institutionalisation as an academic field or even as a discipline having 'science' (encompassing the natural sciences and/or the techno-sciences) as its central object of research, STS began to move out of the fold of the other social sciences. Like science itself, which is often seen as existing outside society, STS strove for a position with privileged analysis, putting itself 'above' or even 'outside' the other social sciences.[24] The price for having established a multi-disciplinary and relatively open scientific field was the severing of many previously existing links to general sociology, organizational sociology, studies of higher education, sociology of medicine and of technology, history, not to speak of that branch of economics which analyses the economics of research and innovation.[25] By adopting the concept of science as understood by scientists, and notwithstanding evidence to the contrary from its own highly diverse micro-sociological case studies, STS continued to treat science as if it were a coherent social institution. It lost sight of many developments and processes which had occurred on a macro-sociological, economic and political level, developments which jeopardize what is understood by 'science' or at least point in the direction we have indicated as Mode 2 knowledge production. The reasons for this new mode of production having emerged at the present time are not hard to find. In the first place, Mode 1 has been eminently successful. However, over the years the number of graduates trained in the ethos of research together with some specialist skill has become too large for them all to be absorbed within the disciplinary structure of academic life. As a consequence, the number of sites where competent research can be carried out has increased. These individuals and the organisations they work in constitute the intellectual resources for, and social underpinnings of, Mode 2. Further, the development and rapid diffusion of the new information and communication technologies have created a capability which allows these sites to interact. They have set the stage for an explosion in the numbers of interconnections and possible configurations of knowledge and skill. The result can be described as a socially distributed knowledge production system, in which communication increasingly takes place across existing institutional boundaries. Even more important may be the fact that knowledge production in this societally distributed system takes place *in the context of application,* thereby also increasing its openness (Gibbons et al. 1994).

It is obvious that the social sciences could profit from science studies, just as we claim that STS has to re-establish its links to the social sciences and especially to social theory. What STS has to offer to the other social sciences is above all the understanding it has gained of *how knowledge is produced, maintained, stabilised, closed, contested or negotiated* – in short, it can bring its analytic insights to a domain which still works with the assumption that the concept of knowledge is largely unproblematic and uncontested, because it is treated as given. Many social scientists still operate along an invisible divide that takes 'the social' as malleable and weak, while 'the rest', be it the natural world or the artificially constructed one, is assumed to be given and hence can safely be left out of social analysis.

The obstinence of this divide has deep historical roots, some of which have recently been illuminated by Stephen Toulmin. He locates the separation as one that took place more than three hundred years ago, when 'reason', seen as embodied solely in formalised and deductive knowledge, took on the hegemonic lead and was split from 'reasonableness', the knowledge provided by the *sciences morales* about the standards and circumstances under which 'reason' could be implanted and applied in the world of institutions and relationships, none of which can function without human judgement (Toulmin 2001). But the most pernicious of the distinctions still overshadowing the social sciences' awareness of their own modes of knowledge production is that between 'hard' and 'soft'. By accepting the verdict that their knowledge is inevitably imprecise, subjectively distorted and inherently unable to make predictions, many social scientists remain trapped in their own inferiority complex. The only escape routes open then are either to 'harden' their knowledge by imitating the natural sciences or to insist on their 'special' status, thus reinforcing the alleged correctness of the divide.[26]

The espousal of a thorough STS perspective would dissolve both the masochistic ritual of re-asserting the necessity to 'harden' one's knowledge and open the door to a vast field of productive insights in the social sciences and humanities. From statistics to transcripts to the actual practices of social surveys as one of the main methodological instruments in use, from the conduct of clinical trials to how regulations come about and to other legal and administrative practices – there is hardly a sustained awareness of the processes and practices through which the social sciences produce their knowledge and how they shape their objects of research. "The sociologists' level of discussion about sociology is not yet sociological," writes Kieserling (Kieserling 1999, 405), to which we could add: their (the social sciences') level of discussion has not yet reached the level offered by STS. It is to the credit of the *Sociology of the Sciences Yearbook* that it has explicitly taken up the challenge; not all of its potential audiences, however, have yet been reached.

There are three important issues to be considered in deciding what a restructuring of the social sciences – and of the humanities[27] – from an STS perspective would imply, and which pertinent insights and findings it could generate. *First,* most of the definitions of what the social sciences are rest upon untenable assumptions. The differences from the natural sciences are usually reduced to the idea that the social sciences deal mainly with human beings, i.e. entities whose special features, qualities, behaviour or values make them distinct from all other existing entities. It is highly doubtful that such a view can be maintained, both in the light of the current knowledge being created in the neurosciences and other life sciences, in primatology as well as on the molecular genetic level. However controversial this new knowledge may be (e.g. cultural anthropologists object vociferously to primatologists applying the term 'culture' to primate behaviour), there can be no doubt that the lines of what it means to be human are currently being re-drawn with vigour. New knowledge is continuously emerging about cognition and language abilities in humans and animals, and the interest in discovering the origins of life is being matched by new questions and new findings about how our early human ancestors evolved and lived.

Science studies in turn have shown over the years that the drawing and re-drawing of such boundaries always occurs for certain purposes, which need to be carefully examined in each case.[28] They are also keenly aware of the implications which the divide between humans and non-humans has for any comparison and distinction between social science knowledge and natural science knowledge, with different functions assigned to each in specific circumstances. Last but not least, a closer look at what the social sciences actually do, in contrast to what they claim they do, reveals a great diversity in what it means to be 'human'. The diversity of the 'human' entities dealt with by the social sciences and humanities is at least as great in range of qualities, behaviours, values and other special features assigned to them, as that attributed to the gap held to separate humans from non-humans. Moreover, in actual practice and in theoretical understanding, 'humans' must be deconstructed, if such different phenomena as the functioning of the economy, the behaviour of firms, voting patterns, collective and individual action, family ties, or innovation are to be properly analysed and explained. To move away from generic descriptions such as the 'human factor' or 'verstehen' as a unique domain of the humanities would not only result in a more adequate and refined picture of what social scientists actually *do,* but it would also further their self-understanding and help them to escape the trap of believing that they merely represent the less developed natural sciences. Perhaps even more importantly, the problem of what it is to be human would be integrated into the epistemological basis of the social sciences. Instead of insisting on an essentialist position which can never (and is never) followed in practice, the social sciences would have to confront the *question* of what it *means* to be human in specific circumstances, and of what follows from it.

The *second* issue is closely related to the first. Much of social theory still operates without an elaborated concept of how to deal with the realm of the *non-human,* be it the environment as a general sociological term separating the social world from the rest, whether artefacts and technologies of different kinds or an entity like 'nature' as used in environmental discourse.[29] This is not to say that what is *non-human* is not taken into account, but it is often done either in an ad-hoc manner, treated as a given boundary condition for social life and action which does not invite closer analysis, or as an often recalcitrant factor or condition which resists and reacts to human action. An STS perspective would have much to offer by showing that it is impossible to refer to any 'environment' or to the materiality of the non-human world without realizing that the established conceptualisations have originated in science and are shaped and altered by science. The power of defining the natural world still lies with the natural sciences. They retain a monopoly on scientific conceptualisations and safeguard the access routes to understanding and explanation, even if it is granted that this monopoly is not complete. For all practical purposes, there is no direct route from the social world to a *non-human* world, be it 'natural' or artificially constructed. The social sciences, if they want to understand the growing complexities and expansion of the materiality in which humans live and act, have necessarily to pass through a language of description and explanation, including the meanings attached to them, as originated in and imprinted by concepts and practices from science and engineering. This is not to acknowledge that these techno-scientific repertoires have to be accepted *tel quel.* STS teaches us that scientific conceptualizations are historically and culturally bounded. They are hence unstable and – within limits – open for contestation. If the social sciences were seriously to reconceptualise the materiality of their social practices and of the environment in which humans act, they would gain, as in the first question of what it means to be human, an understanding of the – flexible and malleable – meaning of this materiality and of the manifold processes and forms of interaction with it.

The *third* issue is the development of a much longer historical perspective for the social sciences, one in which the role played by science and technology has its integrated place along with other institutional and epistemic transformations connected to the making of modernity. Due to the pioneering work of Wittrock and others, the multiplicity of modernities is today coming to the fore in a world-wide context, together with a better understanding of the emergence and spread of democratic nation-states, liberal market economies and a relatively autonomous science based in research-oriented universities (Wittrock, Heilbron and Magnusson 1996; Wittrock 1998). It is no longer sufficient to analyse the major transformations in the West as being premised on industrial-technical and on political revolutions, without including the epistemic and institutional transformations and novel practices which underlie them. The emergence of distinctly modern key concepts, such as the discovery of 'society', the role and function of historical enquiry, of language and new collective identities, needs to be situated in particular institutions and in intellectual practices. This forms part of a long-term agenda upon which the social sciences have embarked. In her work on the origins of knowledge, religion and politics in modernity, Rivka Feldhay has questioned the narrative in which the constitution of each as an autonomous sphere is usually told, highlighting instead the variable figurations they have formed over time. The idealisation of science tends to portray it as a prototype of objective discourse capable of producing value-free, neutral and factual types of knowledge. This kind of knowledge becomes disembedded. What is generally overlooked is that public forms of knowledge are mediated in complex ways through language and the construction of trust. They depend and are embedded in a wider arrangement consisting of other facts, values and interests (Feldhay 2001).

Far from remaining separate institutional entities, the embeddedness of scientific practices, their moral economy and ethos in the societal fabric of the day would become more visible, but so would the specific contributions and the functions of scientific and technical knowledge in the military and colonial endeavours of the nation-state and, more generally, the role played by science and technology in inventing, sustaining and changing the project of modernity in the West and its 'entangled history' with the rest of the world. The STS perspective could be extended backwards in time, in close interaction with historians, historians of science and technology, economic historians, sociologists and political scientists.

THE PLACE (AND A PLEA) FOR SOCIAL THEORY IN STS

Mertonian sociology of science had never attempted to move into the substantive core of science but remained instead on the organisational-structural side and focused on the functioning of science as a sub-system. When Merton turned away from the Parsonian system to theories of the middle range, he nevertheless left room for partial theories still modelled on Western society in general. With the emancipation of STS from Mertonian sociology, such carefully delineated ambitions and assumptions were also left behind. *More studies on 'science' meant fewer on 'society'*. This went so far as to lead to the term 'social' being declared outmoded (Latour 1993).

But the problem of situating the substantive core of science, of assigning it a place and a time in an environment, which retained a whiff of the social, still remained. It was solved by radically and thoroughly contextualising all scientific practices to be studied. 'Context' would from now on become a concept replacing everything previously thought to represent social structures or hierarchies, social differentiations or functions. Since the level of analysis of STS had also taken a decisive turn towards the micro-level, context became even more ubiquitous and useful, since it could now be applied to everything surrounding the micro-objects under study, from instruments to the beliefs of those who operated them, from institutional arrangements to patterns of conflicts or co-operation among scientific practitioners.[30] The emancipation from philosophy of science largely meant a re-focusing on individual case studies and brought with it the 'thick descriptions' taken over from anthropology. Issues of epistemology, if treated at all, were now also subject to this kind of thickening. No new epistemological patterns emerged. At most, a higher resolution of (old) philosophical generalities had taken place.

The double shift to case studies on the micro-level and the general substitution of society by a varying and particularistic 'context' created a theoretical vacuum which was, however, rapidly filled with a stream of new concepts, languages and vocabularies. Attempts at systematising this sprawling but also highly stimulating conceptual mix finally succeeded. Actor-network-theory became *the* (only*) theory* at hand, widely accepted within STS and attracting followers even from outside. The ANT approach worked beautifully and was extremely helpful in connecting humans with non-humans, in showing the heterogeneity of network associations and, more generally, in providing a view which allowed many old, uncritically reproduced assumptions about science and its implications to be swept aside. In its focus on the immediate and the near-by, however, it also eclipsed some of the more important regularities, as well as attention to changing processes of societal development. The problem with ANT is that it works on two levels which are at opposite ends of the scale along which most social sciences operate: the ontological-metaphysical level and the level on which local and particular contingencies are at work. The theories of the middle range, which might connect these opposites, are not yet in place.

In a memorable exchange during the EASST conference in Bielefeld some years ago, social theory, personified by Niklas Luhmann, the grand master of systems theory, was invited into the inner halls of STS for a debate with Bruno Latour, grand master of STS and spokesperson for the 'excluded things', the actants. Luhmann patiently insisted on communication being the fundamental element that constitutes society. For Latour, the world could not be more different. For him, it was the materiality of things and their interconnectedness with the social world which constituted the specificity of science and technology, leaving no room for differentiation of any kind. What was most stunning was the missed opportunity of an encounter between such radically different but equally sophisticated theoretical descriptions of the world and to explore what followed from each. What would have happened if the world of 'things' and 'actants', in all their rich heterogeneity, messiness and fleeting transformation into ever forming and dissolving networks, had taken seriously a world where distinctions have to made all the time in order for communication to occur and observation to be possible?[31]

STS has been able until now to get on with this theoretical anomaly, in which it has de facto only one theoretical position, largely because it has continued to see one of its main tasks as – to use an old-fashioned expression – 'falsifying' traditional notions of what science is, does or claims to be. As already noted, there appears to be an unlimited supply of such notions and representations, coming from many different sources all of which continue to replenish themselves. STS's success consists in showing that '*it was not so*'. There is nothing wrong with such a self-imposed task. Moreover, it fits well into the tradition of the Enlightenment, with a small concession made for a bit of subversion and a whiff of heresy.[32] This leaves one question open, however: if science is *not* what everybody, including the scientists and the public, thought it was – then what is it?

It is here, we believe, that a place must be made for social theory. What counts as 'theory' in STS is usually tailored to the case at hand: it remains particularistic. Systematic attempts at comparison of case studies and their theoretical content are extremely rare, nor are collaborative efforts in such projects encouraged by the existing reward structures. Theory in STS is often understood in the sense of 'critique', as a way of questioning the arrangements one has found rather than of coming up with propositions which would allow one to find some regularities to explain different kinds and realms of order or disorder. If STS were to take seriously the *processes of contextualisation* impinging upon and transforming the production of scientific knowledge today, it would inevitably have to deal with social theory (Nowotny, Scott and Gibbons 2001).

This is not to be taken as a plea for any particular theory, but a general call for integrating social theory into STS, just as we have argued above for a much stronger integration of STS into the social sciences. At this point it might be objected that the age of grand theories, whatever their provenance, is over, or that STS, having successfully emancipated itself from sociology, should not fall back into a state in which theory is largely synonymous with the work of one great old (or dead) man. Our conception of social theory is different. While we acknowledge that the theoretical advances in sociology, to which we shall limit ourselves, have been made by a small number of individuals, we maintain that their impact on the field has only been possible through a widespread, collective effort. Moreover, most of these theoretically-minded individuals have been acutely conscious of the need to continuously ground, nourish and complement, as well as to repeatedly test, their own theoretical work with empirical data of different kinds, whether produced by their collaborators, themselves or, as in most cases, through the work of those in the field who have been inspired by their theories. This is as true for the founding fathers of sociology, Weber, Marx and Durkheim, as it is in the case of the more recent theoretical contributions made, for instance, by Bourdieu[33], Boltanski and Thévenot, Elias, Luhmann or Wallerstein. We will concentrate on these last four theorists, since we believe that their work shows many, as yet unexplored, links to STS that could be followed. Although Luhmann and Boltanski/Thévenot are considered important theorists in their own countries, their international influence has been remarkably low. Although we can only briefly allude to the kinds of questions their work in social theory raises for STS and although we do not maintain that their perspectives, insights and problem definitions are unique, we nevertheless believe that they exemplify the potential that social theory can bring to STS.

Norbert Elias was associated with the *Yearbook* from its beginnings until his death. Starting with the seemingly naïve question of how the authority structure of the sciences, manifest in the position and function of scientific elites and in the pervasive hierarchies of disciplines, affects the production and development of knowledge, Elias moves on to show numerous interdependencies in long-term historical processes with planned but unintended outcomes (Elias 1982). Scientific elites, he maintains, first had to obtain their relative autonomy from other kinds of power elites. They had to strengthen the tendency to see one specific type of scientific knowledge as the only valid one. They differ from other power elites, however, since innovation and the production of new knowledge cannot be rendered routine and must remain open to being tested in non-authoritarian ways. Moreover, knowledge production is seen as a long-term, collective, intergenerational process whereby scientists build upon work done by others. Yet maintaining control and/or a monopoly over the kind of knowledge produced by science also implies highly variable figurations of the 'established' and the 'outsiders' with reciprocal forms of dependence. Elias was very concerned with showing that scientific knowledge, as controlled and monopolized by the scientific establishment, is one special kind of knowledge. Knowledge, for Elias, is a nexus of human-made symbols used as a means of orientation and communication. It is acquired socially as much as it is learned by an individual. It has roots in biology, but is strongly social in the process of acquisition and in its progress. Knowledge advances by attaining ever higher levels of synthesis, such as the theory of (biological) evolution, which has moved beyond the state of law-like theorising towards process models and process-oriented theorising. Having reached a higher level of synthesis also means that the fit has increased between a symbolic ('inner') representation made by humans and the 'outer' world.

What Elias' theory of knowledge (or 'symbol theory', as he also called it) offers is one of the rare attempts to bridge a natural science perspective with a resolutely social perspective of the *longue durée,* in the course of which the mutual chains of interdependence between humans lengthen while their capacity for evolving new means of orientation (knowledge) also grows. Scientists (including social scientists), as a special case of an established power group that is capable to create and maintain a monopoly over the kind of knowledge it produces, therefore mediate between the two. Elias pleads for a sociological type of enquiry, one capable of working out process models of the development of knowledge, fitting into but not reducible to models of the long-term development of human societies. One cannot shuffle, he maintains, 'forms of thinking' into one compartment, the 'objects of thinking' into another, human experience into a third and the social groups within which standards of orientation and thinking develop into a fourth. The actual, observable development comprises all these factors closely interwoven. What will then become clear is that the social value – and hence the power ratio – of groups of scientists is related to the cognitive value of what they produce. Sociologists, according to Elias, would be well advised to reconsider their own position within the hierarchy of scientific monopoly-holders. If the relative autonomy of a scientific group is highly connected to the relative autonomy of its subject matter, they have to avoid becoming squeezed between two power blocks: between the physicists (and the hegemonic claims of 'nature') on one side, and the ideologies of the political world on the other.

Niklas Luhmann was one of the most proliferate and elaborate theorists social theory ever has seen, but he remains little recognized in the Anglo-American literature. For STS, Luhmann promises to be a productive partner in dialogue. He devoted considerable analysis and refined articulation to his theory of science as one of the subsystems of society (Luhmann 1992). His work displays parallels with, but also striking contrasts to, STS. If one takes the observation of 'science' to be one of the tasks STS has set itself, one is quickly confronted with one of the basic assumptions of Luhmann's theory according to which each societal subsystem can only observe another in its own language. A primary and privileged point of observation does not exist. Reflexivity enters through second-order observation, when the system observes its own observations. (Contrary to much 'reflexivity' in STS, reflexive communication for Luhmann only concerns what can be publicly communicated. Luhmann denies that private experience or emotions can ever be made the content of reflexive communication.) All observations are made from a specific location within a system and remain within the system. This theoretical position strips the reflexivity debate of all its moralism, but also of the agonies of continuous self-doubt and questioning. According to Luhmann, there is nothing special about the position of social (or other) scientists. They only follow the communication rules of their own subsystem, and their analysis thus remains confined to what these rules allow them to do.

This position derives from Luhmann's theory of functional differentiation, which stood in such sharp contrast to Latour's position when the two publicly confronted one another in Bielefeld. Modern society is characterised by ongoing processes of functional differentiation, resulting in each subsystem's developing its own code or way of distinguishing between what belongs to the system and what is outside, i.e. in its environment. The code for science is the distinction between true and false; for law it is right/wrong, for politics ruling party/opposition. All decisions within a subsystem have to follow the distinctions upon which the code rests. This is important for science studies since Luhmann claims that it is impossible for a system to work with another code. If the legal subsystem has to decide on scientific questions, it will do so by means of the code of law. Another example is that of environmentalism. Luhmann diagnosed very early that environmentalism can only cause friction and irritation in other subsystems, since it lacks a code with which to operate. Moral content or ethical claims are not sufficient for setting up a code. Science therefore cannot absorb environmentalist ideas, except when it can subsume them under its own code of true/false. The same holds for law, where the idea of assigning a special value to 'nature' cannot be integrated except by transforming it into its code. Luhmann predicted as early as 1986 that the environmental movement would be fully integrated into the existing subsystems, thereby losing all its distinctive features (Luhmann 1986).[34]

Environmental disputes also figure prominently in the work of Luc Boltanski and Laurent Thévenot, and have been further expanded in a comparative mode in the work of the latter.[35] In their 'theory of justification', Boltanski and Thévenot elaborated several 'regimes' of justification which are applied in modern societies in the case of conflict. The regimes are the following: the 'market' performance regime, the 'industrial' efficiency regime, the 'civic' equality regime, the 'domestic' local and personal-ties regime, the regime based on the 'inspiration' expressed in creativity or religious grace and the regime of 'renown', based on fame. In contrast to Luhmann, these regimes of justification are not constrained to operate only in certain subsystems, but can be transferred from one to another. Since all of these regimes (later also called 'cultural repertoires') can be applied in principle, their actual use depends on the relative valuation accorded them in specific conflicts. What is more, they are all in conflict with one another and are all-encompassing, since every one of them is linked to a definition of what humanity should be. Thus it is possible to denounce one's opponents with a full regime, while one's opponent can do the same with another regime. For Boltanski and Thévenot, the outcome of conflicts depends on the valuation of different regimes and not, if we compare theirs to Luhmann's position, on the fit with the code of a subsystem where they are employed. This allows them to observe the relative worth of different regimes of justification in different cultures or different times. It would also be possible to empirically investigate whether some regimes prevail in some parts of society or not, and thus it would be possible to observe whether there are different regimes used than those in use in law, for example. It is also possible to ask whether there emerge new regimes of justification, such as a 'green' regime (Moody and Thévenot 2000, Thévenot, Moody and Lafaye 2000). If such a regime has emerged, as is claimed in Thévenot's most recent work, then we could observe a shift from solely 'humanity'-based regimes of justification to 'nature'-based regimes of justification. If this is empirically observable, as Thévenot claims, this would be a serious challenge to the humanistic grounds of this theory. It would probably also hint at a special kind of scientistic worldview, one which puts the objects of scientific enquiry into the position of regimes. This would mean that what was usually hidden as expert 'truths' are now turned into a regime of justification. Thus it would introduce a naturalistic fallacy as a resource in disputes.[36]

Immanuel Wallerstein might at first sight be an unlikely choice of theorist with potential interest for STS. Over the last years he has made crucial steps away from his earlier Marxism, with its neglect of interpretive questions. In contrast to the fragmented and scattered empirical work of STS, Wallerstein still believes that a systematisation of social science research is possible and that the macro-structures of the *longue durée* can and need to be analysed. His world-systems theory not only encompasses the past, but also reaches out into the future. For the period from 1945 until the nineties he diagnoses a Kondratieff cycle and claims that the downswing of the cycle in the nineties correlates with a dissolution of the power of the state (Hopkins and Wallerstein 1996). What is of potential interest to STS here is the link Wallerstein construes between a decline in state power and a decrease in scientism as part of a macro-sociological shift.[37] Science and scientism were intimately tied to the idea of the state as a machinery kept in operation by science as the major means of guaranteeing its efficiency. Wallerstein's emphasis on scientism raises an interesting question about the emergence of STS as a research field and its own position vis-à-vis the state as a declining power. It could be argued that STS itself is part of the puzzle and that it co-evolved at a time when the authority and legitimisation of the state and, linked to it, of science, were already on the decline.

EPILOGUE: LE NOUVEL ESPRIT DU STS[38]

Let us now perform a little thought experiment. Imagine that not every new discovery and development in the biomedical sciences or other rapidly developing branches, such as nanotechnology, were to result in the call for setting up yet another ethics committee. Instead, public uneasiness and scientific uncertainty would be channeled to STS for consultation and advice. The STS community would have set up a number of highly competent committees with flexible membership, including scientists and others from outside the STS orbit, in order to guarantee a broad range of expertise grounded, nonetheless, in an STS perspective. Of course, these committees would be thoroughly familiar with the writings and research of Brian Wynne, Sheila Jasanoff, Peter Weingart and others, and would possess the knowledge and communicative skills to gauge the anxieties of the public as much as its as-yet unarticulated desires. Scientists would be extremely relieved, since they would no longer feel constrained by the ever-changing or overly vague guidelines set up by the old ethics committees, guidelines which all-too-eager legislators and regulators have also sought to translate into the binding form of laws and regulations. Instead, they would delight in a newly won freedom of self-regulation and self-enlightenment. Having learned to partake in STS reflexivity, they would know how to make their own hidden assumptions explicit, and therefore could move on to a higher form of reflexivity, one which leaves room for comparing their assumptions with those of other actors. Eventually, in a true spirit of collaboration with the other members of the committee, it could be shown how different repertoires or ways of describing the world also affect the world. The committees would now be in a position to give genuine policy advice, since they would have learned the answer to the question 'What follows?'

Let us continue with another little thought experiment. Imagine further that STS, in a move reminiscent of the strategy originally devised by Emile Durkheim to introduce sociology into the French system of higher education, would have succeeded in getting STS courses as mandatory requirements into the curricular training programme of all high school teachers. In order to set up such a subject in secondary education, two preliminary but crucial conditions would have had to be fulfilled. First, in order to promote its view of science, STS would have had to become much more pragmatic and give up some of its sophistication. Particularly the artful jargon practiced in its writings would have had to be toned down to a language which not only high school teachers but also their students could understand and follow. It would also have had to adjust its research agenda, at least to the extent of making room for studies which would allow it to evaluate its own efficiency in communicating, as well as its ability to absorb lay notions. The second condition consists in realizing that science teaching in high schools and universities is, despite serious efforts in science education as a research field, not very professionalized. In many countries science teachers are recruited into high school teaching after having undergone a scientific training themselves with a pedagogical layer on top. After gaining access to the teaching infrastructures at high schools, colleges and universities, STS would have focused on developing curricula for high school students and the education of their teachers. The excellent text books developed by the Open University would have served as additional incentive.

The success of this strategy would have become evident when STS became an obligatory rite of passage for all those concerned with science education. Since practically all highly industrialized countries would be facing a severe shortage in scientifically and technically trained man- and woman-power by now, the problem of attracting more good students into science and engineering could finally be solved thanks to the professionalised and responsible activities of STS in this domain. But this itself would not have been sufficient. The ultimate and intrinsic reward for STS would have come with a market test. STS had always claimed that if science students and their teachers only had a better (i.e. an STS-inspired) understanding of science, then the new generation of students could also make a difference by producing scientific knowledge which showed greater social sensibility. In our thought experiment the market test would have confirmed the expected results. Thanks largely to the STS-inspired curricula with which the new science graduates would have been trained, they would have turned out to be experiencing less difficulty in crossing academic disciplines. Their fungibility would have been considerably enhanced.

Another example in our thought experiment concerns science policy. As is obvious from the above, future policy advisors would also have undergone an STS-inspired training which would have taught them modesty and critical reflection upon their own role. In addition, they would have learned a great deal from Niklas Luhmann about how organisations function. Thus they would know, for instance, "that in organizations crucial structures are being built up that function to deal *ex post* with decisions taken *ex ante*, respectively that serve to provide for future possibilities of the retrospective management of decisions that have to be taken now," and would act accordingly (Luhmann 1988, 167). Long gone would be the days when it was possible to get a position in any of the numerous institutions which dealt with research policy on the national, supra- or international level without showing one's STS credentials. STS would have learned from environmental studies its practical lesson in how to professionalise itself. It would now have come to the realisation that multi-layered links had to be set up, linking potential clients seeking advice and consultation for the practical problems they faced. In providing professional services to them, STS thereby would make its own knowledge an indispensable ingredient, diffusing into new laws and regulations and infiltrating into business contracts about exchanging, buying and selling of materials, organisms and other 'actants' needed for the functioning of the research enterprise. STS would advise, but it would also help to regulate intellectual property-rights agreements which would have become even more important now that the funding patterns of fundamental research were able to come from any source, public, private or both, with any number of hidden or overt strings attached. Science and research policies now would be thoroughly in the hands of STS, which would have generated a need for the expertise it had to offer while educating the experts in demand and by expanding even into private consultancies to employ those experts. While the rhetoric about knowledge societies would persist, STS would not only take advantage of it in an opportunistic mood, but would make the issues it raised an integral part of its research agenda. Far from renouncing its critical stand, it would pursue an intelligent policy of professionalisation.

The last example of our thought experiment takes us back into academia. Although the university structures, management and leadership would have changed considerably, STS would have kept its home there after a somewhat difficult period of transition in which university officials would have been more inclined to favouring business and management studies and an economics-of-innovation approach. Imagine, therefore, STS having discovered that it not only *had a future outside academia,* through its many professionalised links, but *also inside.* Surprisingly however, its base within units devoted to further research in STS as understood previously, would have been kept small. Its main research activities would now be distributed all over the university, even in the most unlikely places and research groups, all of them keen on having an STS person as a member of their team. This would have occurred on the basis of genuine mutual respect for the knowledge which each of them, including the STS person, would have brought to their commonly defined problem. STS knowledge production would now be challenged to leave its critical stance behind (often practiced as critique in the past) and become constructively engaged in a joint effort. STS could no longer complain that its findings were 'wantonly disregarded by their audiences'. It would have to attempt to satisfy demands and expectations of more audiences than it could always manage. Perhaps surprisingly – but the demand would have been so huge and the problems posed so intellectually exciting – the split that some had feared before between a 'critical' and a 'technocratic' strand would not have occurred. It would have been understood in time that such a distinction was arbitrary, operating according to the moral standards of an STS person as an individual, rather than being the collective responsibility of the entire community and its collective vision of the field. Of course, there would remain sufficient room for critical debates, even for controversies, but critique would now no longer be directed at debunking what science was and what scientists did. Rather, STS would be constructively engaged in co-operation with other scientists to help build a world in which science and society met under one roof.

You may now ask how the thought experiment ends. The answer is rather simple and, we hope, not all too disappointing. After David Edge and others had rightly complained about the potential audiences of STS not listening, an anonymous sponsor would have offered a substantial and generous grant for research to look into the problem and to come up with a solution. Initially, some ignominious squabbles about who should carry out this research and receive the grant would have occurred. The sponsor would finally have put an end to them by insisting that an equally anonymous but highly prestigious consultancy firm be hired. Its diagnosis would have boiled down to a two-fold verdict: either no market existed for STS and its products, or STS was behaving like a company which had failed to adapt to changes in its environment. In the latter case, it deserved to go bankrupt and vanish. In the first case a market had to be created and, well, you have just read our recommendations as to how to go about that process.

Swiss Federal Institute of Technology Zurich, Switzerland

ACKNOWLEDGMENTS

The authors want to thank their colleagues Rainer Egloff, Christian Pohl and Monika Dommann for their helpful comments.

NOTES

[1] Although STS and the 'new' history of science have moved much closer together, we do not explicitly include work in the history of science here. For a thorough assessment of the future directions of the history of science see the chapter by Michael Hagner and Hans-Joerg Rheinberger in this volume.

[2] We will use the two terms 'field' and 'discipline' almost interchangeable. We believe that it is currently not possible to tell whether STS is a discipline or not. If we follow the definition of Turner, then "the discipline must be called a discipline, and the name [has to be] shared and used" (Turner, 2000, 47). Obviously this does not really apply to STS since in many places in the world there exist different names for what is called STS here. But STS is also not only a 'field' in the sense of Bourdieu or even Whitley, since it shows many characteristics of what commonly disciplines are said to contain: a shared body of literature, methods and theories as well as degree granting units and a career-structure of its own that enables indoctrination of the canon. It is also important here to notice that this text itself is part of a disciplining exercise, trying to strengthen STS' disciplinarity. Thus the term field will rather be used with reference to internal struggles whereas the term discipline will be used with reference to and issues of professionalization institutionalization.

[3] A number of such textbooks have recently appeared, for example Cutcliffe (2000), Felt, Nowotny and Taschwer (1995), and Hess (1997).

[4] For a good introduction to these debates, see Pickering (1992).

5 For some analyses of the 'science wars', see for example Edge (1999), Segerstråle (2000) and Jurdant (1998).

6 'The' scientists largely came from theoretical physics in the US and over-proportionally from biology in the UK, while scientists from continental Europe remained largely indifferent, at least in public.

7 Among other joys, hear about the ones described in the song with the same title by The Artist Formerly Known as Prince in the album *graffiti bridge*.

8 One should bear in mind that the Handbook was officially sponsored by 4S. Thus it represents the 'official' history of STS (Jasanoff et al. 1995, XI).

9 A special version of this history of progress is the one embodied by the Paris School. After STS has shown that all science is context-dependent – surprise, surprise, enter the Paris School! They are not content with merely cleaning the jungle. They will now show that STS itself is infected by social realism too, being merely the other side of a coin called natural realism. Therefore, a new currency needs to be introduced. It is called ANT, and has quickly spread throughout the tribal lands. See Callon and Latour (1992) and Callon (1995).

10 For a concise but short attempt on a social history of the early developments, see Pestre (2001, 190 ff.). He hints at an important change of political and epistemic positions: "Au cours des deux derniers siècles, la gauche (au sens politique) a plutôt été positiviste et scientiste; ce sont ses ennemis (dont la droite catholique) qui ont insisté sur la caractère construit et toujours limité des savoirs scientifiques" (Pestre 2001, 189). As prime examples of such right wing constructivists he names Pierre Duhem and Michael Polanyi.

11 For an attempt at a sociological explanation why STS never succeeded in France, see Freudenthal (1990).

12 Barbara Beigun Kaplan even claims: "STS as a subject of study, shares [with women's studies] this unique situation of having its origins in a social movement outside academia." (Kaplan 1991, 2), as quoted in Edge (1995, 11). It is an open question whether this statement does not apply to most of the social sciences.

13 In the 3 August 2001 issues of *Science*, one can read the following announcement: "Science's next wave, www.nextwave.org. Global: Saving the Planet. Our August feature explores careers in the environmental sciences – where opportunities encompass not only research in climate change, bioremediation, and biodiversity, but also careers in environmental advocacy, lobbying, and writing and editing."

14 For an early prediction and analysis of this problem, see Luhmann (1986).

15 It should perhaps also be mentioned that the anti-globalisation movement and its grievances have so far been of little interest to STS.

16 In this respect, STS resembles the even more extreme situation of cultural studies in the U.S., where the ritual invocation of political slogans in academic texts is taken to be a substitute for political action. For a stronger critique of this practice, see Gitlin (1997). For an overview of political and interventionist science studies and his own programme, see the last chapter of Hess (1997).

17 This even holds for studies of mathematics or scientific modelling. Among the exceptions, see Heintz (2000) and Restivo (1992).

18 And if so, it seems that historians had to clear the trail. See also the Yearbook volume XV as one of the first of its kind (Wagner, Wittrock and Whitley 1991). For recent attempts of STS scholars to study the economy, see Callon (1998b), Mackenzie (2001), and Knorr Cetina and Bruegger (2000); for an attempt to study the law, see Latour (forthcoming).

19 If one continues the parallels with anthropology and if 'strangeness' is one of the criteria of choice, it follows that interest in a once unfamiliar field must decline once it has been described. This was famously so in the heyday of British social anthropology, as related by Kuper when remembering his education in Cambridge in the early sixties: "An extreme but not exceptional view held if a Malinowskian had worked in that region – or even in the same country – then it had been 'done' and one had best go somewhere else" (Kuper 1999, 20). The idea of returning 'home' probably also stems from the sense that all other fields or societies have been 'done'.

20 For a discussion of how engineering ceased to be a problem in STS see Downey, Donovan and Elliott (1989). For a recent attempt to re-integrate medicine in STS, see Berg and Mol (1998).

21 On the relationship between science in the Western world and science in the South and on the characteristics of dependent knowledge, see Goonatilake (1993).

22 For a recent overview and restatement of this position, see Flyvbjerg (2001).

23 Looking at technology studies, the overwhelming impression is that technology is equated with high-tech, that is, tangible things and artifacts designed by people who are either scientists or close to them. Intangible things such as accounting procedures are left to social historians. Simple things, like self-designed pinboards or the Zimbawe bush-pump, are usually absent. For examples to the contrary, see Mol and de Laet (2000) and Ingold (2000).

24 This is one of the reasons why the attacks coming from the 'science warriors' have been genuinely hurtful. It was extremely sobering to discover suddenly the extent to which the self-image of STS differed from how it was perceived by its 'most significant other', the 'real' (natural) scientists. STS had believed that it had transformed itself into something different from the other social sciences and humanities, only to discover that in the eyes of the natural scientists it was indiscriminately lumped together with the weirdest exemplars of 'postmodern nonsense'.

25 For a recent attempt at a new economics of science, see Dasgupta and David (1994).

26 For a similar discussion, see Latour (2000).

27 In an interesting project on "Multiple Modernities, Multiple Antiquities", Michael Werner and Wolf Lepenies propose research about the reception that the different images of antiquity has had on the tradition and emergence of the humanities in Europe. There hardly existed a project of 'modernity' not underpinned and accompanied by images, representations and constructions of the past, nor by a reconstruction of the antique past as practised by the humanities, which did not include imaginations and projections of the present and future. The project, as far as we understand its intentions, will not contend itself to analyze written texts only, but will include 'hardware' produced by the humanities, at least as executed or realized in accordance with their specifications: monuments of different kinds, buildings like museums or parliaments, urban space, but also material in every-day use (private communication).

28 See Gieryn (1999), Haraway (1989) or Latour (2000).

29 For a recent discussion of these issues in the history of environmental sociology, see Gross (2001).

30 For a discussion of the 'problem of context' in anthropology, see the reader edited by Dilley (1999). For a comparison of the emergence and crisis of 'context' as the main explanatory factor in STS and media studies, see Schlecker and Hirsch (2001).

31 For another account of this event, see Wagner (1996).

32 For a summary of this task, see the article by Pestre, where he states five "contre-propositions" against the "discours dominant en Occident au cours de ces derniers siècles à propos des sciences" (Pestre 1995, 494).

33 We will not discuss the work of Pierre Bourdieu here, because we assume that it is fairly well known, even in the Anglo-American world. But it remains to be observed that Bourdieu has almost disappeared from STS, quite contrary to his role in the other social sciences. For some literature specific to STS see Bourdieu (1991, 2000).

34 A further discussion of Luhmann's theory and its relation to STS can be found in Wagner (1998). Wagner concentrates especially on Luhmann's stance that society consists of communication and its alleged exclusion of objects from society.

35 For the main statement see Boltanski and Thévenot (1991); for a short introduction in English see Boltanski and Thévenot (1999).

36 Latour objects to this claim on grounds of his political philosophy (Latour 1998).

37 One of the most profound thinkers about the shifting relationship among science, politics and the state is undoubtedly Yaron Ezrahi. In his earlier work, he masterfully showed the crucial importance for the rise and functioning of a new conception of politics in the framework of liberal democracies of scientific and technological models of 'witnessing' and of rendering visible the efficiency of the machine. In his later work, Ezrahi returns to the emergence of a notion of politics as the exercise of human freedom and the issue of restraint it raises as putting limits on arbitrary political power. Scientifically certified 'reality' in the form of the necessary laws of nature, helped to create the public space in which this new balance could be played out. Scientific knowledge and its institutions constituted a principal source of apolitical authority with the capacity to appear as setting limits and imposing discipline on political power. In contemporary democracies, these long-term developments are now becoming eroded (Ezrahi 1990, see also this volume).

[38] We borrow the title from the recent book of Luc Boltanski and Eve Chiapello (2000).

REFERENCES

Berg, Marc and Annemarie Mol (1998), *Differences in Medicine. Unraveling Practices, Techniques, and Bodies*, Durham: Duke University Press.

Biagoli, Mario (2000), 'Introduction. science studies and its disciplinary predicament', in Mario Biagoli (ed.), *The Science Studies Reader*, XI–XVII, New York: Routledge.

Boltanski, Luc and Laurent Thévenot (1991), *De la Justification. Les Economies de la Grandeur*, Paris: Gallimard.

Boltanski, Luc and Laurent Thévenot (1999), 'The sociology of critical capacity', *European Journal of Social Theory* 2 (3): 359–77.

Boltanski, Luc and Eve Chiapello (2000), *Le Nouvel Esprit du Capitalisme*, Paris: Gallimard.

Bourdieu, Pierre (1991), 'The peculiar history of scientific reason', *Sociological Forum* 6 (1): 114–36.

Bourdieu, Pierre ([1975] 2000), 'The specificity of the scientific field and the social conditions of the progress of reason', in Mario Biagoli (ed.), *The Science Studies Reader*, New York: Routledge, pp. 31–50.

Callon, Michel (1995), 'Four models for the dynamics of science', in Sheila Jasanoff, Gerald E. Markle, James C. Petersen and Trevor Pinch (eds.), *Handbook of Science and Technology Studies*, London, Thousand Oaks and New Delhi: Sage, pp. 29–63.

Callon, Michel (1998a), 'An essay on framing and overflowing. Economic externalities revisited by sociology', in Michel Callon (ed.), *The Laws of the Markets*, Oxford: Blackwell, pp. 244–69.

Callon, Michel (1998b), 'Introduction: The embeddedness of economic markets in economics', in Michel Callon (ed.), *The Laws of the Markets*, Oxford: Blackwell, pp. 1–57.

Callon, Michel and Bruno Latour (1992), 'Don't throw the baby out with the bath school! A reply to Collins and Yearley', in Andrew Pickering (ed.), *Science as Practice and Culture*, Chicago: Chicago University Press, pp. 343–68.

Cutcliffe, Stephen H. (2000), *Ideas, Machines, and Values. An Introduction to Science, Technology, and Society Studies*, Lanham, MD: Rowman & Littlefield.

Dasgupta, Partha and Paul A. David (1994), 'Toward a new economics of science', *Research Policy* 23 (5): 487–521.

Dennis, Michael A. (1997), 'Historiography of science. An American perspective', in John Krige and Dominique Pestre (eds.), *Science in the Twentieth Century*: Harwood, pp. 1–26.

Dilley, Roy (ed.) (1999), *The Problem of Context*, New York: Berghahn.

Downey, Gary L., Arthur Donovan and Timothy J. Elliott (1989), 'The invisible engineer. How engineering ceased to be a problem in science and technology studies', *Knowledge and Society* 8: 189–216.

Edge, David (1995), 'Reinventing the wheel', in Sheila Jasanoff, Gerald E. Markle, James C. Petersen and Trevor Pinch (eds.), *Handbook of Science and Technology Studies*, London, Thousand Oaks and New Delhi: Sage, pp. 3–27.

Edge, David (1999), 'Editorial postscript', *Social Studies of Science* 29 (5): 790–9.

Elias, Norbert (1982), 'Scientific establishments', in Herminio Martins, Richard Whitley and Norbert Elias (eds.), *Scientific Establishments and Hierarchies*, Dordrecht: Reidel, pp. 3–70.

Ezrahi, Yaron (1990), *The Descent of Icarus. Science and the Transformation of Contemporary Democracy*, Cambridge: Cambridge University Press.

Feldhay, Rivka (2001), 'Knowledge, religion and politics in modernity. mutualities, tensions, and paradoxes', in Helga Nowotny and Martina Weiss (eds.), *Jahrbuch 2001 des Collegium Helveticum der ETH Zürich*, Zürich: vdf, pp. 223–42.

Felt, Ulrike, Helga Nowotny and Klaus Taschwer (1995), *Wissenschaftsforschung. Eine Einführung*, Frankfurt a.M.: Campus.

Flyvbjerg, Bent (2001), *Making Social Science Matter. Why Social Inquiry Fails and How it Can Succeed Again*, Cambridge: Cambridge University Press.

Freudenthal, Gad (1990), 'Science studies in France. A sociological view', *Social Studies of Science* 20: 353–69.

Gibbons, Michael, Camille Limoges, Martin Trow, Helga Nowotny, Simon Schwartzman and Peter Scott (1994), *The New Production of Knowledge. The Dynamics of Science and Research in Contemporary Societies,* London, Thousand Oaks and New Delhi: Sage.

Gieryn, Thomas (1999), 'The U. S. Congress demarcates natural science and social science (twice)', in Thomas Gieryn (ed.), *Cultural Boundaries of Science. Credibility on the Line*, Chicago: Chicago University Press, pp. 65–114.

Gitlin, Todd (1997), 'The anti-political populism of cultural studies', in Marjorie Ferguson and Peter Golding (eds.), *Cultural Studies in Question*, London: Sage Publications, pp. 25–38.

Goonatilake, Susantha (1993), 'Modern science and the periphery. The Characteristics of dependent knowledge', in Sandra Harding (ed.), *The Racial Economy of Science. Towards a Democratic Future*, Bloomington, IN: Indiana University Press, pp. 259–67.

Gross, Matthias (2001), *Die Natur der Gesellschaft. Eine Geschichte der Umweltsoziologie*, Weinheim: Juventa-Verlag.

Haraway, Donna (1989), *Primate Visions. Gender, Race and Nature in the World of Modern Science*, London: Routledge.

Heintz, Bettina (2000), *Die Innenwelt der Mathematik. Zur Kultur und Praxis einer beweisenden Disziplin*, Wien: Springer.

Hess, David (1997), *Science Studies. An Advanced Introduction,* New York: New York University Press.

Hopkins, Terence K. and Immanuel Wallerstein (1996), *The Age of Transition. Trajectory of the World-System 1945-2025,* London: Zed-Books.

Ingold, Tim (2000), *The Perception of the Environment. Essays in Livelihood, Dwelling and Skill,* London: Routledge.

Jasanoff, Sheila, Gerald E. Markle, James C. Petersen and Trevor Pinch (1995), 'Introduction', in Sheila Jasanoff, Gerald E. Markle, James C. Petersen and Trevor Pinch (eds.), *Handbook of Science and Technology Studies*, London, Thousand Oaks and New Delhi: Sage, pp.. 3–27.

Joerges, Bernward and Therry Shinn (2001), *Instrumentation Between Science, State and Industry*, Dordrecht, The Netherlands: Kluwer Academic Publishers.

Jurdant, Baudouin (ed.) (1998), *Impostures Scientifiques. Les Malentendus de l'Affaire Sokal,* Paris: La Découverte.

Kaplan, Barbara Beigun (1991), 'STS, women's studies, and the transformation of the undergraduate curriculum', *Science, Technology & Society* 86/87: 1–11.

Kieserling, André (1999), 'Die Selbstbeschreibung der Soziologie', *Soziale Welt* 50: 395–412.

Knorr Cetina, Karin D. (1992), 'Zur Unterkomplexität der Differenzierungstheorie', *Zeitschrift für Soziologie* 21 (6): 406–19.

Knorr Cetina, Karin D. and Urs Bruegger (2000), 'The market as an object of attachment. Exploring postsocial relations in financial markets', *Canadian Journal of Sociology* 25 (2): 141–68.

Kuper, Adam (1999), 'Postmodernism, Cambridge and the Great Kalahari Debate', in Adam Kuper (ed.), *Among the Anthropologists. History and Context in Anthropology*, London: Athlone, pp. 15–35.

Latour, Bruno (1993), *We Have Never Been Modern*. Cambridge: Harvard University Press.

Latour, Bruno (1998), 'To modernize or to ecologize? That's the question', in N. Castree and B. Willems-Braun (eds.), *Remaking Reality. Nature at the Millenium*, London/New York: Routledge, pp. 221–42.

Latour, Bruno (2000), 'When things strike back. A possible contribution of science studies to the social sciences', *British Journal of Sociology* 51 (1): 105–23.

Latour, Bruno (forthcoming), *Dire le Droit. Une Ethnographie du Conseil d'Etat*, Paris: La Découverte.

Luhmann, Niklas (1986), *Ökologische Kommunikation. Kann die moderne Gesellschaft sich auf ökologische Gefährdungen einstellen?* Opladen: Westdeutscher Verlag.

Luhmann, Niklas (1988), 'Organisation', in Willi Küpper and Günther Ortmann (eds.), *Mikropolitik. Rationalität, Macht und Spiele in Organisationen*, Opladen: Westdeutscher Verlag, pp. 165–85.

Luhmann, Niklas (1992), *Die Wissenschaft der Gesellschaft*, Frankfurt a.M.: Suhrkamp.

Mackenzie, Donald A. (2001), 'Physics and finance: S-Terms and modern finance as a topic for science studies', *Science, Technology, & Human Values* 26 (2): 115–44.

Mol, Annemarie and Marianne de Laet (2000), 'The Zimbabwe bush pump. Mechanics of a fluid technology', *Social Studies of Science* 30 (2): 225–63.

Moody, Michael and Laurent Thévenot (2000), 'Comparing models of strategy, interests, and the public good in French and American environmental disputes', in Michèle Lamont and Laurent Thévenot (eds.), *Rethinking Comparative Cultural Sociology. Repertoires of Evaluation in France and the United States*, Cambridge: Cambridge University Press, pp. 273–306.

Nowotny, Helga (1990), 'Knowledge for certainty. poverty, welfare institutions and the instistitutionalization of social science', in Peter Wagner, Björn Wittrock and Richard Whitley (eds.), *Discourses on Society*, Dordrecht: Kluwer, pp. 23–41.

Nowotny, Helga, Peter Scott and Michael Gibbons (2001), *Re-thinking Science. Knowledge and the Public in an Age of Uncertainty*, Cambridge: Polity.

Pestre, Dominique (1995), 'Pour une histoire sociale et culturelle des sciences. Nouvelles definitions, nouveaux objets, nouvelles pratiques', *Annales* 3: 487–522.

Pestre, Dominique (2001), 'Etudes sociales des sciences, politique et retour sur soi', *Revue du MAUSS* 17: 180–96.

Pickering, Andrew (ed.) (1992), *Science as Practice and Culture*, Chicago: Chicago University Press.

Restivo, Sal (1992), *Mathematics in Society and History. Sociological Inquiries*, Dordrecht: Kluwer.

Schlecker, Markus and Eric Hirsch (2001), 'Incomplete knowledge: Ethnography and the crisis of context in studies of media, science and technology', *History of the Human Sciences* 145–156 (1): 69–87.

Segerstråle, Ullica (ed.) (2000), *Beyond the Science Wars. The Missing Discourse about Science and Society*, Albany, NY: State University of New York Press.

Thévenot, Laurent, Michael Moody and Claudette Lafaye (2000), 'Forms of valuing nature: Arguments and modes of justification in French and American environmental disputes', in Michèle Lamont and Laurent Thévenot (eds.), *Rethinking Comparative Cultural Sociology. Repertoires of Evaluation in France and the United States*, Cambridge: Cambridge University Press, pp. 229–72.

Toulmin, Stephen (2001), *Return to Reason*. Cambridge, MA: Harvard University Press.

Turner, Stephen (2000), 'What are disciplines? And how is interdisciplinarity different?', in Peter Weingart and Nico Stehr (eds.), *Practising Interdisciplinarity*, Toronto: University of Toronto Press, pp. 46–65.

Wagner, Gerald (1996), 'Signaturen der Wissensgesellschaften – Ein Konferenzbericht', *Soziale Welt* 47: 480–4.

Wagner, Gerald (1998), 'Niklas Luhmanns "die Gesellschaft der Gesellschaft" und ihre Bedeutung für die Wissenschafts- und Technikforschung', *Rechtshistorisches Journal* 17: 574–88.

Wagner, Peter, Björn Wittrock and Richard Whitley (eds.) (1991), *Discourses on Society. The Shaping of the Social Science Disciplines*. Vol. XV, *Sociology of the Sciences. A Yearbook,* Dordrecht: Kluwer.

Whitley, Richard (2000), *The Intellectual and Social Organization of the Sciences,* 2nd ed. Oxford: Oxford University Press.

Wittrock, Björn (1998), 'Early modernities. Varieties and transition', *Daedalus* 128 (3) 19–40.

Wittrock, Björn, Johan Heilbron and Lars Magnusson (1996), 'The rise of the social sciences and the formation of modernity', in Johan Heilbron, Lars Magnusson and Björn Wittrock (eds.), *The Rise of the Social Sciences and the Formation of Modernity. Conceptual Change in Context, 1750–1850*, Dordrecht: Kluwer, pp. 1–34.

Zilsel, Edgar (1942), 'The sociological roots of science', *American Journal of Sociology* 47: 544–62.

POSTSCRIPT

CHAPTER 13

BERNWARD JOERGES

REFLECTIONS ON THE MILLENNIUM, CALENDARS, AND THE GREGORIAN HEGEMONY*

Ideology is the other side of chronology
Wolf Lepenies

All essays in this volume take their timeframe from the millennium, using year 2000 as a major marker. A few years into the new millennium, the turn of the twentieth century into the twenty-first continues to organize historical memory and foresight, even if the urgency to couch almost everything in terms of this chronology has dwindled and given way to more meaningful markers – not the least, as many have said, the 'true' beginning of the new century: September 11, 2001.[1]

With a little distance from the great event, with all its accompanying anxieties and hopes and one world-one-mankind hypocrisies, I shall offer here some reflections, or should I say musings, on ways the millennium, and the Gregorian calendar which defines it, frame matters in and outside science. This will be a highly selective exercise, drawing on various themes that have captured my imagination during the time of my association with the *Yearbook* since 1989.

I will start off by looking back at the millennial craze and its embeddedness in the Gregorian *Zeitrechnung*. I will then turn to ways the Gregorian regime has organized the history of science and technology. Two authors, the historian of society Eric Hobsbawm and the historian of technology Thomas Hughes, are drawn upon to demonstrate how calendars order the timescapes of social science too. In this context the 'The End of Science'-discourse as a *fin-de-siècle* phenomenon in debates about science is touched upon. I will then change perspective from calendars *of* social studies of science and technology and turn to calendars *in* social studies of science and technology. I will point out the peculiar hegemony of the Gregorian calendar and its potential cultural ramifications. Finally, as a way of looking ahead, I will ask, in an STS vein, for the place of calendar science/technologies in social studies of science and technology.

B. Joerges and H. Nowotny (eds.), Social Studies of Science and Technology: Looking Back Ahead, 261–291.

THE GREGORIAN TIMESCAPE AND THE MILLENNIUM CRAZE

'In our culture', the *longue durée* measured by calendars is calibrated according to the Gregorian standard for years and begins with year 1 A.D. (or C.E., for Common Era). It is this calibration that allows us, since the Gregorian reform, to frame the past, the present and the future in terms of years, centuries and millennia before and after Christ.

> The Gregorian reform: Pope Gregory XIII had anno 1582 skipped ten days of the (then Julian) calendar in order to bring the Christian calendar in line with the astronomical situation. It seems that his astronomers made a mistake, though: the incongruities that had accumulated since Cesar's Roman time were only compensated back to around 320, one of the reasons why the Gregorian is no good in astronomical contexts. Astronomers today use a simple counting of days elapsed since noon, Monday, January 1, 4713 B.C. on the Julian calendar. The number of days elapsed is called the Julian day number or the Astronomical day number. Why this should be so I do not know ...

Some Western nations have adopted the Gregorian chronological scheme in the twentieth century only, and many nations have other calendars written into their constitutions. In many cultures the Gregorian calendar coexists with other calendars, and the majority of humankind manages their daily lives under different, non-Gregorian time-regimes. And yet, as the turn of the millennium has amply demonstrated, the Gregorian chronology has come to dominate private and public discourse on a truly global scale.

Fixing Time the Gregorian Way

I will not even begin to try to dismantle the deep structure of the Gregorian calendar. Be it enough to point to some of the ways it is embedded in everything.

On the *most mundane level* it is embedded in all machinery around the world that uses electronic circuits. Every operation performed on or by humans as well as non-humans with the help of electronically controlled equipment using time-of-day markers is automatically geared to this structure. Many of these operations have a legal aspect, that is they *must* be performed correctly. One cannot change one's birthday or the date of a credit line as one pleases without incurring trouble. Try to apply the Jewish calendar in these matters. At the *most ceremonial level* the calendar serves to organize historical memory and future vision. Every commemoration of past events, whether in pride or in grief, comes in the numbers the Gregorian count provides and sets in perspective.[2] Many of these occasions again *must* be observed by every respected member of a group or people, at the peril of losing their communal identity: think of all the anniversaries, from one's own to one's nation's, one performs or joins in the life-long work of identity building.

In-between the mundane and the ceremonial we have, among many other layers, the entire gamut of territorial, property and inheritance rights that are codified or contested in calendric terms. Again, this regards personal, national or corporate time-dependent contracts on various chronological scales.

All this applies for any elaborate calendar, and it would be (and is) cumbersome and costly to convert all calendrical references from one *Zeitrechnung* into another. But with the advent of computer networks, it is in principle quite unproblematic to switch calendric markers or maintain some double or triple reckoning at will.[3] The problem lies with the translations of meanings that are not *in the numbers* – in the sense that everyone who can count can see them – but nevertheless are so strongly coupled to the numbers (or more precisely: a position relative to a numerological scheme) as to need them for their evocation, or, the other way around, as to get lost when numbers are switched. This can be demonstrated for all the levels of the calendric scale – hours (high noon), days (lucky Friday), weeks (first advent, recurring weekly festivals), months, years, decades, centuries, millennia...

Here I am interested in scales from years upwards. Years are structured by recurring dates related to recurring rituals – religious, political, personal/local... In some dates, all of these participate, Easter for instance.

> Beyond this, Easter in a way epitomizes the millennial puzzle, because it sets the stage for Christ's second coming which has always been debated in terms of the millennium. – I have never understood the complexities of the Easter-riddle. The debate on the problems of calculating the correct dates for Easter fills libraries, because in determining Easter one somehow has to reconcile both great cosmic cycles, the lunar *and* the solar, both maddeningly fractional: Easter falls on the Sunday after the first full *moon* after the venal *equinox*.

At the time the Gregorian calendar was introduced, the yearly cycle was predominantly shaped by the liturgical calendar which has no significant event corresponding with the turn of the Year. New year festivities were rather played down as pagan. This began to change with the growing importance of a more secular reckoning of decades and centuries.

Reckoning in decades and centuries seems to be a feature of most calendars – the practice of periodisizing history along a scale of magic round numbers is not unique to Western calendars. But the practice of routinely representing past and future events in terms of decades and centuries, and of publicly celebrating the turning points, seems to be closely linked to the growing supremacy of the Gregorian time order.

> In this long and uneven process, protestant countries lagged behind Roman Catholic ones. Sweden, for instance, has a peculiar history: It decided to make a gradual change from the Julian to the Gregorian calendar. By dropping every leap year from 1700 through 1740 the eleven superfluous days would be omitted and from March 1 1740 they would be synchronous with the Gregorian calendar. So 1700 which should have been a leap year in the Julian calendar was not a leap year in Sweden. However, by mistake it seems 1704 and 1708 became leap years. This left Sweden out of synchronization with both the Julian and the Gregorian world, so they decided to go back to the Julian calendar. In order to do this, they inserted an extra day in 1712, making that year a double leap year. So in 1712, February had 30 days in Sweden. Later, in 1753, Sweden changed to the Gregorian calendar by dropping 11 days like everyone else had.

According to Peter Stearns, the Harvard/Carnegie Mellon social historian, the use of century designations became routine after 1600 only. The first time the end of an old and the advent of a new century excited much public rhetoric and celebration was 1700. Since then a considerable culture of centennial celebrations and *fin-de-siècle/debut-de-siècle* framing of historical downs and ups has emerged. Contemporary historiography operates mightily with the figure of centuries. More obviously meaningful turning points like the French revolution or WW II are often aligned with centennial schemes, and not the other way around.

The 'triumph of the century' (Stearns 1996) since 1700 surely reflects many cultural shifts. The best cue here may be the word – *siècle* – itself: 'secular' change (in the sense of 'this-worldly' used since the thirteenth century), the moving away from tradition in all spheres of life, had become the predominant sense of history for intellectuals and political leaders in Europe. Both rapid change and the stabilization (*Verstetigung*) of its driving forces in 'modern' institutions, especially science and commerce, contributed to a growing need to fix definite units of past and, especially, future history. Forward looking long range calendars became, maybe for the first time, as important as reckoning in terms of great past events. The trading of events (with past dates attached) for dates (looking for future events), which has been so characteristic of the recent turn-of-the-millennium rhetoric, became feasible only when literacy and the use of printed calendars and almanacs became common since the seventeenth century.

But the spread of calendars and almanacs also triggered millennistic movements, as opposed to secular thinking and historical attitude. In this sense, reckoning in terms of millennia also is very much a form that came with the Christian calendars. Although millennialist thinking can be found in Islam or Old Chinese *Zeitrechnungen* too, it is not strongly linked to their calendars. It is more messianistic/apocalyptic than millennarian in terms of calendric date. Presumably, numerological millennialism is particularly strong in Western civilization because of its deep biblical roots: 'A thousand years is but one day in the eyes of God.' God created the Universe in 7 days. His creation recapitulates divine history. History will be fulfilled after 7000 years therefore. (We have now entered the third day...). This is the original connection of the millennium as a Judeo-Christian construct and the calendar meanings of 1000, 2000, 3000...

The Three Strands of the Millennium Craze

The big calendars of high cultures operate on a day/into weeks/into months/into years/ into decades/into centuries (or at least dynastic periods)/into millennia basis. As Lepenies has pointed out for the Gregorian period (and as could probably be shown for other great calendar schemes too, although he makes this contingent on a precise the notion of 'century'), the metaphors of age and aging, anthropomorphically taken from the yearly cycles of human existence, are powerfully at work in thinking about centuries and millennia too (Lepenies 1999). The magic of 'round numbers' is apparent in all these cycles and the way followers of the Gregorian have experienced the bi-millennium mirrors the meanings conventionally attributed to (simplifying to three cycles) yearly, centennial and millennial markers and turns.

Years are full of recurring life-time events, both secular and religious. The year-cycle is very old and quite universal, although most calendars will place the beginning of a new year somewhere around Easter. Celebrating the New Year at the beginning of January slowly came with the Gregorian calendar.

Centuries are historical-calendrical and secular. Europe went secular-calendrical after 1200, and definitely since the last three centuries.

Millennia are messianic/apocalyptic and religious in a broad sense. Associating apocalyptic visions with a thousand years in a calendar is again an achievement of the

Gregorian calendar.[4] A new apocalyptic millennialism came with New Age, in combination with a mystification of the Maya calendar whose turning point in 2012 was easily synchronized with the Gregorian 2000 (what is 12 years in view of the New Age?). Messianism has great traditions outside Christianity in Judaism and Islam, but at least in Islam it did not acquire millennistic character (according to Stearns 1996; see also Oliver and Steinberg 1998; Stowasser 2000). On the other hand, centuries did not become important units of ordering time in calendars of other civilizations.

The bi-millennium powerfully *fused* these three strands in organizing history around the round numbers of the Gregorian calendar. This fusion of yearly reckoning, centennial periodization and oftentimes millennistic/messianic distinctions of millennial eras in bi-millennial rhetoric made for the cultural interest of the event. But can it be shown to have had any definite effects on the course of history? In an earlier paper on the dynamics of the millennium I had hypothesized that, despite its essential arbitrariness, this great date in search for great meanings would set off *great projects* in search of legitimations that would otherwise not be engaged with (Joerges 1990). But most of the large-scale projects that were indeed designed over the last decade of the twentieth century have become shelved or diminished, not so much for lack of money but for lack of time. This was true, for instance, for the grand projects that were on the plate in Rome, the mother of all 2000 metaphors and stories. Or in England where the London Dome was indeed built but bankrupted before the new millennium gathered momentum. It looks then as if the *economic* mobilization the year 2000 effected was largely exhausted in big parties – splash economics, as it were.

Potential long-range *cultural* effects of 2000 seemed to me, before the turn, to merit interest in close connection with 2000 projects and programs. All of these, whether they have been realized or not, finalized in time or not, came with characteristic symbolisms celebrating and propagating particular versions of communal, national or global (future) history. A quasi-universal discourse of ending something and beginning anew orientated the projec(tion)s of many public spokespeople and thereby presumably conditioned the world-views of many more ordinary people. From this, one might have expected ubiquitous cultural effects of the millennial turn that carry a long way into the future. I will come back to this point in the context of what I call the Gregorian cultural hegemony. But in fact, the millennium has not occasioned the critical culture debates, let alone militant resistance, I felt at the time it would.

CALENDARS AND PERIODIZATIONS *OF* SOCIAL STUDIES OF SCIENCE AND TECHNOLOGY

Exploring how calendars organize historical material – why not begin at home? The calendric hegemony of the Gregorian *Zeitrechnung* – in the sense of its imposing a particular periodization on cultures with their own *Eigenzeiten* and calendars – applies to the periodizations and chronologies of the sciences and of science studies as well. In a broad sense, all sciences use periodizations in structuring and comparing their narratives and the history of their narratives.[5] Depending on how students of science and technology choose to construct the historicity of their objects of study (i.e., other sciences, and thereby also the historicity of the objects of scientific and technological knowledge), different timescapes organize their narratives.

No wonder then that the sense of an ending and of a beginning which the turn of the millennium infused into everything was hard to escape in science & technology studies too.[6] And of course, towards the end of the twentieth century, the history of science was massively written as synchronic history: the ups and downs and progressions of science being aligned with grand history of ideas and political history which in turn is inexorably tied up with our *Zeitrechnung* and its symbolic codes. In the following I will draw on two examples – one from social history open to science/society issues, one from the history of technology – in order to show how certain centennial and millennial interpretations have impregnated the history of science & technology 'in the twentieth century'.

Science in the Short Twentieth Century

Eric Hobsbawm has given a rare account of twentieth century science, in the penultimate chapter of his 'Age of Extremes', before he looked 'Towards the Millennium' in his final chapter.[7] How did he represent science (and for that matter technology – he calls the respective separate chapter 'Sorcerers and Apprentices', where the apprentices are the non-scientific users of science)?

Hobsbawm chose two epigrams for his chapter on science. One by Lévi-Strauss who said in 1988 that there is still a place for philosophy in today's world, but that science has changed the rules of the game for good, 'it has revolutionized the rules by which the intellect operates.' The other is by Margaret Jacobs who pointed out that

> the standard text in gas dynamics written by his author while on a Guggenheim fellowship has been described by him as having had its form dictated by the needs of industry. Within this framework, confirming Einstein's theory of general relativity came to be seen as a critical step toward improving 'ballistic missile accuracy by accounting for minute gravitational effects.' Increasingly, post-war physics narrowed its concentration into those areas thought to have military applications (Jacobs 1993, 66–7).

Science between Philosophy and War. Hobsbawm seemed to signal right from the outset, that the world-changing powers of the sciences will be framed on the one hand as intellectual history, the history of ideas one might say; on the other as the history of the dominance of nations, as political history.[8] And he ended his analysis on a sceptical and (I think) ironic note: The 'world-changing powers' to which the scientists held the key in the twentieth century, protected them, for these powers

> appeared to depend on allowing an otherwise incomprehensible and privileged elite – incomprehensible, until late in the century, even in its relative lack of interest in the external signs of wealth and power – to go its own way.

For this reason, all nations adopted relatively non-interfering science policies. But governments are not interested in scientific truth, continues Hobsbawm. They are interested in the instrumental potential and the symbolic uses of science, that is, national prestige, 'in which the pursuit of Nobel prizes preceded that of Olympic medals and still remains more highly valued.'

> Such were the foundations on which the triumphant structures of scientific research and theory were erected, by which the twentieth century will be remembered as an age of human progress and not primarily human tragedy (Hobsbawm 1995, 557).

A curious gloss indeed on Mertonian and older notions of the autonomous constitution of scientific institutions.

In between such beginnings and endings Hobsbawm has a highly nuanced story. In the social history of science, as pictured by Hobsbawm, the sciences play three crucial roles: they are considered a major force in societal development, thanks to their splendid if politically opportune autonomy; they are seen, in particular, to drive economic development, both production and consumption; and they are seen as jeopardizing the human environment. Not surprising, of course, but the interesting thing is the specification of the mechanisms that make for science's relations with overall history, economic change, and the natural environment. The fundamental narrative of both the history of science and social history tends to remain the history of nations and empires. In the twentieth century this is the struggle between democratic nations and fascist/communist regimes. The resulting frame is 'the short twentieth century' 1914/18 – 1989/90. Twentieth century history of science and technology is synchronized with the overall chronological scheme of Hobsbawm's history of the short century. This is the basic plot, and the drama of science is staged as a drama of world-historic struggles for supremacy.[9]

May it be added that there are more than a few twists. First of all, the scientists did not know what they were used for. Hobsbawm talks about a basic paradox 'the historian of the century must grapple' with, namely that no period since Galileo's recantation has been more dependent on scientists, and yet no period has been less at ease with them. He does not resolve this one, but leaves us with a sibylline prophecy: the twentieth century will be remembered, thanks to science, not as the age of human tragedy but of human advance... More importantly, however, Hobsbawm's story is impressively nuanced and belies, all in all, the world-historic simplification. Read closely, the events marking various turns in the development of science and technology do not correspond too well with grand political history.

Hobsbawm presents science and technology as very important, but nevertheless as a subsystem whose history can be looked at separately, only to demonstrate its synchronicity with a social history which is in turn linked to the great politico-ideological changes of the century. Science is almost exclusively treated separately from the rest of the story, the interrelation of science (as distinguished from technology – *nota bene* production technology, not the technology of science) with general history is added after the fact of writing down the social history of the autonomy of science. A peculiar mix of externalist and internalist interpretations.

Technology in the American Century

Similarly, the historian of technology Thomas Hughes has set out the basic narrative of the development of large technical systems in stages that are variously synchronized with a centennial historical frame, allowing for ordering and making sense of events across many systems and system types. Over the years, one can discern three forms of periodization in Hughes' work.

The first, developed in Networks of Power, reconstructing (energy-)time from 1870 to 1930, had three stages. The three steps, which Hughes had established studying the American, British and German electricity generation and distribution systems, comprise 'invention and development'; 'innovation and competition'; 'consolidation and rationalization': a script for a clearly upward and outward reaching movement resulting, if successfully played out, in mature large technical systems of high momentum in post-war Western economies. In an evolutionary metaphor, most historians and sociologist of technical systems following Hughes have posited an orderly, directed (if not always planned and sometimes crisis-prone) trajectory from small (young) and local to large (old) and national systems (Hughes 1983). Here we have, in many of the stories told in the literature on large technical systems, for the last time, harking back to the eras of heroic system-builders of the Gründerzeit and of enthusiastic post-war reconstruction, interpretations of system development as the story of an inventive beginning, a tempestuous development and a stable state of maturity.

But this form of periodization proved not entirely defensible. Hughes gradually developed it further in his second big study on the development of large-scale technological systems, published 1989. *American Genesis: A Century of Invention and Technological Enthusiasm* leads up into the 1970s. Around the same time, Renate Mayntz published her *On the Development of Technical Infrastructural Systems*, concluding that at least in continental European nations

> . . . the evident affinity between centralized government. . .ruling principles and the organizational imperatives of communication and transportation net-work technologies probably played an important role: This has encouraged both government interest in the development of these LTS and, conversely, their development into large monopoly organizations.

She goes on to argue that the historical specificity of this congruency may have brought the heyday of large technical systems, organized on a strictly centralized basis, to an end towards the end of the century. Both as a result of technical developments, which are less dependent upon the traditional network set-up and can thus be more de-centrally organized, and because the predominance of hierarchical social orders are possibly disappearing (Mayntz 1988, 275f.).

Mayntz and Hughes abandon here the stage model and its implicit success story in favour of a story of the rise and fall of large technical systems, in tune with the American self-interpretations of decline so characteristic of the 1980s. The structural-analytical concept of the 'momentum' of large technical systems following the electrical pattern yields to a parable of cyclical generation and decay. Now Hughes sees large technical systems as culturally threatened, obsolete and doomed to extinction by the counterculture emerging in the 1970s, with its vision of a decentralized, soft technology. He was impressed that substantial thinkers among the counterculture elite identified large-scale systems as a salient negative aspect of contemporary culture. Herbert Marcuse, Paul Goodman, and Theodore Roszak are cases in point. Oppressive technological – not political – systems were their major concern.

And he maintains that out of the counterculture came post-modern thinking as well, especially in architecture. 'In part, post-modern architecture is a counter response to the order and control of large-scale modern systems' (Hughes 1997, 187).

As a consequence, Hughes has tended since to concentrate on 'younger' systems like 'management systems' and the technically hybrid telematic systems of the more recent past. At the same time, he seems now to interpret 'older' systems resembling grid-based energy systems as 'modern' and the non-grid based, hybrid project-, management- and information-systems of the 1970s and 1980s as 'post-modern' (Hughes 1998). The secular metaphor has also been adjusted: the implied epochs are now distinguished into a modern era, beginning approximately with the pre-last turn-of-the-century, and a post-modern era, strongly drawing on a pervasive and persuasive *fin-de-siècle* rhetoric.[10]

The End of Science – Fin-de-Siècle and Millennialism in Science Studies

Toward the end of Hobsbawm's 'short twentieth century' and Hughes's 'American century', the idea to take distance and to look anew at science and technology has led to various attempts to place science & technology research in a broader framework of historical concerns. A 'sense of an ending' (Frank Kermode), the notion that something irrevocable had happened to good old modern (enlightenment) science took hold of much writing about science. True, the idea of 'the end of history' which has had a short half-life in the late 1980s had already become largely history itself; still, an overwhelming sense that science and technology are part of a wider historical sweep best described as the end of an era was widespread.

The End of Science thesis came in many guises: completeness, insolvability, disinterest, de-institutionalization, uselessness, perversion (Elvee 1992). John Horgan, for example, a former *American Scientific* science writer, had in the 1990s hailed the end of science because, he claimed, prominent practitioners from all kinds of fields have told him that all important problems in science have been solved (except for TOE, and that would be settled soon, too). Remaining problems are said to be unsolvable (Horgan 1996; see also McGinn 1996; Schick 1997, on some scientist's philosophical naïveté; Maddox 1998, against Horgan; Mittelstraß 1998, on the philosophical issues). Horgan's point was adopted by epistemological ironists who claimed that scientific theories were not believed any more by their proponents to be capable of validation and that decisions as to their validity were not arrived at by controlled experiment but *via* aesthetic considerations.

Whatever one made of such claims, in the 1990s, received public representations of Science clearly were progressively at odds with much current scientific practice and emerging unofficial representations from within and without. One way of interpreting this was that practice had run away from its venerable symbolic façades. Representing the scientific enterprise could for a long time play a self-referential game, without much reference to what scientists did, but not forever. A crisis of reference occurred because one had to realize that official concepts for science and technology did not fit anymore; holding fast to old representations and institutional forms made the thing itself, i.e., actual research practice, vanish. Another way of interpreting it worked in the opposite direction: emerging representations of science and technology urgently needed to be translated into corresponding practice. The way science and technology was still being done badly lagged behind. Again, the thing itself was missing behind its ceremonial façades. This view seems to me behind much of 'mode 2' discussions (Gibbons et al. 1994).

Whichever way: assuming that this powerful movement, the *Aufklärung*, has ever happened (and who am I to say it has not), some analysts also began, towards the end of the century, to see that the enlightenment process had to be, and indeed was, rewritten the same way the processes of the division of labour and the processes of the advance of high science & technology had to be rewritten. The implication was that it had happened slightly differently than we were taught in school all along – and that it would probably continue for a while to do so. Part of it was that it seemed hazardous to prophesize how it would go on, in this new millennium, after the great divide in historical time. No way of knowing, not because times had become so turbulent in the twentieth century but because they always were. In this sense, no end of science, or at least no epochal rupture but rather haphazard continuity.

The story of scientific and technological advances in the twentieth century, as told by Hobsbawm or Hughes and others, was quite straightforward: through the successful application of science and technology as solution to social (including military) problems more problems were created worldwide than solved. The public façades and legitimations of science (including its claims to state support) did not allow for admission of this outcome. Consequently, the façades crumbled. New hypocrisies were wanted and were missing badly, new practices looked for their proper hypocrisies. For analyses of science and technology primarily interested in the performance of the façades, this made look science and technology indeed entering into a deep crisis, if not vanishing. If and when the assumed reference out there had changed to a degree of becoming obviously non-functional and even illegitimate – for instance because scientists' spokesmen themselves had largely changed their language – science & technology studies tended to observe a crisis of science & technology that was in fact a crisis of science & technology studies.

The plausibility of the crisis image owed much to a particular framing of the question then. The history of science and technology, much like social history at large for that matter, continues to be framed as part of enlightenment history, and enlightened history. But, to borrow a phrase: we need here to 'rethink, not un-think' the enlightenment (Elkana), and science and technology studies have a prominent role to play in this project. In this sense, wholesale talk about the ultimate crisis and in a way the end of modern science at the eve of the century was unproductive. Notions of radical ruptures in science and technology tended to be devoid of specific analyses of scientists' practices.

In sum: the debates around the themes of crisis, rupture and 'the end of science' could hardly have occurred outside a *fin-de-siècle/fin-de-millennium* framework. As Diana Relke justly put it at the time:

> The End of Science belongs to a rapidly growing body of *fin-de-millennium* literature that feeds our need to attach meaning to the meaningless calendrical fact that the end of the second millennium is nigh. Horgan, one of the most eloquent science journalists in the business, pursues the thesis that science is doomed because all the Big Questions have been answered--all, that is, except the biggest one, the imminent answer to which will ring down the final curtain on the scientific enterprise (Relke 1998, 23).

If today we feel that science has lost its boundaries, that the project of a universalistic science has been abandoned in many places, that scientific and research institutions have lost autonomy and public authority, that the course of scientific advance has become unpredictable and so forth, then this may not be attributable so much to epochal ruptures and turns in the history of science but to a lucky turn in science studies: We begin to understand that on-going science and technology always had uncertain boundaries, precarious autonomy/authority, unpredictable outcomes, localized jurisdictions, multiple and shifting time-frames, and so forth.

Yet, just as toward the end of the nineteenth century *fin-de-siècle* rhetoric did, *fin-de-millennium* rhetoric *a forteriori* represented, at the end of the twentieth, a sort of chronological *Kampfbegriff* used by ideological camps in order to 'influence with their politics of time the politics of their times' (Lepenies 1999, 7). The same for science policy?

Heroic Periodization

Scholars in the history of both science and technology tend to impose on these fields the heterodox timescapes of grand history. In the face of such heroically periodisizing stances, a wealth of social studies of science & technology of the 1980s/1990s have produced much evidence for highly contingent (indirect, mediated) links between the tides and ebbs of science and technology and great politics respectively. Hobsbawm's and Hughes' monumental pictures do not agree too well, I feel, with the views cultivated in much of recent empirical work on science and technology. Apart from the fact that these studies rarely relate their subjects to social history in the standard meaning of the term, the periods and turning points identified in this work correspond little with overall historical phasing.

In turn, an important reason why historians tend to disregard social studies of science and technology may be that their respective periodization schemes differ. Hobsbawm and other historians relate loosely linked social developments to grand epochal changes in world history, in the relations of nations. Wars and the rise and fall of political regimes mark the phases and periods of things. Of course there are good reasons for this. Economic change is more or less closely linked to grand history, and military politics no doubt have major impacts on the ways of science and technology. Yet this seems to me to be grossly overdone. The development of the stuff of particle physics or microbiology is not, even by externalists, seen as something related to big politics, or rarely so. (The fact that the findings are couched in terms of possible military applications does not prove the contrary.)

If, in other words, the social history of this century were rewritten as a detailed social history of science, technology, work and the environment, the links between overall great history and social history would be loosened. One would discover more synchronization between scientific-technical movements and social history and less synchronization with political history, for instance, and wholesale heroic periodizations would lose much of their meaning.[11] Interestingly, Steven Weinberg, who contributes the chapter 'The Expansion of Physics' to the recent *Oxford History of the Twentieth Century*, in refusing to frame change in physics in terms of political periods and events, ends up with not much more than a list of names and dates: chronological time has no historical time here, except that he finds it impossible not to go way back into the nineteenth century to explain what happened in the twentieth (Weinberg 1998).

On a much larger scale, the history of the 'war of civilizations' also tends to be written heroically and inscribed into the timescapes of its winners and survivors. With the exception of William McNeill, all contributors to the *Oxford History of the Twentieth Century* write basically European history, innocently draped in Gregorian clothes.[12]

METROPOLITAN TIME, OR: CALENDRIC HEGEMONY

The millennial turn occasioned less economic and cultural projects with lasting impacts than one could expect. And yet, the ominous date 2000 may have been of more intrinsic cultural interest. A thought experiment: Imagine yourself a self-respecting post-colonial public intellectual, educated and fully conversant in the traditions of your own non-Western (you would see it in fact as pre-Western) culture, sad or disturbed about the spiritual havoc created by Western civilization. And imagine that you were annoyed by the way the politicians in your own society, your intellectual interlocutors (if any) in other societies, the media you are exposed to daily (and of course your bank, your TV-program and your computer) continuously pushed on you the upcoming turn of the millennium as a major historic marker. Would you not have been tempted very strongly to use this date (and the parties that you could not avoid anyway) as a platform for raising a few issues in cultural politics?

You would understand, after all, that the fixation of the birth of Christ as year 1 of the Christian era, a process beginning sometime in 526 A.D. of their counting, was, at the time, introduced only for purposes of forward reckoning, the years B.C. still be counted *ab urbe condita*. Christian time was to follow a different count from Roman time. Eventually this resulted in the imposition on the West of the Gregorian time order in 1582, and in due time on your own civilization. All this was of course set in motion by a clash of civilizations in the first place: the desperate attempt of the 'Western' Pope to regain some legal supremacy over Eastern Christendom.

> The Christian emperors had retreated to Constantinopolis, the West had been taken over by the heathen Germans and Goths. The Pope in Rome availed himself of a smooth public intellectual, Dionysius Exiguus, who in 526 A.D. (at that time year so and so after Diocletian, the prosecutor of Christians) fixed the Easter dates for the next hundred years to come, thus allowing the pope to successfully arbitrate in the Easter controversy in the East – every Eastern diocese used another Easter date. The same public intellectual then had the idea that it was inappropriate for Christians to count their years after 'the godless emperor' Diocletian and persuaded his Pope to institute the birth of Christ as year 1 . . . We know now, that the little monk got it quite wrong, relying on a mix of assumptions about the age and date of birth of Christ which are considered off the mark, according to present informed guesses, by 4 to 7 years. We are already farther into the next millennium after Christ. Lepenies for instance dates the birth as August 30, 0002 B.C.

The new calendar was then, right from its beginnings, the outcome of a cultural confrontation, a means to manage and legitimize political meanings.

Would you not be tempted to take the opportunity of the millennial turn to make the case that the millennium craze epitomizes a vast cultural hegemony imposed through the Gregorian time order? I felt you would, and began, around 1998, looking at things through this frame.

Local Calendars/Global Calendars and the Colonization of Time

By placing the mark 2000 on just about everything in public life, establishing as it were a black hole in time from whose attraction no event coming into its vicinity before or after would ever escape, the Gregorian calendar has become paramount in the symbolic representations of the affairs of humankind. There was, in particular, one symbolism in innumerable millennium-projects that tied in with the calendric hegemony of Western Time: the image of 'One Mankind, One Planet, One Fate' which came with many programs couched in terms of the millennial turn. The icon of the globe/earth/planet was probably the most common visual element in texts evoking the grand event:

The 2000 UN conference on cities, to give only one example, was called Urban 21. Like the one before in Istanbul 1997, called Habitat, the problems of urbanization and life in 'the city' were very strongly coupled to issues of global ecology and the future of modern civilization, that is the future of mankind at large. The life of the myriads of city dwellers from the Bronx to the favelas of Sao Paolo or the shanty towns of Mexico City and their equivalents (?) in Cairo, Timbuktu, Lhasa, Vladiwostok, Heyderabad, Kyoto (you name it) was symbolically aligned to the great watershed before 2000/after 2000. But while these dates were packed with meanings for most who did the talking at Urban 21, one may assume that (beyond a short moment on the evening of Dec. 31, 1999) they were quite meaningless for those who do the living in these cities.

In other words, the alignment of local histories with global 2000-history may not have made much sense for those who do not speak and think and dream English or Russian or Spanish in these cities. Because (disregarding the deracinated for the moment) people outside the Western sphere have their own calendars.

> On January 1, 2000 (Gregorian), Mayans would have written: 12.19.6.15.0, French revolutionaries: Décade II, Duodi de Nivôse de l'Année 208 de la Révolution, Muslims in Kairo: 24 Ramadan 1420, Muslims in Iran: 11 Dey 1378, some Jews: 23 Teveth 5760, businessmen following the proposed ISO-standard: 6 of week 52 of 1999, Ethiopians: 23 Takhsas 1993, Copts: 22 Kiyahk 1716, Old-style-Chinese: Cycle 78, year 16, month 11, day 25, and so on and so forth.

All these calendars are filled with their own symbolisms for the representation and legitimation of the daily practices of those following them. They deal with their problems at hand by relating them to overarching periodizations, memorials and turning points, much like everybody does in every culture. Only theirs are different ones, more or less unrelated and incapable of assimilation to code 2000.

It may seem that, beyond historiographic issues of how it all came about, the Gregorian order is quite inconsequential: people just take it for granted, it has been in effect for a long time and nobody apart from some calendar experts has a conceivable interest in changing it. But then again this may not be the case. Generally speaking, I can think of four constellations where things become more complicated.

The first constellation is when two or more *Zeitrechnungen* present in a cultural sphere *compete* in the regulation of the same process and a person or a group has to negotiate which one should apply to which element. As a Russian-orthodox Serb living in catholic Croatia one has to negotiate ceremonial time quite a bit around Christmas and Epiphany. The second is when a reigning *Zeitrechnung* is contested or *revolutionized* by some groups within a society in order re-write its history. If one were a Jacobean in 1794 one would live in year 2 of the First French Republic which attempted to erase all memory of the *ancien régime.*

> Revolutionary change almost as a rule is accompanied by attempts to overthrow reigning calendars. Stalin tried it, the SS tried it, Khomeini tried it. The *Jacobin* Revolution of 1793–94 was the Year 2 of the Revolution (Year 1 having begun with the establishment of the First French Republic in September 1792). Of course the *ancien régime* memory lingered, and was mightily refreshed at the occasion of the 1992 French bicentennial, not the least by historians who erased versions á la Michelet. Interestingly, the Jacobins instituted two commissions in 1792, one for the total revision of the metrologies of space and mass, one for the revision of time-metrology, including calendar-chronology. Both revisions were undone by Napoleon, but while the first was in the larger run perfectly successful, the chronological revolution never captured popular imagination, or undid peoples' religious imagination. The reason why the decimalization of space measurement succeeded may have been that restructuring space was loosely coupled to religious representations, whereas chronology was intimately linked to religious life (Michel Vovelle in Reichardt and Schmitt 1988, p. 66). Accordingly, the short-lived revolutionary calendar only took root in metropolitan areas and larger cities (Meinzer 1992, 23ff.).

The third constellation is when a dominant and basically uncontested time order is considered to be partially impractical and in need of *reform* (see below, on resistance from within). The fourth is when one *Zeitrechnung* is imposed from the outside on a cultural sphere whose *Eigenzeit* and its symbolic codes are very different from the hegemonic calendar. If one were a non-assimilated Jew in Eastern Europe a while ago one resented the alienating pressure of the Christian calendar hegemonically imposed on one's working life and one's possibility to openly perform one's own memorial events.

Code 2000 was/is metropolitan. The term echoes of course the old *dependencia*-theorists and third-worldists who attributed underdevelopment to the economic relationships between center and periphery, both in a global view and in a fractal view of ever-repeated center-periphery asymmetries as one moved down the scale from global society to the last village. This is thoroughly out of fashion in most post-colonial theories today, and the globe has indeed become far too perplexing to understand economic matters in these terms (Larrain 1989). But World time and the Western calendar might be the most successful metropolitan hegemony ever in precisely this sense. Successful because, obviously, it was not a burden but a resource for those new in power, no matter at which level. The survival, and in some circumstances revival of competing *Zeitrechnungen* is part and parcel of this: it serves to assuage and manage the cultural irritations that came with joining the post-colonial global order or disorder.

The invasion of codified time-orders outside the Western World by the Western order goes on and is carried forward by all kinds of fully legitimate institutions within non-Western societies. The seemingly most unobtrusive institutions are probably media entertainment and computerized economic operations – TV-time and bank time is Western Time. But the same holds for most other time-regulated activities requiring link-up with international communication networks. If there is one colonial achievement that splendidly survives the age of colonization and imperialism, it is the world-wide implantation of the Gregorian calendar.

Resistance from Within

The reign of the Gregorian has not gone unchallenged in its proper domain.

> A glorious feat of resistance against the United Gregorian Calendar Experts & Scholars was of course that everyone has celebrated the Event at New Year's eve 1999, although according to the calendar the new millennium began on New Year's eve 2000: the first year of the new age is 2001, because the monk failed to account for the year 0 at the time. On the other hand, the same scholastic calendar scientists who keep pointing this out will also tell us that the millennium already started in 1997 (or such), because Gregory and his astronomers mis-dated the birth of Christ by an uncertain number of years. We were all duped and missed the 'true' turn . . .

In a more serious vein, the reform of the Gregorian Calendar was and is on the agenda of diverse groups and interests. Big global business was never pleased with it. In the 1920s, Eastman propagated a calendar based on Comte's positivistic calendar reform scheme (the so-called Cotsworth calendar).[13] ISO (the International Organization for Standardization) has not sanctioned an alternative to the Gregorian Calendar currently used by most of the world; but it has provided standard notations for recording Gregorian dates, one of which dispenses with months.

> The ISO dating-by-weeks notation can be found in the technical standard referred to as ISO 8601. Dating by weeks is common in European countries; and it is becoming increasingly more common world-wide. All 52 or 53 weeks of the year receive a number, and days of the week are numbered 1 through 7, beginning with Monday. So a typical week date notation in accordance with ISO 8601 would look like this: 1995-W03-6. That notation indicates the sixth day of the third week of 1995. The 'W' in the notation is optional, but the '0' should precede week numbers less than 10. All week numbers have two digits. The ISO Week Date of the day I began research on this essay (27.08.1998) was 1998-W35-3. It was week 35 of year 1998, day 3 of the week, and day 238 of the year.

Global reform was discussed in 1954: the 'World Calendar', considered at the time by the UN, had the 365th day as dateless and weekless, but was not adopted. Similarly, the US congress has considered a 'perpetual' calendar, which aligned the first day of the four, annual quarters to always fall on a Monday.

As the millennium approached, a great number of rationalizing calendar reform schemes were put forward. There was another initiative to have the UN vote in a more 'rational' New Millennium Calendar.

> Here is a bit of the rhetoric from the website arguing their case for a clean-up: "Today, economic, political & cultural globalization appears substantial. However in several years, at the current rate of change, conditions of today may, in retrospect, then appear relatively primitive. Standardized systems are critical to such development, which in turn is critical to realization of great potential benefits. The calendar is a vital fundament of all infrastructure. But, one country has just returned to the lunar calendar, in 1998; world-wide, a number of systems still coexist at this time. Our current Gregorian regime is pa tently not worthy of being a standard, with ever

shifting day/dates and variable months, underpinned by a far from optimal week-system ... Many major enterprises cannot, effectively, use today's calendar for planning; in practice using notional 28 or 30 day months instead. Planners, like all of us, cope with the awkward vagaries knowing no other system; much as business coped without fax/modems for a hundred years – despite their predating telephony. Calendar reform was last on the UN agenda to facilitate business and government planning, then fixing just four days of the year, the first of each quarter to fall on a Monday. But ALL 365 days of the proposed nmc year are perfectly fixed." Now, with the proposed NMC system, complete rational symmetry at last. Monday, from Norse 'Monandaeg' or 'moon's day', symbolizes the lunar connection of the old Gregorian system, who's lunatic features end with it (there are no Mondays in the six day NMC week).

Another scheme, with links to public intellectuals like Noam Chomsky, was called the ABC (A Bioregional Calendar), propagated by the following rhetoric:

> At present human society is dominated by a single calendar system. This is a reflection of a culture of dominance permeating humane society. A Bioregional Calendar (ABC) is an alternative that provides humanity with the ability to organize while encouraging autonomous cultural expression.

In this proposal, B.C./A.D. was meant to be substituted by a new Year Zero in 1945 (Gregorian), and designated B.H./A.H. (before Hiroshima/after Hiroshima). The ABC-calendar advanced a syncretistic symbolism supporting multiculturalist aspirations. Yet another scheme, The Global Era Calendar, put forward in Oregon in the name of humanity, converted 2000 into Year 0 and added leap year adjustments supporting globalist values. Richard Landes, Professor of Medieval History at Boston University, is said to have called the calendar 'a gem' and a 'call to mindfulness'.

Most of these schemes have been somehow addressed to the United Nations, all of them are meant to simplify, rationalize and un-enculture/re-enculture a Gregorian calendar considered both messy and encumbered with historical ballast. They can be considered rationalist-universalist cults: they assume a superior rationality of the Gregorian calendar as compared to other chronological schemes that have been worked out in other cultures, but look for more perfection and less cultural bias (except for the obsession with rationalizing and perfecting of course).

The drive for perfecting the Gregorian calendar is fed by rapid advances in the perfection of the second, which used to be defined astronomically, representing a standard movement in the sky, but has long since been dissociated from its heavenly context by the European introduction of mechanical clocks, and lately by atomic clocks. The standard today is the caesium standard, and fractions of seconds can be calculated and, especially, used for calibrating machinery, to many points after the comma. Working up one's way from the continuously improved second to other units of time and then to standard chronologies produces more and more imperfections in the calibration of the machines called calendars. These remain inescapably (it seems) tied up with models of the skies, i.e., astronomical calculation. The intellectual challenges this creates for the astro-time schemes mentioned above are beyond my understanding. As to the challenges it creates for more mundane requirements to synchronise valid calendars with new, caesium-based time-standards for global computer-networks requiring exact calibration of same-time in whatever time-zone of world-time they operate: I have not worked out those.

In a Weberian vein, the task of a sociology of calendars would be to point out the ways in which the Gregorian calendar incorporates more of the criteria of occidental rationality than alternative calendars and for that reason has managed to put the reckoning of macro-time into a global iron cage.[14]

In what sense could the Gregorian be said to be more rational than Non-Western calendars? According to the mediaevalist Hans Maier, the Gregorian calendar is the only one that has a zero point, thus allowing the projection and the retrojection of time-intervals in *both* directions, the future and the past, indefinitely. All other great calendars take as their base some foundational date, like the Roman *ab urbe condita* or the biblical day of creation.[15] As Maier notes,

> (i)t is one of the ironies of history that the same *Zeitrechnung* which moved Christ into the center of time finally succeeded in the age of enlightenment – that is in a time which began to break with many Christian traditions. But the practical advantages of a chronological reckoning from a fixed point could not be denied even by critics of Christendom and the Gregorian era (Maier 1989, 42).

In other words, the seemingly christocentric base B.C./A.D. of the Gregorian makes this calendar the best candidate for a scheme acceptable to everybody on 'rational' grounds: because, as many internal reform schemes show, the zero-point is arbitrary, we could decide to exchange B.C./A.D. by, say, B.H./A.H. without disturbing the chronology as such, whereas agreeing on a global *ab urbe condita* seems utterly hopeless. Whose City?

Resistance from Without

What then about resistance from without, in the eyes of the Other? The hegemony of the Western time zone, the magic reinforcement this dominant frame has received from its major time-mark 2000, the unassailability it has achieved outside its context of origin and cultivation are truly remarkable historic processes. And because this is so, one would expect dissenting voices. Imagine yourself a self-respecting post-colonial public intellectual... Looking into this issue as the millennium approached, my problem was that I did not know any such intellectuals personally and that they did not publish or talk in the media I had access to. At the time, most people I told of my hunch – that somewhere out there some people would seriously question the significance of the millennium for their worlds – refused out of hand the notion that this might happen on any scale that would make it interesting in terms of longer-range cultural effects. They may have been right, and in this case Samuel Huntington should really not be so worried.

On the other hand, even if such voices existed, our chances to hear them would not be good, because (to put it in shorthand) they would not argue in English. In fact, the situation must be expected to be more complicated: In Arabic literature, for instance, there are today books that are written in Arabic for translation into foreign language. These books are not successful with the local connoisseurs, it seems. The books in Arabic they prefer are written in a genre which is difficult to translate and has no local connoisseurs over here. Also, paradoxically, it may be too early, because the accumulated effect of the millennium beyond the big party is still to come.

The Internet, this quasi-global amplifier of preferably undisciplined murmurs from any conceivable debate that attracts people with computers, is the first medium one would have turned to in order to find out more (even if it is dominated completely by English, followed by Hebrew and a few other European languages). Probing the Internet in 1999 first of all pointed to a phenomenon which was very much to be expected from general millennium-theory: that the grand event had amazing synchronizing power. Web sites devoted to empathetic development aid and world-improvement, and (less surprisingly) esoteric web sites, draped their programs in millennial language. The great date was exploited to underline and dramatize whatever message for the future was transmitted, and all kinds of apocalyptic/messianic urges naturally fed on the millennium. Islamic web sites too couched indigenous messianic messages in millennistic terms. Also to be found were attempts to reinterpret the religious and secular meanings of Christmas and New Year in terms of fixed dates and seasons in non-Gregorian calendars.

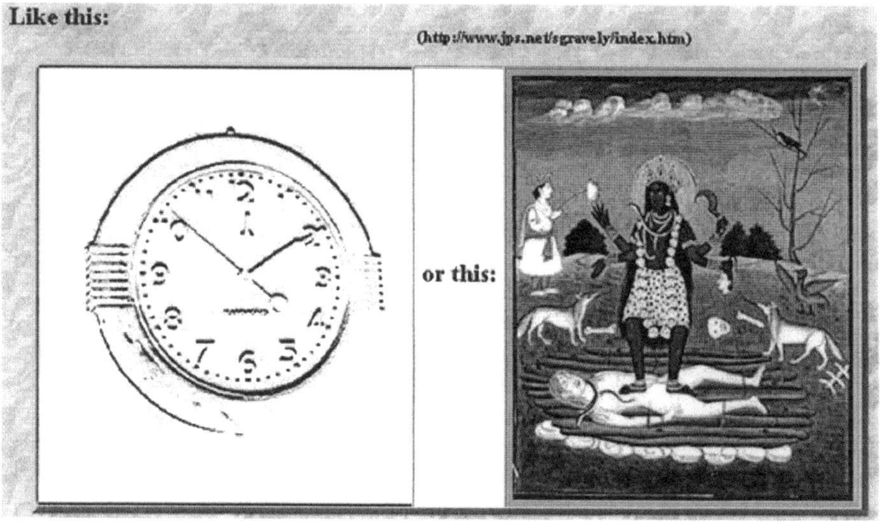

In any event, probing the Internet around the Millennium turned out signs of consent rather than dissent. Careful distancing, like adding C.E. (common era) to B.C./A.D.-dating in learned and p.c. discourses, remained token reverences to a time order one cannot refuse. As far as I could see, overt problematization was absent from this medium.[16] Also, stressing the hegemonic element could easily been refuted by pointing, as Lepenies does, to some alleged, computer-based universal time regime:

> The critique of the eurocentrism and christocentrism of the present millennium-panic – a worldend-final definitely following European rules – does not capture the point, though, because in the age of the Internet, the microchips and global networking of computers have long since shaped an ecumenical time regime, whose universality on our planet has no historical precedent (Lepenies 1999, 7).

However, Lepenies seems to me to disregard here that the planet is not 'ours' and that, in his own words, "(i)n view of this 'long' twentieth century [referring to a long Schumpeterian century couched in economic terms as opposed to Hobsbawm's short political century, B.J.], the globally increasing inequality represents the central problem of our time." Today this inequality can best be measured by differential access to the computer-matrix. In any case, my attempt to look at the millennium in hegemonic terms yielded scarce results. As my correspondent from the *Muslims Student Association* said: "The only ones who have brought this issue (non-Western protest) to our attention are people in Christian countries like you," indicating that the Other is mostly an internalized instance.

So much for the Internet. And yet, "the millenarian theme ... continues to occupy many minds among the faithful," as the Arabist Barbara Stowasser concedes. According to her, the year 2000 produced an abundance of books and pamphlets in the Arabic world which tied the coming bi-millennium of the Gregorian calendar to the final defeat of the West. Often predicated on old prophesies about the coming fall of Byzantium (which indeed occurred), she says, this literature predicted the coming fall of Byzantium's latter-day incarnations, Europe and America. And she quotes one author whose book with the remarkable publication date 'before 1999' castigates Arab and Muslim intellectuals for their

> almost total disregard of traditions about the Antichrist and the bloody battles at the end of time ... [while] today the members of Western civilization themselves speak of nothing else but the day of doom, the end of history, the clash of civilizations, the atomic Armageddon. How can intelligent men not see that the enemies have fixed their bayonets and closed their ranks, while we talk about culture and civilization and development ... mere intellectual prattle ... [which is, itself] a sign of the Hour, when intelligent minds become twisted and almost none [still sound] are to be found (Salim (before 1999), 20; quoted after Stowasser 2000, 18).

Hegemony, Competition, Revolution, Reform?

The time-zone constellations of competition, revolution, reform and hegemony set up different questions as to the relationship between *Eigenzeiten* and *Fremdzeiten*, own time and other's time. While elements of all these constellations can be found in people's time management at all times in all places[17], I still hold that the four situations combine today in a more dramatic fashion in the time management of people *outside the Western civilization*. But the now receding event – the turn of the millennium – did not have, over and above its many absurd and short-lived side-effects, the interesting and potentially far-reaching cultural significances I suspected.

All in all, the experience of the millennium's turn has demonstrated splendidly that the Gregorian is well in place on a planetary scale. It has demonstrated that there are many ways of living with two or more other calendars, even if negotiating them has its irritations. There may have been attempts here and there to contest it, but they remained largely symbolic. As, presumably, with space travel in the future: there will be a need to establish interplanetary calendars, but they will for a long time remain merely technical. The Gregorian will then for a long time to come be the main chronological currency to which other currencies tend to be coupled unilaterally. TV stations and banks will make sure of it.

But is not that what hegemony means? The Gregorian calendar comes with all kinds of recurring fixed dates whose meanings are not empty, from Sundays, to Christmas and New Year eves to turns of centuries and millennia. The meanings of these are variable but not entirely changeable. They can be mapped, trans-scribed onto other time-charts, but not without changing the meanings previously inscribed into them. The trans-scribed meanings will change too, relative to the time of origin, but there will be little or no traffic of meaning in the other direction.

MACRO-TIME AND CALENDARS *IN* SOCIAL STUDIES OF SCIENCE AND TECHNOLOGY

What then is the place of calendars and chronologies in the social sciences, in particular social studies of science and technology? Let me bring in a few quite general considerations about calendrics before I turn to the neglected subject of macro-time in STS.

More Observations on Time Scales and Calendrics

The general interest of calendrics as a scientific and technological field results from its strange and ineluctably awkward position in the canon of knowledge. It saddles the opposites of natural, formal, cultural, and theological sciences. For that reason, its epistemological status remains forever precarious or outright ideological.[18]

A strange scientific field indeed: it includes mathematicians and astronomers, astrologers, theologians, scholars in cultural studies, historians, scholars in legal studies, New Age gurus, and many more. The symbolisms of day and night, sun, moon, planets, stars and their conjunctions, changing seasons, equinoxes and eclipses, comings and goings of comets tie calendars to both religious experience and mundane know-how, science and mythology. Both our practical requirements (making contracts, calculating the right time) and our quests for deeper meaning (the quest for cosmologies) seem to contribute to an incontrovertible need to resort to *numerical regularity*. Distinct from time understood as movement in history, this gives us *chronologies* as a means to find order in a confusing and ever-changing world.

As sorting devices for ordering events past, present and future, all calendars are somehow based on the three great natural cycles everybody observes and understands (and in this sense alone might be called an evolutionary universal). Expressed in Galilean terms: the earth circles the sun in a year, the moon circles the earth in a month, and the earth turns around itself in a day. The rest of the major calendric cycles – weeks, seasons, decades, centuries and millennia – are more difficult to map on astronomical or generally natural observations, although it has been tried of course.

> Nobody seems to know for sure how the almost universal standard of the 7-day week evolved. The obvious reference to Genesis only begs the question: where does Genesis have it from. The Babylonians already had the 7-day week (allegedly based on and named after the five planets then known plus the sun and the moon). It became a binding standard under the Romans, and later, worldwide, with the British Empire. In purely astronomical terms, the 7-day week makes little sense, except as an approximate marker of four moon phases. As for seasons, again 13 weeks approximately make up one season in our solar calendar.

Looking at the world of time in terms of centuries and millennia especially turns out to be as far from 'natural' as possible: it resulted from highly contingent historical struggles and culture wars.

In a nutshell, the problem arises from 'God, the bad mathematician': calendars and their turning points cannot elegantly be anchored in nature and her harmonies. Repairing this defect is behind any calendar invention and interpretation. As Steven Jay Gould said in his chatty and charming little book on the Millennium:

> If God were Pythagoras in Galileo's universe, calendrics would never have become an intellectual subject at all. The relevant cycles for natural timekeeping would all be nice, crisp, easy multiples of each other – and any fool could simply count. We might have a year (earth around sun) with exactly ten months (moon around earth), and precisely one hundred days (earth around itself) ... But God ... did not fashion a very regular universe after all. And we poor sods of his image are therefore condemned to struggle with calendrical questions till the cows come home, and Christ comes around again to inaugurate the millennium (Gould 1997, 134).

The unpleasant truth: the regularities in nature are not numerical, God must have been a lousy mathematician when he created. The earth's rotations around itself and the sun, and the revolution of the moon cannot be expressed as even multiples of each other. Nature offers only fractionality to infinite decimal places here.

With years and days it is not so bad: only about a quarter day off the mark, which has given us the Gregorian calendar's leap year (requiring, if I have this correctly, its own correction by dropping the leap year every 100 years for a while, but another correction was due and 2000 has been a leap year after all ...). So the solar year has 365 days, 5 hours, 48 minutes and 45.96768 ... seconds – what did He think? With the moon it is far worse: the moon travels around earth in 29.53059 days. The closest one can get from this to a 'regular' solar year is dividing it into 12 lunations, which gives 354.367706 days, close to 11 days short of the solar year. So we have our 30./31. monthly alterations, plus the twenty-nineth of February every 4 years. There have been other solutions to this: both the Maya and the Jacobins came up with 12 months at 30 days plus 5 days at the end (which were considered unlucky by the Maya and turned into a big party by the French, with an added sixth day of feasts every four years . . .).

Of the many ingenious solutions that have been devised for this problem, 'ours' is the predominantly solar Gregorian calendar. Civilizations who opted for lunar calendars rather (Babylonian, Imperial China, Judaism) bring them in line with the solar year by alternating full and deficient months (30 or 29 days) and by inserting leap months every now and then, according to another rough astronomical regularity called the Metonic cycle.

After the Greek astronomer Meton, but the cycle was discovered independently at various times in human cultural history. The Metonic cycle denotes the shortest sequence of years that brings lunar and solar calendars into near-alignment: it runs 19 years and requires the insertion of one leap month in any 7 of these 19 years. The Jewish calendar for example inserts a thirteenth month of 30 days into the third, sixth, eleventh, fourteenth, seventeenth and nineteenth years of a Metonic cycle.

Apart from leap months and leap years, we are seldom reminded of the artificiality of our calendars, except if we partake in two or more different time orders. Thus, the negotiation of the Judaic and the Gregorian calendar in daily life must be known to everyone who celebrates Chanukah: it tends to slip back through the Gregorian calendar about 10 days every year, only to hop forward into late December just before slipping into November. Its the Metonic cycle. The Islamic calendar, also predominantly lunar, does not use the Metonic correction.

Iran is the only country in the Middle East that uses the 'Islamic solar calendar' which originated before the beginning of the Persian Empire (550 B.C.). After the conversion of the vast majority of Iranians to Islam, the ancient Iranian calendar was adjusted to begin with the year of Prophet Muhammad's immigration to Medina (the hijrah), equivalent to A.D. 622 in the Gregorian calendar. To convert from the Islamic solar calendar to the Gregorian, 621 or 622 years, depending on the time of year, are added to the Iranian year. (For example, the Iranian year 1361 began on March 21, 1982, and ended on March 20, 1983). Noh-ruz (New Year's Day) falls on March 21, the venal equinox. For religious observances, Iranians use the Arabic lunar calendar.

This is why Ramadan and other fixed dates keep moving back in our predominantly solar Gregorian calendar all the time. By the same token, 'their' count of years runs away from 'ours' – a thousand years are over 31 years plus a few weeks faster for a Moslem.[19]

The Study of Macro-Time

Few people apart from calendar specialists would be concerned with such matters if the millennium, as a date fixed in time which almost nobody could ignore, had not drawn a measure of attention to the politics and pleasures of studying macro-time:

> If we regard millennial passion in particular, and calendrical fascination in general, as driven by the pleasure of ordering and the joy of understanding, then this strange little subject – so often regarded as the province of drones or eccentrics, but surely not of grand or expansive thinkers – becomes a wonderful microcosm for everything that makes human beings so distinctive, so potentially noble, and often so actually funny (Gould 1997, 158).

This seems, as Gould conceded himself, 'a risibly inadequate stab at lyricism,' but still: part of the fun is in the variety of solutions to an unsolvable problem and of the confusion resulting from having to live with several solutions at once. This pleasure may be diminishing in the long run, however, under the pressure of calendric hegemony.

Yet, chronological macro-time has been scarcely studied from a social science perspective.[20] Not that time and its orders has not been analyzed from broadly sociological perspectives, of course. A while ago, David Landes has presented a pioneering if ethnocentric attempt to write the cultural, economic and scientific-technological history of clocks (Landes 1983).[21] Since then, time disciplines conventionally determined and organized with the help of clocks have been intensely researched in all kinds of cultural contexts, including the scientific-engineering. These time studies have – much like other fields of social and cultural studies in the past decades – produced their most interesting insights by going local (if not native). On the way, some rather doubtful theories about time frames being essentially local-cultural, in the sense of their trans-cultural incompatibility or un-translatability, were produced.[22]

On the other side of studying micro-timescapes, the largely unexplored white spot of all-inclusive *Zeitrechnungen* waits for examination in a sociology-of-science and sociology-of-culture perspective.[23] No attempt comparable to Landes' has yet been made to open the field of calendrics to social studies.

History has seen an amazing variety of calendar systems. What are their rules and how do they serve their complex functions? The double intellectual challenge all working calendars present is indeed to understand how they negotiate the amazingly complicated astronomical (and in their wake arithmetical) problems and how various religious and other powerful traditions are affected by the solutions to these problems. This requires a perspective that combines science and technology studies with studies of the manifold uses of calendars in organizing social life.

One reason why *Zeitrechnungen* have scarcely been subjected to sociological analysis may be that chronology tends to be considered as distinct, almost external from the notion of social time. Luhmann for instance argues that the experience of temporal movement and change in itself does not yet constitute social time. Social time, he suggests, should be understood as 'the interpretation of reality in view of a difference of past and future' (Luhmann 1990, 124). And he shows that the notion of an irreducible plurality of social time is well ingrained in sociological approaches to time since Merton, Sorokin and Gurvitch have laid the foundations.[24] Chronology, he argues, can then be understood as 'standardized scheme of movement *and* time' (Luhmann 1990, 124, emphasis Luhmann). Luhmann, and I suspect the majority of time-sociologists, insists on this difference between chronology and time.

Luhmann then turns to the differences between modern (functionally differentiated) societies of the West and less complex societies with respect to the form the relation between future and past takes. He is not interested in the role chronologies play in this, although he has an interesting analogy:

> Not time, as Aristotle thought, but chronology creates distance. Chronology functions as an evolutionary universal which combines very simple rules with highly complex functions – like money (Luhmann 1990, 125).

Typically, Luhmann grants social time its plural status, but uses chronology in the abstract singular, not looking into the 'very simple rules' making up this medium. Calendric issues are relegated to a footnote.[25]

Another reason for the dearth of studies in calendrical time may simply be that sociological classics have not articulated these issues, adopting naively the typically whiggish and entirely Eurocentric historical timescapes of contemporary historians. Max Weber for instance, that great theoretician of rationalization by numbers (*Rechenhaftigkeit*), has never concerned himself with *Zeitrechnung*, except – marginally – in his studies of the economic ethics of world religions.[26] Thus in his volume on *Konfuzianismus und Taoismus* he distinguishes the Eastern 'chronomantical' orders – interested in the prediction of happy and unhappy dates and events – from Western 'chronological' order (Weber 1991).

Apart from this, there is just one place in his monumental work in cultural history where Weber discusses calendric matters: in the volume on *Das antike Judentum* (Weber 1972). Here he observes that the institution of the *Sabbath* originates in the Babylonian moon cult. Whereas this day remained strictly coupled to the lunar cycle in Babylon (i.e., did not fall on every seventh day), in Israel the Sabbath was routinely counted as every seventh day, 'disregarding the moon-phases,' as Weber notes (Weber 1972, 160). Weber's interpretation follows his overall rationality hypothesis: The Babylonian time order was determined by the 'aristocratic astronomical knowledge of priests,' to which economic interests remained subordinated. The change to a chronological time order in Israel results 'simply from the stronger preponderance of agricultural interests and habits in Palestine, oriented to the local city markets.' Peasants and city-dwellers had a vital interest in a regular recurrence of a market days: 'the eventual imposition of a through-numbered Sabbath is due to the emergence of a market economy' (Weber 1972, 161).

It is remarkable that Weber did not follow through this thought – that time orders became detached from their celestial foundations for mundane reasons and were based on some purely numerological procedures – in the direction of an analysis of the emergence and imposition of Western calendars. Consider that Weber's sociology of religions is based on a broad evaluation and absorption of the literature in the sciences and humanities of his time. It is quite puzzling, for that reason, that he did not peruse the broad literature on *Zeitrechungen* available to him.[27] But Weber's disinterest in rational time orders is matched by Weber's many interpreters and exegetes. As far as I can see, nowhere in the very substantial secondary literature on Weber can leads be found to possibly implicit discussions of time orders in his *oeuvre*, and nowhere are reasons for Weber's disinterest explored.

One answer to this may lie in the circumstance that of course the catholic countries followed Gregorian papal orders much earlier than protestant ones. In Weber's logic of the protestant ethic and the 'spirit' of capitalism, the Gregorian may not have been a proper candidate for the most rational calendar. Interestingly, England, the country that was first in the industrial revolution, reluctantly adopted continental Gregorian conventions long after most other countries in the Western world, although many of the aborted attempts at calendar reform in European history originated in England. But by 1582 England had parted company with Rome and resisted innovations from continental Europe in a way that remains familiar. Besides, there was a strong French enlightenment tradition to resist the Gregorian calendar on account of its irrationality and missing decimal elegance (Maier 1991, 128, referring to the entry *ère* in the *Encyclopédie)*.

Be that as it may, post-classic sociologists seem to have experienced a certain postmodernist disdain with chronological issues, possibly in reaction against the powerful ceremonial façade of historical periodizations in 'modernist' science.[28] But no one seems to be able to resist the centennial and millennial calendrical frames and to strictly conceive of local and lateral events in terms of their *Eigenzeit* – a sense of historical time that needs not chronopolitically be reduced to one calendrical hegemony with its epitomical time mark, the millennial turn.

How then is macro-time made and managed and how are local times and global times, sacred times and mundane times reconciled and mutually articulated? Why have some cultures been satisfied with calendars that just sort days into months (and why is 'our' calendar used for this purpose, most of the time)? Why have others extended their chronological institutions to lunar or solar or lunar-solar years? Why have some cultures, past and present, remained fixated in ritualized old chronological systems, while others invested in highly politicized schemes of perfecting their chronologies? Why have some cultures chosen to subordinate astronomical reckoning to reckoning in terms of dynastic succession, economic expediency or religious (astrological) orthodoxy? What drives calendar reconstruction – science, technology, economics, politics, religion - or *some other thing*?

Calendars as Big Time Machines

Langdon Winner, the American philosopher and sociologist of technology, begins his much-quoted essay *Do Artifacts Have Politics?* with the observation that certain details of form in our landscapes – bridges, streets and roads – habitually are taken to be meaningless, 'and yet power relations are literally built into them' by design (Winner 1980, 123). With this he leads up to his often retold story about Robert Moses' low bridges on Long Island, New York, through which he is said to have prevented the poor and the black from frequenting Long Island beaches. While the parables of bridges and similar small-scale, massively tangible artefacts serve didactic purposes better than analytical ones, Winner's approach extends to large-scale intangible artefacts, for instance zoning laws and their technologies.[29] Not only landscapes, timescapes too have their zoning devices and indeed zoning laws. In drawing attention to the large-scale timescape of grand historical periodizations and turning points, and the calendars that are used to chart them, I have in fact asked a Winnerian question: Do calendars have politics?

Calendars and the time zones they set up can be seen as large-scale intangible, yet powerful cultural artefacts for the management of historical meaning.[30] Adopting Winner's approach for chronological artefacts as embodying political *designs*, however, would be like saying that the building of hours into clocks *ipso facto* forced people who use them to do certain things and prevented them from doing something else, *and* that this is why clocks are built this way. The argument sounds more exotic than it is – many control theorists in social studies of technology have in fact advanced similar theses. Applied to the Gregorian calendar this would translate into: Living in this *Zeitrechnung* ipso facto forces us to organize our memories in a certain way and prevents us from organizing them differently. And there must be political reasons for this – if not literally somebody, which would be too tall a claim, then some more comprehensive historical designs must be at work here.

But the argument from design is far too simple and unnecessarily conspirational. Looking at calendars and the time zones they support as large-scale cultural artefacts for the management of historical meaning rather leads back to the initial question: What drives calendar construction and reconstruction – science, technology, economics, politics, religion – or some other thing? All of the former, obviously. It took a lot of scientific and technical expertise, artistic and literary imagination, economic and political interest and ingenuity, religious and philosophical teachings and propaganda to establishing the Gregorian calendar and any other big time machine (for good overviews see Maier 1991; Stearns 1997). The Gregorian, like any other calendar, is therefore filled to the brim with historic meanings and legitimation claims, and the millennium has epitomized many of these. Turning the argument around in a way, however, one should not forget that the Gregorian *Zeitrechnung* is, like all *Zeitrechnungen*, quite meaningless in itself. At the least, there are many Gregorian calendars. U.S. Christmas is not like German Christmas, the symbolisms of sun and moon which still govern the experiences and emotions of many people are different in Latin and in Nordic cultural spheres, the millennium meant something else for Mr. Bush, Mr. Putin and Mr. Arafat or Mr. Chamenei, or whoever will have replaced them in the coming millennium.

This is precisely the reason why the Gregorian calendar cannot be changed at will, and why revolutionary attempts to erase history by exchanging calendars must fail. Historical meanings require for their institutionalizations calendars. In this trivial sense, calendars have historical meanings. But their power does not reside in their formal properties, the cycles and counts and round numbers they come with. Their meaning is delegated meaning.

And that leads to 'the other thing', the other thing being the opportunity opened up for translation and re-inscription by clever standardization. As far as I can see, no one in social studies of science and technology has studied calendars and similar chronologies in this perspective. Looking at calendars in terms of a sociology of large technical systems, one must beware then to fall into the traps of technology studies. Especially: the power of calendars is not inherent. Their formal properties constrain nothing. It is the authorized meanings that count, and calendars like the Gregorian, with their far from simple rules, are capable, for that reason, to support many meanings.

Overarching time orders established through calendars make then for the translatability of local time orders; they also bring parts of the cultural material of local orders into their always and by necessity more standardized and expert-based realms. They therefore – although they are progressively de-secularized and proceduralized scientifically – will also always remain subject to various kinds of *Ideologieverdacht* (the notion that they are instruments of power and control): accused of colonizing cultural and political intents and/or effects. 'Ideology is the other side of chronology' (Lepenies 1999, 7). By virtue of being *the* overarching time order at present times, the Gregorian calendar will share this with alternative overarching chronologies.

It remains to be seen whether the radical secularization brought to modern chronology via the 'perfection of the second' will eventually, for many technical and economic purposes, somehow replace Gregorian World-Time. Only some properly standardized chronological scheme, based all the way up on the cesium standard, i.e., a scheme dispensing with heavy social symbolisms (except alas the symbolisms or formal perfection), could one day affect its institutional integrity, precarious as it may have been all along. But historically, advances in standardization have always and everywhere tended to occasion veritable explosions of new cultural forms.

Laboratory cesium frequency standard

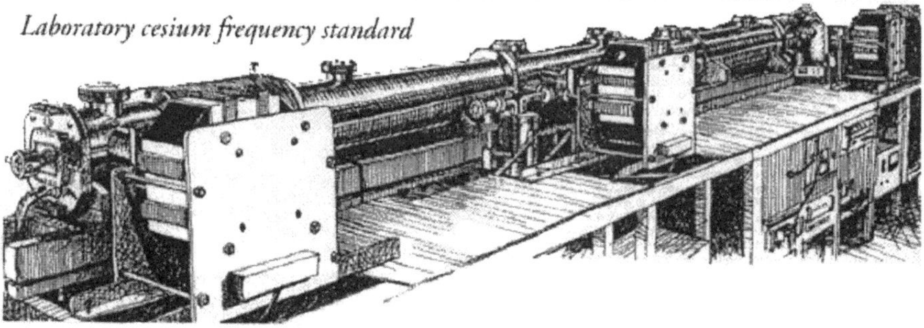

In many European languages, the words for telling (his)stories and for counting can be traced to the same root.

Pointing out that the Latin computare is very close to conter, contar, raccontare, to tell and so on, Arno Borst notes that "the less educated (*Nichtgelehrten*) of Europe have linked the telling of stories and the counting of time." In German the very word for storytelling (*erzählen*) contains the word number (*Zahl*) (Borst 1991, 41, quoted after Maier 1991, 125). This can remind us that standardized chronologies may never be entirely dissociated from *narrative* knowledge and self-knowledge – and there is maybe no other one. Foucault notes that, in the twentieth century, nineteenth century obsessions with timing and periodization have given way to spacing or *heterotopia* (Foucault 1986). In a similar sense, *heterochronia* may become a permanent feature of the timescapes of globalizing societies in the twenty-first century and beyond.[31] By the same token, *Ideologieverdacht* will also remain a permanent feature of uniform world-time - even if the celebration of the millennium has not given much occasion for deep cultural irritations in parts of the world predominantly living in timescapes other then the Gregorian.

Wissenschaftszentrum Berlin für Sozialforschung Berlin (WZB), Germany

NOTES

* Even if I failed to take them fully into account, my thanks for critical comments on consecutive versions of this essay go to Helga Nowotny, Richard Rottenburg, Joerg Potthast, Terry Shinn and Hans-Jörg Rheinberger.

1 Calendar enthusiasts and numerologists immediately began speculating about the significance of that date. Christopher Hitchens advanced the "persuasive explanation that ... it was on Septmeber 11 1683 that the conquering armies of Islam were met, held, and thrown back at the gates of Vienna" (Hitchins 2001), only to be told that the siege really ended on September 12. Another speculation has September 11 1922, the day the League of Nations proclaimed Britain's mandate over Palestine. Others say, it's because the twin towers looked like the number 11 . . .

2 A number in itself marks nothing of course, as Niclas Luhmann has shown, using the example of 'Europe before the potato' and 'Europe after the potato'. In order to make a historical event meaningful, one always has to mark two points in time, or three events (Luhmann 1986).

3 The cost argument would play a major role in all explanations of the success of the Gregorian calendar as a 'market failure' in terms of economic theories of path dependence or lock-in à la QUERTY, Betamax, Microsoft or Macs. It should be clear from this essay that I hold with the debuncers of 'the fable of the keys' (Liebowitz and Margolis 2001) who fail to find any convincing empirical case for technological lock-in. This does not mean that the Gregorian is the best product in the calendar market, but that it must be pretty good.

4 Historically, Islamic apocalypticism is not closely linked to millennialism (see Arjomand 1998). However, enthusiastic students of the millennium, like the group around Richard Landes at Boston University's Center for Millennial Studies, may disagree and at present make much of the apocalyptic dimensions of Bin Laden and 'revolutionary millennialism' of September 11.

5 Not all specialties dealing with historical processes argue on a calendric scale: paleontologists don't usually, nor do neuropsychologists or laboratory-ethnomethodologists. But at least in their self-representations *as* sciences, in writing their own history, they always do.

6 To the extent that they address change of any kind. Granted, social studies of science and technology, as represented for instance in the volumes of the *Yearbook for the Sociology of the Sciences*, tend to be rather a-historical. Of course, there will always be a historical subtext, but I venture that few of these studies explicitly relate scientific/technological change to social history, let alone refer to the work of social historians. As to the strictly history-of-science & technology literature, I feel that in addition to traditional internalist approaches there is a preponderance of studies relating scientific/technological change to *cultural* history (as opposed to social and political history).

7 Hobsbawm 1995. For present purposes one may assume that Hobsbawm's reading of the history of science and technology in the twentieth century and his way of relating it to other parts of the story is not at all untypical for contemporary social historians. (With the exception of his emphasis on environmental problems, perhaps, although he does not enter into a 'history of the environment'.)

[8] Not as a matter of social history in the continental sense, e.g., Weberian themes such as the *Rationalisierung der Lebensführung* or *Säkularisierung der Weltanschauung* or science as a major erosive force in the decline of religious and authoritarian institutions, as *Ersatzreligion* or even – closer to the heart of contemporary social historians – as changing irreversibly the character of work.

[9] In a similar fashion, Clark (1998), for instance, has synchronized, in opposition to formalistic theories, modernistic art in the twentieth century with social and political history. His century, like Hobsbawm's, is a short one, ending the modernistic episode in 1989 with the fall of the Berlin wall.

[10] Be it said that this division of the history of large technical systems into modern and post-modern exemplars loses some of its persuasive force because the old systems rhetoric survives in part: the thinking of systems in terms of complex causal interrelationships and control has not been abandoned, although it doesn't quite fit into a postmodernist rhetoric of uncontrollable contingency and self-organization. For a more extended treatment see Joerges (1994).

[11] This is of course at the heart of the controversy whether scientific-technological development could be understood as strongly coupled to military history. The notion that the Internet must be understood in terms of its early military genealogy is a current example.

[12] But even McNeill has to put the mythical observer from Mars in the same Eurocentric-centennial timeframe: "An intelligent visitor from outer space, observing the earth as a whole in 1900 and again a hundred years later, would surely find the increase in human numbers and urbanization, together with associated alterations in the planet's array of other life forms, the most remarkable change of the century" (McNeill 1998). For a wonderful history of the periodizations and chronologies in the history of *philosophy*, mother of all social and cultural studies, up until the *Neuzeit* see Daniel Garber and Michael Ayer's *The Cambridge History of Seventeeth Century Philosophy* (1998).

[13] My birthday in Comte's Positivist Calendar is Gutenberg 20, for instance.

[14] In terms of an economics of calendar technology it would be to demonstrate that the Gregorian is not a technology imposed by a monopoly on 'locked-in' consumers who cannot rid themselves of an inferior or unwanted product at reasonable cost. Not being an economist, I still feel that this would be hard to show.

[15] Which proved useless even *within* one civilization because at Gregor's time estimates of the date differed by about 3000 years, and around A.D. 1750, 200 competing ways of dating the creation had accumulated (Maier 1992, 42).

[16] "As Muslims, we consider the Moon calendar for historical events and religious activities. Muslims do not know and do not recognize the Sun (or Western) calendar for many major historical events in Islam," says one of my correspondents, Sultan Al-Harbi for the Muslim Students Association of USC. And Jerome C. Glenn, a spokesperson of the many millennium/calendar lists on the net, flatly denies the occurrence of any 'non-Western protest'.

[17] See Nowotny's *Eigenzeit* (1993).

[18] For an introduction in *Kalenderwissenschaft* see for instance Zemanek 1987, on Calendars and social life Zerubavel 1985, for a comprehensive cross-disciplinary book listing of the scientific literature on time http://book.netstoreusa.com/index/bkbsg600.shtml.

[19] In their terms, in ours only 30 something years; have your pick and legitimize it... The Islamic calendar will define its own millennium some time ahead: Muharram 2000 will be in January – February 2562 A.D.

[20] I denote macro-time very extended and institutionally pervasive time-schemes, spanning decades/generations and civilizations.

[21] Typically, Houston University College of Engineering's impressive project of documenting "machines that make our civilization run, and the people whose ingenuity created them," with its approximately 1500 episodes, has 46 entries dealing with clocks, and just 2 with the calendar (one on the problem of inserting leap seconds and one on the Bimillennium).

[22] They also tend to see clocks, 'the original sin' (Hörning) in the realm of time technologies, as embodyments of social order and control. For different approaches see Hörning, Ahrens and Gerhard (1999), Mackenzie (2001).

[23] There is, of course, since Joseph Justus Scaliger (*De emendatione temporum* 1538, *Thesaurus temporum* 1606), no dearth of work on calendars, and the approaching millennium has produced a deluge of publica-

tions. Practically everybody writing on the subject has presented the story in a strictly Judaeo-Christian perspective.

24 In the meantime, a large philosophical and research literature dealing with temporality (as opposed to chronology) and its cultural construction/evolution has emerged (see Nowotny 1993, 1997, 1998).

25 "The primary function of primitive chronology seems to be the integration of recurrent ecological changes and behavior-regulating social norms" (Luhmann 1990, 146, FN 18).

26 My thanks go to Ralf Thies for helping me with researching Weber.

27 Basnizki for example references 8 texts on the Jewish calendar in 1908 (see Basnizki 1986, 56). Ginzel's standard 3-volume *Handbuch der mathematischen und technischen Chronologie. Das Zeitrechnungswesen der Völker* has extensive chapters on the time orders of Indians, China, Japan, Jews, Mohammedans, Egyptians and Romans – exactly the civilisations and religions Weber concentrated on (Ginzel 1906–14).

28 Foucault has expressed it well when he spoke, in order to call for more concern for place and space, of the great obsession of the nineteenth century, namely "history: with its themes of development and of suspension, of crisis and cycle, themes of the ever-accumulating past..." Our epoch, he claimed decades ago, will perhaps be above all the epoch of space and simultaneity (Foucault 1980, 149; 1986, 22).

29 Among others the entire apparatus of spatial mapping and planning, and the technologies of restructuring spatial form. For a critique of Winner's design theory see my *Do Politics have Artefacts?* (Joerges 1999).

30 In characterizing them as intangible I certainly do not mean to say that they are not rooted in all kinds of tangible artifacts; the point is that beyond their paper-form these systems of artifacts remain largely invisible and virtually unknown to anybody in their entirety.

31 Note that I have finally given in to the urge to sort matters on a centennial scale.

REFERENCES

Arjomand, Said Amir (1998), 'Islamic Apocalypticism in the Classic Period', in B. McGinn and S.J. Stein (eds.), *Encyclopedia of Apocalypticism*, New York: Continuum, Vol. 2, pp. 238–83.

Basnizki, Ludwig (1986), *Der jüdische Kalender. Entstehung und Aufbau*, Königstein/Ts.: Jüdischer Verlag.

Borst, Arno (1991[1999]), *Computus. Zeit und Zahl in der Geschichte Europas*, Kleine kulturwissenschaftliche Bibliothek, 28 (letzte überarbeitete Ausgabe dtv 1999), Berlin: Wagenbach.

Clark, Timothy J. (1998), *Farewell to an Idea. Episodes from a history of modernism*, New Haven, CON: Yale University Press.

Elvee, Richard Q. (ed.) (1992), *The End of Science?: Attack and Defense* (The Nobel Conference Series), Landham, MD: University Press of America.

Foucault, Michel (1986), 'Of other spaces', *Diacritics* 16: 22–7.

Gibbons, Michael, Camille Limoges, Helga Nowotny, Simon Schwartzman and Peter Scott (1994), *The New Production of Knowledge: The Dynamics of Science and Research in Contemporary Societies*, London: Sage.

Ginzel, Friedrich K. (1906–14), *Handbuch der mathematischen und technischen Chronologie. Das Zeitrechnungswesen der Völker*, 3. Bd., Leipzig: J.C. Hinrichs.

Gould, Stephen J. (1997), *Questioning the Millennium: A Rationalist's Guide to a Precisely Arbitrary Countdown*, London: Cape.

Hitchins, Christopher (2001), 'Why the suicide killers chose September 11', *The Guardian*, October 3, 2001.

Hobsbawm, Eric (1995), *The Age of Extremes: The Short Twentieth Century 1914–1991*, London: Abacus Edition, Little, Brown and Co.

Hörning, Karl H., Daniela Ahrens and Anette Gerhard (1999), 'Do technologies have time?', *Time & Society* 8(2): 293–308.

Horgan, John (1996), *The End of Science: Facing the Limits of Knowledge in the Twilight of the Scientific Age*, Helix Books.

Hughes, Thomas P. (1983), *Networks of Power*, Baltimore and London: Johns Hopkins University Press.

Hughes, Thomas P. (1989), *American Genesis. A Century of Invention and Technological Enthusiasm*, New York: Viking.

Hughes, Thomas P. (1998), *Rescuing Prometheus*, New York: Pantheon.

Joerges, Bernward (1990), '"Global 2000" – Social science, ecology, and the bimillennium', *Futures* (January/February): 3–20.
Joerges, Bernward (1994), ‚Ein Zeitalter der Energie', in W. Zapf und M. Dierkes (eds.), *Institutionenvergleich und Institutionendynamik*, WZB-Jahrbuch 1994, Berlin: edition sigma, pp. 106–33.
Joerges, Bernward (1999), 'Do Politics Have Artefacts?', *Social Studies of Science* 29(3): 411–31.
Landes, David S. (1983), *Revolution in Time: Clocks and the Making of the Modern World*, Cambridge, MA: Harvard University Press.
Larrain, Jorge (1989), *Theories of Development: Capitalism, Colonialism and Dependency*, Oxford: Polity Press.
Liebowitz, Stan J. and Stephen E. Margolis (2001), *Winners, Losers & Microsoft, Competition and Antitrust in High Technology*, Oakland, CA: The Independent Institute, revised edition.
Lepenies, Wolf (1999), 'Das Altern unseres Jahrhunderts', *Tagesspiegel*, 19. Juni 1999: 7.
Luhmann, Niklas (1986), 'Das Problem der Epochenbildung und die Evolutionstheorie', in H.U. Gumbrecht and U. Link-Heer (eds.), *Epochenschwellen und Epochenstrukturen im Diskurs der Literatur- und Sprachtheorie*, Frankfurt a.M.: Suhrkamp, pp. 11–33.
Luhmann, Niklas (1990), 'Die Zukunft kann nicht beginnen: Temporalstrukturen der modernen Gesellschaft', in P. Sloterdijk (ed.), *Vor der Jahrtausendwende: Berichte zur Lage der Zukunft*, Frankfurt a.M.: Suhrkamp, pp. 119–49.
Mackenzie, Adrian (2001), 'The technicity of time: From 1.00 oscillations/sec to 9,192,631, 770 HZ', *Time & Society* 10(2–3): 235–57.
Maddox, John (1998), *What Remains to Be Discovered: Mapping the Secrets of the Universe, the Origin of Life, and the Future of the Human Race*, New York: Macmillan.
Maier, Hans (1991), *Die Christliche Zeitrechnung*, Freiburg: Herder.
Mayntz, Renate (1993), 'Große Technische Systeme und ihre gesellschaftstheoretische Bedeutung', *Kölner Zeitschrift für Soziologie und Sozialpsychologie* 45 (1): 97–108.
Mc Neill, William (1998), 'Demography and Urbanization', in M. Howard and W.R. Louis (eds.), *The Oxford History of the Twentieth Century*, Oxford: Oxford University Press.
McClellan III, James E. and Harold Dorn (1998), *Science And Technology In World History: An Introduction*, Baltimore: Johns Hopkins.
McGinn, Colin (1996), *Die Grenzen vernünftigen Fragens. Grundprobleme der Philosophie*, Stuttgart: Klett-Cotta.
Meinzer, Michael (1992), *Der Französische Revolutionskalender (1972–1805). Planung, Durchführung und Scheitern einer neuen Zeitrechnung*, München: Oldenbourg Verlag.
Mittelstraß, Jürgen (1998), *Das Undenkbare denken. Über den Umgang mit dem Undenkbaren und Unvorstellbaren in der Wissenschaft*, Berlin: Max-Planck-Gesellschaft für Wissenschaftsgeschichte, Preprint No. 105.
Nowotny, Helga (1993), *Eigenzeit*. Frankfurt a.M.: Suhrkamp.
Nowotny, Helga (1997), *Time: The Modern and Postmodern Experience*, Oxford: Polity Press & Blackwell.
Nowotny, Helga (1998), 'Times of Complexity', in J.T. Fraser, M.P. Soulsby and A.J. Argyros (eds.), *Time, Order, Chaos*, Madison, CON: International Universities Press, pp. 91–107.
Oliver, Anne-Marie and Paul Steinberg (1998), 'The politics of apocalypse in the underground media of the Islamic resistance movement (HAMAS), a position paper for the center for millennial studies', www.mille.org/template.htm (28 August 1998).
Reichardt, Rolf and Eberhard Schmitt (eds.) (1988), *Die Französische Revolution als Bruch des gesellschaftlichen Bewußtseins*, München: Oldenbourg.
Relke, Diana M.A. (1998), 'In celebration of doomsday: The end of science and the beginning of equity', *Vox*, 23 (February): http://www.usask.ca/wgst/journals/voxrelke.htm.
Salim, Fahd (1999), *Asrar al s'a wa-hujum al gharb*, Madbuli alsaghir: Cairo before 1999.
Schick, Theodore (1997), 'The end of science?', *Skeptical Inquirer* March/April 1997: 36–9.
Stearns, Peter N. (1996), *World History: Patterns of Change and Continuity*, 2nd edition, New York: Westview/Harper Collins College Publishers.
Stowasser, Barbara (2000), *A Time to Reap. Thoughts on Calendars and Millennialism*, Washington, DC: Center for Contemporary Arab Studies.
Weber, Max (1972), *Gesammelte Aufsätze zur Religionssoziologie, Die Wirtschaftsethik der Weltreligionen, III: Das antike Judentum*, Tübingen: Mohr and Siebeck.
Weber, Max ([1915–1920] 1991), *Die Wirtschaftsethik der Weltreligionen: Taoismus und Konfuzianismus*, Max Weber Gesamtausgabe, Band 1/19.

Weinberg, Steven (1998), 'The expansion of physics', in M. Howard and W.R. Louis (eds.), *The Oxford History of the Twentieth Century*, Oxford: Oxford University Press.

Winner, Langdon (1980), 'Do artifacts have politics?', *Daedalus* 109: 121–36.

Zemanek, Heinz (1987), *Kalender und Chronologie. Bekanntes und Unbekanntes aus der Kalenderwissenschaft*, 4. Ausg., München and Wien: Oldenbourg

Zerubavel, Eviatar (1985), *Hidden Rhythms: Schedules and Calendars in Social Life*, Berkely, CA: University of California Press.

LIST OF CONTRIBUTORS

Ezrahi, Yaron, Professor Dr., Department of Political Science, The Hebrew University of Jerusalem, Levy Building, Givat Ram, 91904 Jerusalem, Israel

Felt, Ulrike, Professor Dr., Universität Wien, Institut für Wissenschaftstheorie und Wissenschaftsforschung, Sensengasse 8/10, A-1090 Vienna, Austria

Gomart, Emilie, Dr., Afdeling Politicologie, Universiteit van Amsterdam, OZ Achterburgwal 237, NL-1012 DL Amsterdam, The Netherlands

Guggenheim, Michael, lic. phil., Eidgenössische Technische Hochschule Zürich, Collegium Helveticum, Schmelzbergstrasse 25, Sternwarte, CH-8092 Zurich, Switzerland

Hagner, Michael, PD Dr., Max-Planck-Institut für Wissenschaftsgeschichte, Wilhelmstrasse 44, D-10117 Berlin, Germany

Hajer, Maarten A., Professor Dr., Afdeling Politicologie, Universiteit van Amsterdam, OZ Achterburgwal 237, NL-1012 DL Amsterdam, The Netherlands

Hilgartner, Stephen, Associate Professor, Department of Science & Technology Studies, Cornell University, 306 Rockefeller Hall, Ithaca, NY 14853, USA

Jasanoff, Sheila, Professor Dr., Harvard University, John F. Kennedy School of Government, 79 John F. Kennedy Street, Cambridge, MA 02138, USA

Joerges, Bernward, Professor Dr., Wissenschaftszentrum Berlin für Sozialforschung, AG Metropolenforschung, Reichpietschufer 50, D-10785 Berlin, Germany

Maasen, Sabine, Professor Dr., Wissenschaftsforschung/Wissenschaftssoziologie/ Science Studies, Universität Basel, Missionsstrasse 21, CH-4003 Basel, Switzerland

Nowotny, Helga, Professor Dr., Ph.D., Director of "Society in Science: The Branco Weiss Fellowship", based at ETH Zurich and Chair of EURAB, the European Research Advisory Board of the European Commission.Society in Science, ETH-Zentrum HAA, CH-8092 Zürich, Switzerland.

Rheinberger, Hans-Jörg, Professor Dr., Max-Planck-Institut für Wissenschaftsgeschichte, Wilhelmstrasse 44, D-10117 Berlin, Germany

Shinn, Terry, Professor Dr., Maison des Sciences de l'Homme, CNRS – GEMAS, 54, boulevard Raspail, F-75270 Paris Cedex 06, France

Weingart, Peter, Professor Dr., Institut für Wissenschafts- und Technikforschung, Universität Bielefeld, Postfach 100 131, D-33501 Bielefeld, Germany

Whitley, Richard, Professor, Manchester Business School, University of Manchester, Booth Street West, GB-Manchester M15 6PB, UK

Wittrock, Björn, Professor Dr., SCASSS (Swedish Collegium for Advanced Study in the Social Sciences), Götavägen 4, 75236 Uppsala, Sweden

BIOGRAPHICAL NOTES

Yaron Ezrahi earned his doctorate in Political Science at Harvard University in 1972. He has written and published extensively on the impact of modern science and technology on democratic governments and the conduct of public affairs. Noteworthy among this publications is his book *The Descent of Icarus, Science and the Transformation of Contemporary Democracy*, published by Harvard University Press in 1990. He was the co-editor, with Everett Mendelsohn and Howard Siegal, of *Technology, Pessimism and Postmodernism*, published by Kluwer Academic Publishers in 1994. Professor Ezrahi's most recent books are *Rubber Bullets, Power and Conscience in Modern Israel* (Farrar, Straus Giroux New York 1997 (in paperback Berkley 1998), and *Israel Towards a Constitutional Democracy* with Mordechai Kremnitzer and the assistance of Margit Cohen and Eytan Alimi (published by the Israel Democracy Institute 2001 – in Hebrew). He received the National Jewish Book Award for 1997, in the category of Israel and Zionism. Since 1993 he is a senior fellow at the Israel Democracy Institute in Jerusalem where he directs the project "Mass Communication and the Democratic Process."

Ulrike Felt is professor of social studies of science at the University of Vienna, Austria. She has been initially trained as a physicist and is working since 1983 in the field of science and technology studies. During her career she held numerous position as invited professor, mainly in the French speaking countries (France, Canada). Her research interests are in the field of science - publics relations, science policy and gender. She has published widely in these domains. Since July 2002 she is editor of the journal *Science, Technology and Human Values.*

Emilie Gomart receivedg her PhD from the Ecole Nationale Superieure des Mines de Paris in the Sociology of Innovations (*Surprised by methadone: An experiment on substitution)* in 1999. She is now post-doc in the department of Political Science at the University of Amsterdam. Her main research focuses on landscape designers working in political circles in the Netherlands and aims to draw from the ethnography and philosophy of science and technology a coherent approach to politics.

Michael Guggenheim is a PhD student at Collegium Helveticum, ETH Zürich. He currently works about the knowledge production and organisation of environmental consultancies.

Michael Hagner is senior fellow at the Max Planck Institute for the History of Science in Berlin. He has been visiting professor at the universities of Salzburg, Tel Aviv, Frankfurt a.M. and fellow at the Collegium Helveticum in Zürich. He is the author of *Homo cerebralis. Der Wandel vom Seelenorgan zum Gehirn* (1997) and has edited *Ecce cortex. Beiträge zur Geschichte des modernen Gehirns* (1999), *Ansichten der Wissenschaftsgeschichte* (2001), and co-edited an issue of Science in Context on *Mindful Practices: On the Neurosciences in the Twentieth Century* (2001). His is currently finishing a book on the history of brains of geniuses in the nineteenth and twentieth centuries.

Maarten A. Hajer is professor of Political Science and holds the Chair in Public Policy, University of Amsterdam. He studied Political Science and Urban and Regional Planning at the University of Amsterdam, obtained a D.Phil. in Politics from the University of Oxford. After that he worked as assistant professor and later as associate professor at the Institute of Sociology of the Ludwig-Maximilians-Universität Munich, was senior research at the Scientific Council for Government Policy in The Hague (the think tank to the Dutch Cabinet). Among his publications are *The Politics of Environmental Discourse – Ecological Modernization and the Policy Process,* Oxford University Press, Oxford / New York (Hardback 1995, Paperback 1997, reprinted in 2000); *Living with Nature – Environmental Politics as Cultural Discourse,* edited together with Frank Fischer for Oxford University Press, 1999. He is co-editor of *Deliberative Policy Analysis – Understanding Governance in the Network Society,* Cambridge University Press, April 2003 forthcoming (with Hendrik Wagenaar).

Stephen Hilgartner is Associate Professor in the Department of Science & Technology Studies at Cornell University. His research focuses on social studies of science and technology, especially biology, biotechnology, and medicine. His recent book, *Science on Stage: Expert Advice as Public Drama (Stanford, 2000)*, won the Carson Prize from the Society for Social Studies of Science (4S) in 2002. *Science on Stage* explores the processes through which the expertise of science advisors is established, contested, and maintained. Hilgartner has also written articles on such topics as data access, risk, popularization, and the Sokal affair. He is a member of the 4S Council and of the Steering Group of the Section on Societal Impacts of Science and Engineering of the American Association for the Advancement of Science (AAAS). He is currently completing a book on genome mapping and sequencing in the 1990s.

Sheila Jasanoff is Pforzheimer Professor of Science and Technology Studies at Harvard University's John F. Kennedy School of Government. She has held faculty appointments or visiting positions at several universities, including Cornell, Yale, Oxford, and Kyoto. At Cornell, she founded and chaired the Department of Science and Technology Studies. She has been a Fellow at the Berlin Institute for Advanced Study (*Wissenschaftskolleg*) and Resident Scholar at the Rockefeller Foundation's study center in Bellagio. Her research centers on the role of science and technology in the authority structures of modern democratic societies, with a particular focus on the use of science in legal and political decision making. Her books on these topics include *The Fifth Branch* (1990) and *Science at the Bar* (1995). Jasanoff has served on the Board of Directors of the American Association for the Advancement of Science and as President of the Society for Social Studies of Science.

Bernward Joerges, senior research fellow at the *Wissenschaftszentrum Berlin für Sozialforschung* (WZB) where he directs the Metropolitan Studies Group, and Professor of Sociology at the *Technische Universität Berlin.* He has been Resident Scholar at the Rockefeller Foundation's study center in Bellagio and visiting professor at, among others, the universities of Uppsala, Linköping, the LSE and the Sorbonne. Joerges has served as chair of the Sociology of Science and Technology section of the German Sociological Association (DGS) and president of the German Association for Science and Technology Research (GWTF). He has published widely in German, English and French in the fields of science, technology and urban studies. Book publications include *Technik ohne Grenzen* (Suhrkamp 1994, with Ingo Braun), *Technik – Körper der Gesellschaft* (Suhrkamp 1996), *Körper-Technik: Aufsätze zur Organtransplantation* (edition sigma 1997), *Instrumentation between Science, State and Industry* (Kluwer 2000, with Terry Shinn).

Sabine Maasen is Professor of Science Studies at the university of Basel. Her training is in sociology, psychology, and linguistics and her research interests are dynamics of knowledge, inter-/transdisciplinarity, processes of scientification, science management as well as science policy. She has published several articles and books in the sociology of science and knowledge including *Biology as Society, Society as Biology. Metaphors* (with E. Mendelsohn and P. Weingart, Kluwer 1994), *Metaphors and the Dynamics of Knowledge* (Routledge 2000, with P. Weingart), and *Die Genealogie der Unmoral. Therapeutisierung sexueller Selbste* (Genealogy of the Immoral. Therapeutic Constructions of Sexual Selves; Suhrkamp 1998). Most recently she works in the fields of the social construction volition and consciousness (e.g., *Voluntary Action. On Brains, Minds, and Sociality*, Oxford 2002, with W. Prinz & G. Roth) and of science policy (*Politics By Discourse*, work in progress).

Helga Nowotny has been Professor of Social Studies of Science at ETH Zurich since 1996 and Director of Collegium Helveticum until 2002. Currently she is Chair of EURAB, the European Research Advisory Board of the European Commission and Director of the post-doctorate Fellowship programme "Society in Science. The Branco Weiss Fellowship". She has a doctorate in law from the University of Vienna and a Ph.D. in sociology from Columbia University, New York. She has held teaching and research positions in Vienna, Cambridge, Bielefeld, Berlin and Paris and has been a Fellow at the Wissenschaftskolleg zu Berlin. From 1974-1986 she has been Executive and Founding Director of the European Center in Vienna and for seven years Chairperson of the Standing Committee for the Social Sciences of the European Science Foundation. From 1987 she was Professor of Social Studies of Science at the University of Vienna and Permanent Fellow of Collegium Budapest/Institute for Advanced Study before moving to ETH Zurich.

Hans-Jörg Rheinberger (born 1946). Studied Philosophy and Biology in Tübingen and Berlin. Molecular biologist and historian of science. 1978–1990 research scientist at the Max Planck Institute for Molecular Genetics in Berlin, 1990–1994 Assistant Professor at the Institute for the History of Medicine and Science, University of Lübeck, 1994–1996 Associate Professor at the Institute for Genetics and General Biology, University of Salzburg, since 1997 Director at the Max Planck Institute for the History of Science in Berlin. Numerous research papers in molecular biology and history of science. Books among others: *Toward a History of Epistemic Things. Synthesizing Proteins in the Test Tube* (Stanford University Press, Stanford 1997); ed. (together with Peter Beurton and Raphael Falk), *The Concept of the Gene in Development and Evolution* (Cambridge University Press, Cambridge 2000).

Terry Shinn is a Research Director at the CNRS and based at the Paris Maison des Sciences de l'Homme. His recent research includes studies of the transverse impact of instrumentation systems on knowledge as well as intellectual and social organization – *Instrumentation between Science, State and Industry* (with B. Joerges, 2001). Shinn has also published books and articles dealing with the history and sociology of scientific and technical education, science/industry relations, the intellectual and social organization of laboratory work, structures of career mobility, and science popularization. He continues to explore the history and sociology of metrology and has lately initiated research into the dynamics of 'innovation'.

Peter Weingart studied economics and sociology at the universities of Freiburg, Berlin and Princeton, holds a chair for sociology of science at the University of Bielefeld. He was director of the Center for Interdisciplinary Research (ZiF) 1989–1994, Fellow of the Wissenschaftskolleg 1983/84, Visiting Scholar at Harvard University 1984/85, Research Scholar at the Getty Research Institute 2000, since 1996 Visiting Professor at the University of Stellenbosch (South Africa), and currently heads the Institute for Science and Technology Studies at Bielefeld. He has published numerous books and articles in the sociology of science and science policy.

Richard Whitley is Professor of Organisational Sociology at Manchester Business School, University of Manchester and is also Visiting Professor in Comparative Business Systems at the Erasmus University of Rotterdam. Recent books include: *The Multinational Firm: Organizing across institutional and national divides*, *Divergent Capitalisms: The social structuring and change of business systems* and the second edition of *The Intellectual and Social Organisation of the Sciences* (all published by Oxford University Press), and *Competing Capitalisms* (Edward Elgar). He also edited a special issue of *Organization Studies* on The Dynamics of Innovation Systems in 2000. In 1998-99 he served as the Chair of the European Group for Organizational Studies and in 1999-2000 was the President of the Society for the Advancement of Socio-economics. Current research interests include the institutional structuring of innovation patterns and competence development in multinational firms.

Björn Wittrock is University Professor at Uppsala University and Principal of the Swedish Collegium for Advanced Study in the Social Sciences, Uppsala. He has published extensively in the fields of intellectual history, historical social science and social theory. He is a Member of the Academic Councils of the Institutes for Advanced Study in Berlin (Wissenschaftskolleg zu Berlin), and in Budapest (Collegium Budapest). He is European Chairman of the European-American Young Scholars' Summer Institutes, a joint activity of six European and American institutes for advanced study. He is a member of the Science Advisory Board of the Swedish Government. He has published fourteen books, most recently: *Public Spheres and Collective Identities* (with Shmuel N. Eisenstadt and Wolfgang Schluchter) Transaction Press, 2001; *Participation and Democracy* (with Dietrich and Marilyn Rueschemeyer) M.E, Sharpe, 1998; *The Rise of the Social Sciences and the Formation of Modernity: Conceptual Change in Context, 1750-1850* (with Johan Heilbron and Lars Magnusson) Kluwer Academic Publishers, 1998 (paperback edition 2001). In 1999 he was awarded the Torgny Segerstedt Medal (highest award of Uppsala University to a Swedish scholar in the social sciences and humanities). He is a Member of the Academia Europaea.

AUTHOR INDEX

Sociology of the Sciences

1. E. Mendelsohn, P. Weingart and R. Whitley (eds.): *The Social Production of Scientific Knowledge.* 1977 ISBN Hb 90-277-0775-8; Pb 90-277-0776-6
2. W. Krohn, E.T. Layton, Jr. and P. Weingart (eds.): *The Dynamics of Science and Technology.* Social Values, Technical Norms and Scientific Criteria in the Development of Knowledge. 1978 ISBN Hb 90-277-0880-0; Pb 90-277-0881-9
3. H. Nowotny and H. Rose (eds.): *Counter-Movements in the Sciences.* The Sociology of the Alternatives to Big Science. 1979 ISBN Hb 90-277-0971-8; Pb 90-277-0972-6
4. K.D. Knorr, R. Krohn and R. Whitley (eds.): *The Social Process of Scientific Investigation.* 1980 (1981) ISBN Hb 90-277-1174-7; Pb 90-277-1175-5
5. E. Mendelsohn and Y. Elkana (eds.): *Sciences and Cultures.* Anthropological and Historical Studies of the Sciences. 1981 ISBN Hb 90-277-1234-4; Pb 90-277-1235-2
6. N. Elias, H. Martins and R. Whitley (eds.): *Scientific Establishments and Hierarchies.* 1982 ISBN Hb 90-277-1322-7; Pb 90-277-1323-5
7. L. Graham, W. Lepenies and P. Weingart (eds.): *Functions and Uses of Disciplinary Histories.* 1983 ISBN Hb 90-277-1520-3; Pb 90-277-1521-1
8. E. Mendelsohn and H. Nowotny (eds.): *Nineteen Eighty Four: Science between Utopia and Dystopia.* 1984 ISBN Hb 90-277-1719-2; Pb 90-277-1721-4
9. T. Shinn and R. Whitley (eds.): *Expository Science.* Forms and Functions of Popularisation. 1985 ISBN Hb 90-277-1831-8; Pb 90-277-1832-6
10. G. Böhme and N. Stehr (eds.): *The Knowledge Society.* The Growing Impact of Scientific Knowledge on Social Relations. 1986 ISBN Hb 90-277-2305-2; Pb 90-277-2306-0
11. S. Blume, J. Bunders, L. Leydesdorff and R. Whitley (eds.): *The Social Direction of the Public Sciences.* Causes and Consequences of Co-operation between Scientists and Non-scientific Groups. 1987 ISBN Hb 90-277-2381-8; Pb 90-277-2382-6
12. E. Mendelsohn, M.R. Smith and P. Weingart (eds.): *Science, Technology and the Military.* 2 vols. 1988 ISBN Vol, 12/1 90-277-2780-5; Vol. 12/2 90-277-2783-X
13. S. Fuller, M. de Mey, T. Shinn and S. Woolgar (eds.): *The Cognitive Turn.* Sociological and Psychological Perspectives on Science. 1989 ISBN 0-7923-0306-7
14. W. Krohn, G. Küppers and H. Nowotny (eds.): *Selforganization.* Portrait of a Scientific Revolution. 1990 ISBN 0-7923-0830-1
15. P. Wagner, B. Wittrock and R. Whitley (eds.): *Discourses on Society.* The Shaping on the Social Science Disciplines. 1991 ISBN 0-7923-1001-2
16. E. Crawford, T. Shinn and S. Sörlin (eds.): *Denationalizing Science.* The Contexts of International Scientific Practice. 1992 (1993) ISBN 0-7923-1855-2
17. Y. Ezrahi, E. Mendelsohn and H. Segal (eds.): *Technology, Pessimism, and Postmodernism.* 1993 (1994) ISBN 0-7923-2630-X
18. S. Maasen, E. Mendelsohn and P. Weingart (eds.): *Biology as Society? Society as Biology: Metaphors.* 1994 (1995) ISBN 0-7923-3174-5
19. T. Shinn, J. Spaapen and V. Krishna (eds.): *Science and Technology in a Developing World.* 1995 (1997) ISBN 0-7923-4419-7

Sociology of the Sciences

20. J.Heilbron, L. Magnusson and B.Wittrock (eds.): *The Rise of the Social Sciences and the Formation of Modernity.* Conceptual Change in Context, 17501850. 1996 (1998) ISBN 0-7923-4589-4
21. M. Fortun and E. Mendelsohn (eds.): *The Practices of Human Genetics.* 1998 ISBN 0-7923-5333-1
22. B. Joerges and T. Shinn (eds.): *Instrumentation: Between Science, State and Industry.* 2000 ISBN 0-7923-6736-7
23. B. Joerges and H. Nowotny (eds.): *Social Studies of Science and Technology: Looking Back Ahead.* 2003 Hb ISBN 1-4020-1481-3; Pb 1-4020-1482-1

KLUWER ACADEMIC PUBLISHERS – BOSTON / DORDRECHT / LONDON